Lecture Notes in Physics

Edited by J. Ehlers, München, K. Hepp, Zürich
R. Kippenhahn, München, H. A. Weidenmüller, Heidelberg
and J. Zittartz, Köln
Managing Editor: W. Beiglböck, Heidelberg

114

Stellar Turbulence

Proceedings of Colloquium 51
of the International Astronomical Union
Held at the University of Western Ontario
London, Ontario, Canada
August 27–30, 1979

Edited by
D. F. Gray and J. L. Linsky

Springer-Verlag
Berlin Heidelberg New York 1980

Editors

David F. Gray
The University of Western Ontario
Astronomy Department
London, Canada
N6A 5B9

Jeffery L. Linsky
Joint Institute for
Laboratory Astrophysics
National Bureau of Standards
and University of Colorado
Boulder, CO 80303
USA

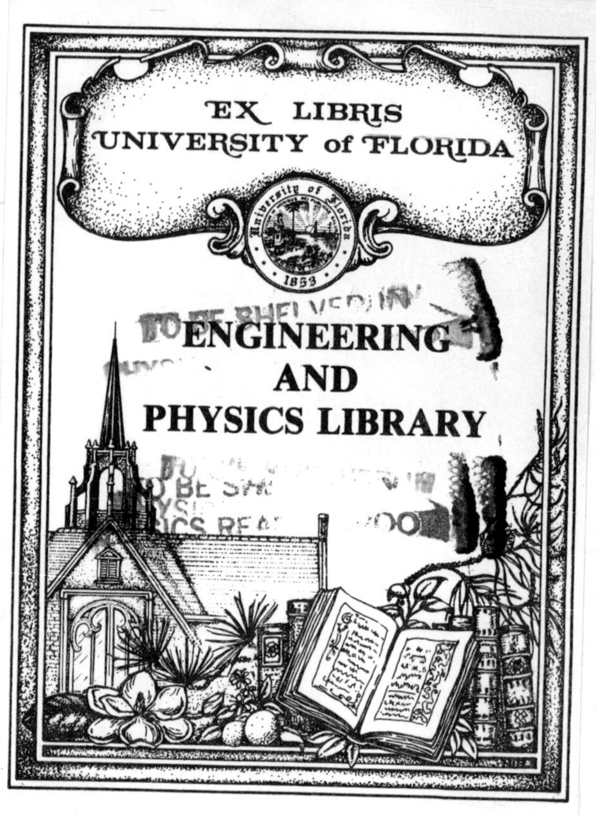

ISBN 3-540-09737-6 Springer-Verlag Berlin Heidelberg New York
ISBN 0-387-09737-6 Springer-Verlag New York Heidelberg Berlin

Library of Congress Cataloging in Publication Data
Main entry under title:
Stellar turbulence.
(Lecture notes in physics; 114)
Bibliography: p.
Includes index.
1. Stars--Atmospheres--Congresses. 2. Turbulence--Congresses. I. Gray, David F., 1938-
II. Linsky, Jeffery L., 1941- III. International Astronomical Union. IV. Title: International
Astronomical Union colloquium no. 51. V. Series.
QB809.S74 523.8 79-27623
ISBN 0-387-09737-6

Printed in Germany

Printing and binding: Beltz Offsetdruck, Hemsbach/Bergstr.
2153/3140-543210

TABLE OF CONTENTS

1. THE PYHSICAL ORIGIN OF TURBULENCE

2. OBSERVED PROPERTIES OF STELLAR TURBULENCE

3. CONCEPTUALIZATIONS OF TURBULENCE

4. SOME EFFECTS OF STELLAR TURBULENCE

List of Participants

Richard C. Altrock, Sacramento Peak Observatory, Sunspot, NM 88349, U.S.A.

Per Andersen, Department of Physics, Brandon University, Brandon MB Can., R7A 6A9.

Lawrence Anderson, Department of Physics and Astronomy, University of Toledo, Toledo, OH 43606, U.S.A.

Hiroyasu Ando, Tokyo Astronomical Observatory, University of Tokyo, Mitaka, Tokyo, Japan.

R. Grant Athay, High Altitude Observatory, P.O. Box 3000, Boulder, CO 80302, U.S.A.

T.R. Ayres, C/O JILA, University of Colorado, Boulder, CO 80309, U.S.A.

Jacques M. Beckers, Multiple Mirror Telescope Observatory, University of Arizona, Tucson, AZ 85721, U.S.A.

Gaetano Belvedere, Osservatorio Astrofisico, Citta Universitaria, 1-95125, Catania, Italy.

Andy Bernat, Kitt Peak National Observatory, P.O. Box 26732, Tucson, AZ 85726, U.S.A.

Erika Böhm-Vitense, Astronomy Department, University of Washington, Seattle, WA 98195 U.S.A.

David H. Bruning, Department of Astronomy, New Mexico State University, Box 4500, Las Cruces, NM 88003, U.S.A.

W. Buscombe, Department of Astronomy, Northwestern University, Evanston, IL, 60201, U.S.A.

C.J. Cannon, Department of Applied Mathematics, The University of Sydney, Sydney 2006, New South Wales, Australia.

Francesco A. Catalano, Osservatorio Astroficiso, Citta Universitario, I-95125 Catania, Italy

R. Cayrel, CFHT Corporation, Waimea Office, Kamuela, HA 96743, U.S.A.

S.M. Chitre, Tata Institute of Fundamental Research, Homi Bhabha Road, Colaba, Bombay 400 005, India.

J.L. Climenhaga, University of Victoria, Department of Physics, Box 1700, Victoria, B.C. Can., V8W 2Y2

Lawrence E. Cram, Sacramento Peak Observatory, Sunspot, NM 88349, U.S.A.

S. Cristaldi, Osservatorio Astrofisico, Citta Universitaria, 1-95125, Catania, Italy.

Lucio Crivellari, Osservatorio Astronomico di Trieste, via G.B. Tiepolo 11, Trieste, Italy.

Y. Cuny, Observatoire de Paris, 92190 Meudon, France

D. Dravins, Lund Observatory, Svanegatan 9, S-222 24 Lund, Sweden.

B. Durney, Sacramento Peak Observatory, Sunspot, NM 88349, U.S.A.

Dennis Ebbets, Department of Astronomy, University of Texas, Austin, TX 78712, U.S.A.

Kjell Eriksson, Uppsala Universitets Astronomiska Observatorium, Sweden.

Nancy Remage Evans, Astronomy Department, University of Toronto, Toronto ON Can., M5S 1A7

Eric Fossat, Departement d'Astrophysique, Universite de Nice, 06034 Nice, Cedex, France

R. Foy, Observatoire de Meudon, F-92190 Meudon, France

Rubens Freire, c/o Dr. Paul Felenbok, Observatoire de Paris, 5 Place Janssen, 92190 Meudon, France.

H.-R. Gail, Institut für Theoretische Astrophysik, Im Neuenheimer Feld 294, 6900 Heidelberg, Germany.

T. Gehren, Max-Planck-Institut für Astronomie, Königstuhl, 6900 Heidelberg 1, Germany.

Peter Gilman, Advanced Study Program, National Center for Atmospheric Research, Box 3000 Boulder CO 80303, U.S.A.

R. Glebocki, Institute of Physics, University of Gdansk, Poland.

Leo Goldberg, Kitt Peak National Observatory, P.O. Box 26732, Tucson, AZ 85726, U.S.A.

David F. Gray, Astronomy Department, University of Western Ontario, London ON Can., N6A 5B9.

Mart de Groot, Armagh Observatory, College Hill, Armagh, BT61 9DG, Ireland.

Hans G. Groth, University of Munchen, Institut für Astronomie u. Astrophysik, Scheinerstrasse 1, D-8000 Munchen 80, West Germany

E.A. Gurtovenko, Chief of Solar Department, Main Astronomical Observatory, Ukrainian Academy of Sciences, 252127 Kiev - 127, U.S.S.R.

E.A. Gussman, Zentralinstitut für Astrophysik, Potsdam, Germany.

L. Hartmann, Center for Astrophysics, Harvard College Observatory and Smithsonian Astrophysical Observatory, Cambridge, MA 02138, U.S.A.

Toshio Hasegawa, Hokkaido University of Education at Asahikawa, Hokumon-cho 9, Asahikawa, Japan.

Alan H. Karp, Palo Alto Scientific Center, 1530 Page Mill Road, P.O. Box 10500, Palo Alto, CA 94304, U.S.A.

K. Kodaira, Department of Astronomy, Faculty of Science, University of Tokyo, Bunkyo-ku, Tokyo, Japan.

H.J.G. Lamers, Space Research Laboratory, Beneluxlaan 21, Utrecht, The Netherlands.

J. Leibacher, Space Science Lab, Dept. 52-12, Bldg. 202, 3251 Hanover Street, Palo Alto CA 94304, U.S.A.

Jeffrey Linsky, Joint Institute for Laboratory Astrophysics, University of Colorado, Boulder, CO 80303, U.S.A.

L.B. Lucy, Department of Astronomy, Columbia University, New York, NY 10027, U.S.A.

P. Maltby, University of Oslo, Institute of Theoretical Astrophysics, P. Boks 1029, Blindern, Oslo 3, Norway.

A. Mangeney, Observatoire de Meudon, Departement de Recherche Spatiale, 92190 Meudon, France.

Pilar Martin, Observatoire de Meudon, F-92190 Meudon, France.

George D. Nelson, Code CB, Johnson Space Center, Houston TX 77058, U.S.A.

Anastasios Nesis, Keipenheuer-Institut für Sonnenphysik, Schöneckstrasse 6, D-7800 Freiburg, Germany.

Å, Nordlund, Nordita, Nordisk Institut for Teoreisk Atomfysik, Blegdamsvej 17, DK-2100 Kobenhavn Ø, Denmark.

Y. Osaki, Department of Astronomy, Faculty of Science, University of Tokyo, Bunkyo-ku, Tokyo, Japan.

Lucio Paterno, Osservatorio Astrofisico, Citta Universitaria, 1-95125 Catania, Italy.

Francois Querci, Observatoire de Paris, F-92190, Meudon, France.

Monique Querci, Observatoire de Paris, F-92190 Meudon, France

Jorge Ramiro de la Reza, Observatorio Nacional, Rua General Bruce, 586, 20921-São Cristovão, Rio de Janeiro-RJ, Brazil.

R.J. Rutten, Sterrekundig Institut, Sterrewacht "Sonnenborgh", Zonnenburg 2, 3512NL Utrecht, Netherlands.

A.J. Sauval, Observatoire Royal de Belgique, Avenue Circulaire 3, 1180 Bruxelles, Belgique.

E. Sedlmayr, Lehrstuhl für Theoretische Astrophysik, D-6900 Heidelberg 1, Im Neuenheimer Feld 294, Germany.

Svein Sivertsen, University of Tromso, Institute of Mathematical and Physical Sciences, P.O. Box 953, N-9001 Tromsø, Norway.

R.C. Smith, Astronomy Centre, Physics Building, University of Sussex, Falmer, Brighton BN1 9QH, Great Britain.

G. Sonneborn, Department of Astronomy, 174 W. 18th Avenue, Columbus, OH 43210, U.S.A.

S.R. Sreenivasan, Department of Physics, The University of Calgary, Calgary, AB Can., T2N 1N4.

R. Stein, Department of Astronomy and Astrophysics, Michigan State University, East Lansing, MI 48824, U.S.A.

Robert E. Stencel, Joint Institute for Laboratory Astrophysics, University of Colorado, Boulder CO 80303 U.S.A.

S.P. Tarafdar, Theoretical Astrophysics Section, T.I.F.R. Colaba, Bombay-400005, India.

G. Traving, Institut für Theoretische Astrophysik, Im Neuenheimer Feld 294, 6900 Heidelberg, Germany.

William H. Wehlau, Department of Astronomy, University of Western Ontario, London ON Can. N6A 5B9.

Peter Wilson, Department of Applied Mathematics, University of Sydney, Sydney, 2006, N.S.W. Australia.

K.O. Wright, Dominion Astrophysical Observatory, 5071 W. Saanich Road, Victoria BC V8X 3X3.

J.P. Zahn, Universite de Nice, Observatoire, 06007 Nice, Cedex, France.

Preface

Stellar turbulence is a fascinating and fundamentally important aspect of stellar atmospheres. I.A.U. Colloquium 51 was organized to bring together both stellar and solar astronomers, to have them share views about the generation, nature, and implications of stellar turbulence.

The program was arranged into several conceptual pieces: the generation of motions by convection, rotation, oscillations, the measurement and observed characteristics of turbulence, modeling and theoretical interpretation of turbulence, and the relation of chromospheres, coronae, and mass loss to the turbulence. This same order of material has been preserved in these proceedings. We chose not to record the questions and dialogue which transpired. This choice was made in part because of the publisher's preference and in part because it leads to greater freedom and intensity of discussion among the participants. Two types of material are found in these proceedings. First, there are the invited papers which are presented in full. Second, there are the contributed papers which are presented in abstract form.

The 73 registered participants came from 19 countries.

The cross-exchange between the stellar and solar domains was very useful. The type of physical goings-on seen on the sun is very informative when trying to puzzle out turbulence in other stars. At the same time, stars give us the conditions where the phenomena comprising turbulence are likely to occur in combinations and strengths quite different from the solar example. Stars also allow us to explore the dependence of turbulence on temperature, pressure, rotation, and similar physical variables - levers not available for the sun alone. Experience has shown that stellar turbulence is not an easy subject to deal with. We need every advantage we can bring to the problem, if progress is to be made. The unification of stellar and solar information, people, and thinking is a very basic step in this direction.

For me personally, I.A.U. Colloquium 51 was a mind expanding and pleasureable experience. I am grateful to those working with me on the organizing committee: J. Beckers, E. Gurtovenko, K. Kodaira, H. Lamers, J. Linsky, L. Lucy, E. Müller, J.-P. Zahn, for their h elp with organizing and running the meeting. I would like to thank the Union and the National Science and Engineering Council of Canada for contributing toward the travel and running expenses of I.A.U. Colloquium 51. I also wish to express my warm thanks to the participants themselves for their constructive and hard working attitudes and for their kindness to me during the meeting.

David Gray
Chairman of the Scientific
Organizing Committee

STELLAR CONVECTION THEORY

Jean-Paul Zahn
Observatoire de Nice
06300 Nice, France

1. INTRODUCTION

This meeting, which deals with turbulence in stars, opens with a review on thermal convection. There is no better way to state from the start that among all instabilities that are likely to arise in stars, it is thermal convection which is the most firmly established as a cause for the turbulence that we observe on their surface. Our confidence in this comes mainly from the theoretical prediction that convective instability sets in whenever the density stratification becomes superadiabatic, as is expected in late type stars whose outer layers are very opaque, due to the ionization of the two most abundant elements, hydrogen and helium. And, in these stars at least, thermal convection occurs close enough to the photosphere to influence, be it indirectly, the profile of spectral lines.

A whole IAU colloquium has been devoted three years ago in Nice to the topic of stellar convection, and one finds in its proceedings an extensive account of what was the state of the problem. Some progress has been accomplished since then, and naturally I will spend most of my time describing recent work, and even work in progess that I an aware of. But on the assumption that some of you are not too familiar with the subject, let me first recall some generalities.

2. A HIGHLY NONLINEAR PROBLEM

Thermal convection is described by a set of well-known equations, which state the conservation of mass

$$\frac{\partial \rho}{\partial t} + \nabla \cdot \rho \underline{V} = 0 \quad , \qquad (1)$$

that of momentum

$$\rho \left(\frac{\partial \underline{V}}{\partial t} + (\underline{V}\nabla)\underline{V} \right) = -\nabla P + \underline{g}\rho + \underline{f}_V \qquad (2)$$

and that of heat (or entropy)

$$\rho T \left(\frac{\partial S}{\partial t} + \underline{V} \cdot \nabla S \right) = \nabla \cdot (K \nabla T) + d_V \tag{3}$$

One can combine these equations to establish the conservation of total energy

$$\frac{\partial}{\partial t} \left(\frac{1}{2} \rho V^2 + \rho H - P \right) + \nabla \cdot \left(-K \nabla T + H \rho \underline{V} + \frac{1}{2} \rho V^2 \underline{V} + \underline{F}_V \right)$$
$$= \underline{g} \cdot \rho \underline{V} \tag{4}$$

I use here the classical notations for the gravity vector ($\underline{g}$) and the velocity field ($\underline{V}$); the pressure (P), the entropy (S), the enthalpy (H) and the thermal conductivity (K) are all known functions of the density (ρ) and the temperature (T). The viscous force ($\underline{f}_V$), the viscous dissipation (d_V) and the so-called viscous flux ($\underline{F}_V$) need not to be explicited for this purpose. The assumption has been made that the medium is optically thick, so that the radiative flux is proportional to the temperature gradient; $H \rho \underline{V}$ is the convective flux, and $1/2\ \rho V^2 \underline{V}$ the mechanical energy flux.

Even a layman realizes that the differential system formed by the first three equations is highly nonlinear, and he may guess that most difficulties in treating the problem are due to this nonlinearity. To stress that point, it suffices perhaps to recall that the most simple version of this system obtained by neglecting the buoyancy and assuming that the density is constant, and thus retaining only the two first equations (which the fluid dynamicists refer to as the Navier-Stokes equations), has no general solution that would be valid over a large range of parameters.

The nonlinearities of the problem have a double role. First they prevent the motions, once the instability has set in, from growing for ever exponentially. Second, they generate a whole set of scales in the velocity field, and they interconnect them. The term responsible for this is the advection term $(\underline{V} \nabla)\ \underline{V}$, and no wonder that this term, plus a similar one in the heat equation, namely $\underline{V}\ \nabla S$, are the most difficult to deal with when attempting to solve the problem numerically.

In fact, the various approaches to the problem that have been devised so far differ mainly in how they treat these advection terms. The most ambitious approach would be of course to consider all scales, and to calculate explicitly those nonlinear terms. However in all cases of astrophysical interest, this is out of question since the scales involved span several orders of magnitude (typically eight or more), and even if the velocity field were homogeneous in space, which it is not, describing all these scales would be well beyond the capabilities of the present computers. Therefore one has always, at one stade or another, to make some simplifying assumption. An impressive portion of the fluid dynamical literature has been devoted to the subject of how to best model convective transport, that is the advection, by prescribed motions, of a scalar or of a vector field. Let us briefly recall how this is done most commonly.

3. APPROXIMATING THE CONVECTIVE TRANSPORT

The simplest situation one can imagine is that of the advection of a passive scalar, ignoring molecular diffusion. The conservation equation for this scalar (which may be a dye in water, a pollutant in the atmosphere or entropy in thermal convection) takes the simple form

$$\frac{\partial s}{\partial t} + \underline{V}\cdot\nabla s = 0 \tag{5}$$

s being the concentration.

Experiments and observations show that in many instances, in particular when the motions are turbulent, the scalar just diffuses away. Thus in that case some local average of s, which we will designate by $\bar{s}$ without further defining it, obeys a diffusion law

$$\frac{\partial \bar{s}}{\partial t} - \nabla\cdot(\chi_T \nabla\bar{s}) = 0 \tag{6}$$

The turbulent diffusivity χ_T is given by

$$\chi_T = l u$$

u being the mean turbulent velocity and l the *mixing length*. The latter has been introduced by Prandtl (1925), and can be viewed as the distance a turbulent element travels before it dissolves in the medium and dumps there the scalar quantity that it has carried to that point.

For this diffusion approximation to be valid, a necessary condition is obviously that the mixing length be smaller than the scale characterizing the spatial variation of $\bar{s}$. Unfortunately, this is not a sufficient condition, as it has been recognized in many cases.

The most recent work on this subject is that by Knobloch (1978), who begins to establish a formal expression for the time derivative of $\bar{s}$

$$\frac{\partial \bar{s}}{\partial t} = \left(\sum_{m=2}^{\infty} \mathcal{K}_m\right)\bar{s} \tag{7}$$

The $\mathcal{K}_m$ are differential operators which can in principle be determined for a given velocity field. If this field is isotropic and varies slowly enough in space and time, the $_m$ are given by

$$\mathcal{K}_m = \eta_2^m \nabla^2 + \eta_4^m \nabla^4 + \ldots + \eta_{m'}^m \nabla^{m'} \tag{8}$$

(where m', an even number, is either m or m-1). The coefficients η_i^m are integrals involving the velocity distribution and its successive moments.

For $m = 2$, one recovers the above mentioned diffusion approximation, with η_2^2 being the turbulent diffusivity. For higher m, the coefficient in front of the Laplacian operator becomes

$$\eta_2^2 + \eta_2^3 + \eta_2^4 + \ldots. \tag{9}$$

and thus the diffusivity is modified, even when the higher order operators play a negligible role. The important fact to notice therefore is that at each step of approximation, as m increases, the coefficient multiplying each differential operator is renormalized.

Moreover, the series is known only to converge if the following dimensionless number is smaller than unity: $u\tau/\lambda < 1$, where τ and λ are respectively the correlation time and the correlation length characterizing the velocity field (see Knobloch's paper for more details). In most cases however that requirement is not met, and this restricts seriously the applicability of the diffusion approximation. Needless to say that the problem becomes even more intricate if it is a vector field that is being advected, if it feeds back on the velocity field, if the latter is no longer isotropic, etc.

At this point we realize that the only reason the diffusion approximation is so widely used is that one has not yet found by what to replace it. And we come to understand that the degree of refinement (I hesitate to say reliability) of a given treatment of thermal convection can be measured by the level at which the diffusion approximation is introduced, as we will see next.

4. THE MIXING LENGTH APPROACH

In the current mixing length approach, which has been applied to stellar convection (in the core of a star) by Biermann (1933), the diffusion approximation is used right from the beginning to model the heat transport. Thus the advection term in Eq. 3, $\rho T \underline{V} \cdot \nabla S$, is replaced by

$$- \nabla \cdot (\chi_T \rho T \nabla S) = - \nabla \cdot (\underline{F}_C) \qquad (10)$$

the quantity in parenthesis being the convective flux, whose magnitude is more commonly stated as

$$F_C = \rho C_P T (u l/H) (\nabla_{ad} - \nabla) \qquad (11)$$

Here C_P is the specific heat at constant pressure and H the pressure scale height, ∇_{ad} and ∇ are respectively the adiabatic and the true logarithmic temperature gradient, with respect to the pressure. The turbulent diffusivity has been replaced by $u l$, where usually the mixing length l is taken proportionaly to the pressure scale height and the velocity u estimated from the conservation of mechanical energy:

$$\frac{1}{2} \rho V^2 \sim g \rho' l \qquad (12)$$

The density difference ρ' between a convective element and its surroundings is evaluated in turn through

$$\rho' / \rho \sim (\nabla_{ad} - \nabla) l/H \qquad (13)$$

All quantities are then defined at any given depth, and the temperature gradient, for instance, depends only on those local values.

Many improvements have been brought to this original form of the mixing length treatment, for instance by Öpik (1950). Vitense (1953) took into account the heat exchanges between the convective elements and the surrounding medium, which enabled her to model a convection zone up to the surface of a star, thus linking its atmosphere with the interior. More recently Shaviv and Salpeter (1973), followed by Maeder (1975), endeavoured to transform the local mixing length treatment into a non-local one, which would be capable of also describing penetrative convection. Another line of action has been to refine the evaluation of the diffusion coefficient χ_T, by introducing the effect of rotation and magnetic field; I will mention here only the most recent contributions from Gough (1978) and from Durney and Spruit (1979), since P. Gilman will cover this subject in more detail in his review.

All these improvements however cannot hide the main weakness of the mixing length approach, namely that it evluates the heat transport through the diffusion approximation. Some progress should therefore be expected when abandoning this approximation, as it is done in the so-called hydrodynamical treatments.

5. HYDRODYNAMICAL APPROACHES

In such approaches, a great effort is made to solve the original fluid dynamical equations, at least for those scales of the velocity field that transport most of the convective flux. The diffusion approximation cannot be avoided entirely, but it will be used only to evaluate the momentum transport by the smaller scales. This is possible in principle if only the larger scales are driven by buoyancy, the smaller scales being generated through the nonlinear interactions that have been mentioned earlier.

By large, we mean here the scales that in the laboratory are of the order of the thickness of the unstable layer. In a star, several of such scales seem to be present, as indicated by the solar observations: those of granulation and supergranulation are well established, but others may also exist, as we will learn from J. Beckers later in this meeting. These scales are certainly related to the vertical structure of the convection zone, although we do not know yet precisely how, especially when the hydrogen and helium ionization regions are merged in the same unstable zone.

The numerical techniques available to solve the nonlinear system (Eq. 1-3) fall into three categories. First, you may choose to stay in the physical space; you divide it in a mesh as fine as your computer allows it, you transform the differential equations in finite differences equations, and you solve those by an appropriate scheme. Its transparency makes this method very appealing, but it remains restricted to rather mild convection, which does not require too high spatial resolution. Alternatively, since you are not interested in each detail of the various fields, but merely in their statistical properties, you might prefer to work in the Fourier space. This is done indeed by most people who study developed turbulence, but the method is only well suited for homogeneous fields.

Most calculations of stellar convection are based on a third method, which combines the main advantages of the other two. Postulating, as it seems very natural, that the various fields are homogeneous on horizontal surfaces in some statistical sense, you take a Fourier expansion (or something similar) in the horizontal directions, but you stay in the physical space for the vertical dimension, in which you are certain to encounter much more structure. This procedure is known as that of *modal expansion*; let us sketch it below when applied to a scalar field, say the temperature:

$$T(x,y,z,t) = \bar{T}(z,t) + \sum_i f_i(x,y)\, \Theta_i(z,t) \tag{14}$$

The temperature is split into its horizontal average $\bar{T}$, which depends only on the vertical coordinate z and time t, and the fluctuation around this mean, which is expanded into a series of horizontal planforms $f_i(x,y)$, the modal amplitudes Θ_i being again functions of only z and t.

The horizontal planforms used most commonly, mainly for simplicity, are those which separate the variables in the linear limit of the problem; they obey the harmonic equation

$$\left(\frac{\partial^2}{\partial x^2} + \frac{\partial^2}{\partial y^2} \right) f_i + a_i^2 f_i = 0 \tag{15}$$

where a_i is the wavenumber that characterizes the horizontal structure of this mode. Fourier modes of the same wavenumber can be assembled to describe cells of various shapes: horizontal rolls, prisms of rectangular base, of hexagonal base, etc., which may differ stronlgy in their nonlinear properties.

Many other possible choices exist for these planform functions; one has recently been explored by Depassier and Spiegel (1979), which needs only one single mode to render the high contrast that is often observed between rising and falling motions.

6. RECENT DEVELOPMENTS

The modal expansion procedure has first been applied to the *Boussinesq approximation*, which is the form the fluid dynamics equations take when the thickness of the considered layer is very small compared to the density scale height, as it is the case in laboratory convection. In this approximation, the only nonlinear terms are - not surprisingly - the advection terms $(\underline{V}\nabla)\underline{V}$ and $\underline{V}\cdot\nabla S$, and much can be learned by studying this simplest example of thermal convection. The parameters defining a given experiment reduce then to only two: the Rayleigh number, which measures the strength of the instability, and the Prandtl number, which is the ratio of the viscosity to the thermal diffusivity. It is therefore relatively simple, in principle, to explore the behavior of convection, both experimentally and theoretically, in the two-dimensional space of these parameters.

For a thorough discussion of Boussinesq convection we refer to the review paper by Spiegel (1971). The most extensive calculations, with the modal expansion procedure, have been performed by Toomre *et al.* (1977); their results are in satisfactory agreement with the laboratory experiments. One difficulty with the modal method is the selection of the planforms that are used in the expansion, especially when the available facilities limit the number of modes to one or two only. Numerical experimentation of the kind just quoted then offers an indispensable guidance on how to make such choices.

a. Strong density stratification

The Boussinesq approximation is of course not well adapted to describe stellar convection, which takes place in general over several density scale heights. Therefore some of the recent works have dealt with more realistic situations and have focussed on the effects due to a strong density stratification. Another approximation is then used very often, namely the *anelastic approximation*. It consist mainly in filtering out the acoustic waves, which probably contribute very little to the energy transport. To be applied, it requires that the convective velocities be small compared to the speed of sound (see Gough 1969), which is generally the case.

The simplest example to consider is that of a thick layer of perfect gas, in which the dynamical viscosity and the thermal conductivity are both constant. In the absence of motion, the stratification is that of a polytrope, with $\ln \rho = n \ln T + \text{constant}$. As is well known, the stratification is convectively unstable when n, the polytropic index, is less then 1.5 (for either a monoatomic or a fully ionized gas).

The linear stability problem has been investigated by Spiegel (1965) and by Gough *et al.* (1976); steady nonlinear solutions have been constructed by Massaguer and Zahn (1979). The main result of those studies is that the pressure forces play a very important role, in contrast with Boussinesq convection, and the reasons for this can be easily understood.

In the anelastic approximation, the equation of momentum conservation can be written

$$\bar{\rho} \left(\frac{\partial \underline{V}}{\partial t} + (\underline{V}\, \nabla)\, \underline{V} \right) = - \nabla P' + \underline{g}\, \rho' + \underline{f}_V \qquad (16)$$

where P' and ρ' are the fluctuating values of the pressure and the density, with respect to their horizontal mean (of which only that of the density, $\bar{\rho}$, enters explicitly in this equation). As before in Eq. 1, $\underline{f}_V$ is the viscous force. One sees that the pressure force, on the right hand side, has been split in the contribution of the fluctuating pressure, $-\nabla P'$, and that of the mean pressure, $\underline{g}\rho'$, which is usually referred to as the buoyancy force.

In Boussinesq convection, the buoyancy force is always acting upwards on a rising fluid element in the unstable region. A first consequence of the density stratification is that *the buoyancy changes sign* in the upper part of the unstable domain. To under-

Figure 1

A cross-section through steady cellular convection

a. The pressure fluctuations are positive where the flow diverges, negative where it converges.

b and c. Variation of the relative fluctuations of the pressure ($P'/\bar{P}$), the temperature ($T'/\bar{T}$) and the density ($\rho'/\bar{\rho}$), along the vertical line A B, where the fluid is rising.

b. In the limit of vanishing density stratification (Boussinesq case), $P'/\bar{P}$ is negligible and thus $\rho'/\bar{\rho}$ is the mirror image of $T'/\bar{T}$.

c. With a strong stratification, $P'/\bar{P}$ is of the same order as the other fluctuations, and therefore $\rho'/\bar{\rho}$ (and thus the buoyancy force) changes sign in the upper part of the domain.

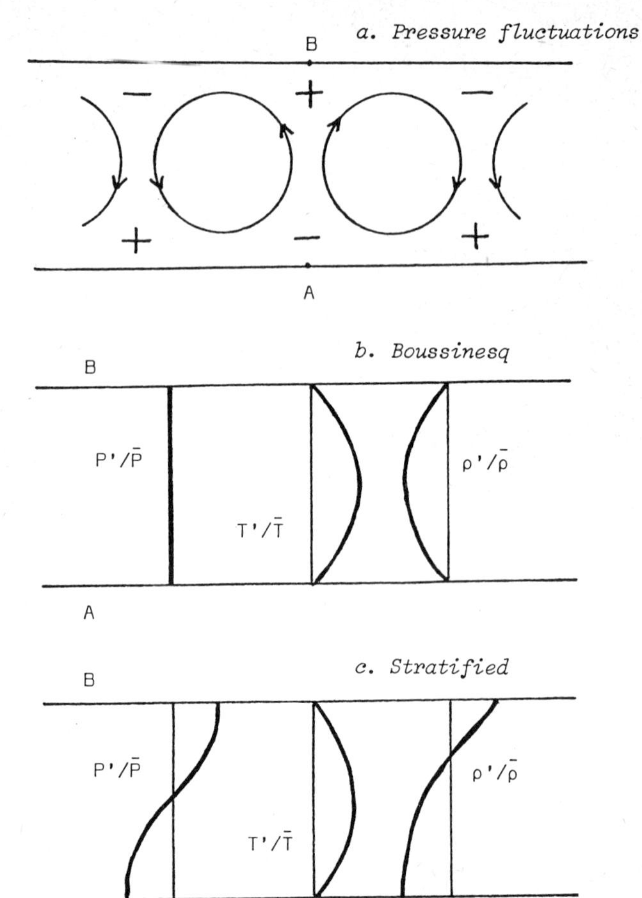

stand this, let us consider Figure 1a where we have sketched a cross-section through a convective layer. It makes no difference here whether the cells are rolls or take another shape: since it is the fluctuating pressure which is responsible for turning a vertical flow into a horizontal one, it must have a maximum in each horizontal plane where the flow diverges, and a minimum where it converges. This obliges the vertical gradient of the fluctuating pressure to be predominantly in the direction of the vertical flow, as depicted in Figure 1c. The temperature fluctuations, on the other hand, are positively correlated with the vertical velocities, since they originate from the advection of heat in a superadiabatic temperature gradient (in the language of the mixing length theory: a rising fluid element is hotter than its surroundings).

Now in Boussinesq convection the pressure fluctuations are vanishingly small compared to the temperature fluctuations: the equation of state therefore imposes that the density fluctuations then just be the mirror images of these temperature fluctuations, as shown in Figure 1b. But in stratified convection, where the fluctuations of the pressure may be of the same order as those of the temperature (depending on the thickness of the convective layer as compared to the density scale height), the buoyancy force reflects the horizontal variations of both the temperature and the pressure, as indicated in Figure 1c. This is again imposed by the equation of state, which in the anelastic approximation takes the simple linear form

$$P'/\bar{P} = \rho'/\bar{\rho} + T'/\bar{T} \qquad (17)$$

with our assumption of a perfect gas. The profiles sketched in Figure 1 are valid for non penetrating convection, with the temperature fluctuations vanishing at top and bottom, but the results can easily be extended to other cases.

Another consequence of strong density stratification is *the modification of the energetics* of thermal convection. The work done by the fluctuating pressure force, per unit time and over the whole considered domain, is

$$E_P = - \iiint \nabla P' \cdot \underline{V} \, dv \quad , \qquad (18)$$

which can be transformed into

$$E_P = - \iint \nabla \cdot (P'\underline{V}) \, ds + \iiint P' \nabla \cdot \underline{V} \, dv \quad . \qquad (19)$$

With the usual boundary conditions on $\underline{V}$, the surface integral is zero and only remains the volume integral, which also vanishes in the Boussinesq limit since there $\nabla \cdot \underline{V} = 0$. But one sees that in a stratified medium, where the conservation of mass imposes that $\nabla \cdot \underline{V} \neq 0$, the fluctuating pressure too contributes to the work done on the fluid. There are many cases, as shown by Massaguer and Zahn (1979), where this pressure integral is larger than that representing the work done by the buoyancy force; there are instances where the latter work becomes even negative, due to the reversal of the buoyancy that has been discussed above.

We should perhaps remind ourselves at this point that in the mixing length approach the pressure fluctuations are generally ignored and that the convective velocities are estimated from the work done solely by the buoyancy. The application of this procedure should therefore be restricted to the Boussinesq limit, where the pressure fluctuations are indeed negligible and only the buoyancy force produces net work. But the mixing length approach is widely used to describe very thick convection zones, where these conditions are no longer realized, and this is another reason for not taking its predictions too seriously.

The properties of non Boussinesq convection that have been presented above in the case of a polytropic atmosphere are entirely confirmed by more realistic calculations, such as those performed by Latour, Toomre and Zahn with application to the envelope of an A type star; a first report on this has been given in Toomre *et al.* (1977). The medium is treated as a real gas, the variations of the thermal conductivity and of the turbulent viscosity are taken in account. The computational domain goes from above the photosphere to a depth of 28,000 km, thus encompassing both convection zones (that due to the ionization of hydrogen and the first ionization of helium, and that due to the second ionization of helium); the density varies by a factor of 60 between top and bottom. The star has an effective temperature of 8000 K and a gravity of $1.15\ 10^4 \mathrm{cm\ s^{-2}}$.

A typical solution is displayed in Figure 2 for one choice of the horizontal planform of the single mode that has been retained in this calculation: it corresponds to cells of hexagonal base with the flow directed mainly upwards along their centerline. The horizontal size of the cell is comparable to the thickness of the lower convection zone. Notice that the pressure fluctuations are of the same order as those of the temperature; as in the simple polytrope, the buoyancy force is controlled by the pressure fluctuations in the upper part of the domain, whereas the temperature fluctuations dominate in the lower part. One verifies also that the fluctuations of these thermodynamical variables do not exceed 8%, which justifies *a posteriori* the use of the anelastic approximation, in which only linear terms in these variables are retained.

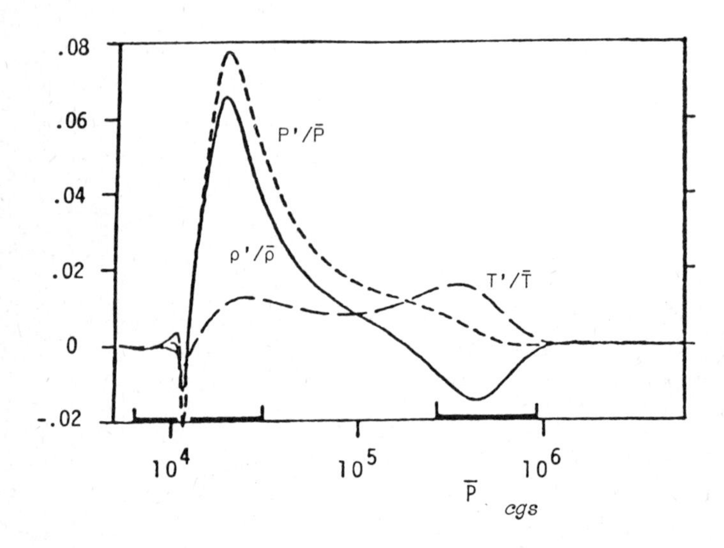

Figure 2

Fluctuations of the thermodynamical variables in the envelope of an A type star

The relative fluctuations of the pressure ($P'/\bar{P}$), the temperature ($T'/\bar{T}$) and the density ($\rho'/\bar{\rho}$) are shown versus the mean pressure, taken as measure of the depth. The location of the unstable zones is indicated in heavy lines on the pressure scale. The results refer to a single mode solution, and the fluctuations which are displayed are the modal amplitudes. The horizontal planform is hexagonal and the vertical velocity is predominantly upwards in the centerline of the convective cells, whose horizontal dimension is of the order of the thickness of the lower convection zone.

b. Penetration

But the most striking result of this calculation is that the convective motions, which are driven in the lower convection zone, penetrate through the stable region above and the upper convection zone up to the surface of the star. This appears clearly in Figure 3, where the two components of the velocity are shown for the same single mode solution as above. The strong horizontal flows which develop in the atmosphere are organized in cells whose dimensions are large compared to the local density or pressure scale heights; one is therefore tempted to identify this with some kind of supergranulation. In an earlier work by Toomre *et al.* (1976), there was already an indication for such an extended penetration, which has the consequence of connecting the two convection zones; according to the current mixing length theory, these zones should be well separated.

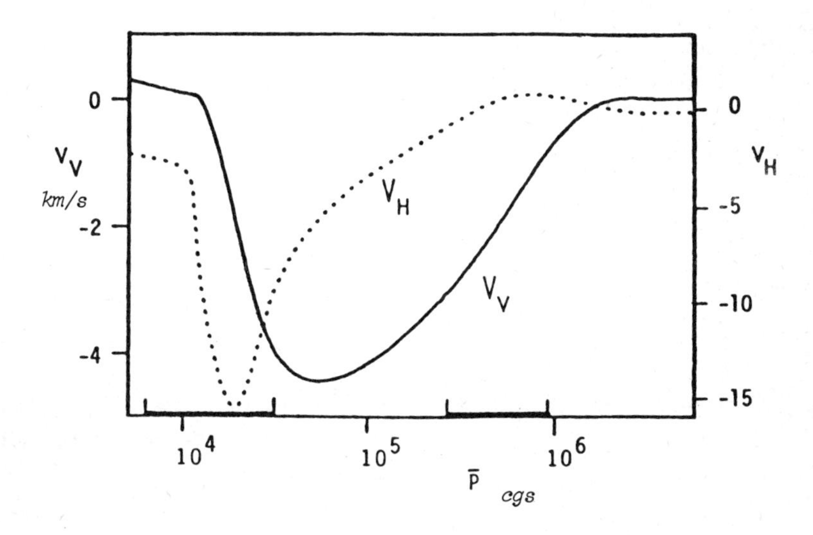

Figure 3

Convective velocities in the envelope of an A type star

The modal amplitudes of the vertical and horizontal velocities, designated respectively by V_V and V_H, as functions of the mean pressure. This single mode solution is the same as that of Figure 2. Over most of the domain, the vertical velocity has a negative sign, implying that it is directed upwards in the centerline of the convective cells. Notice the ample penetration from the lower convection zone all the way up to the surface of the star, and also the large horizontal velocities which occur there.

Convection in smaller scales has not been investigated in this calculation, because the radiative transfer has been treated for simplicity in the diffusive approximation, which is not applicable in the optically thin limit. Nelson and Musman (1977) and Nelson (1978a) have tackled this problem, with application to the solar granulation. They treat the transfer in the Eddington approximation, whose validity has been established by Unno and Spiegel (1966) (at least when the mean radiation field is uniform). The mean stratification is imposed, and the amplitude of the convective motions is limited through a nonlinear process which is similar to a turbulent viscosity. Only the upper part of the solar convection zone is considered, above optical depth $\tau = 25$; although the boundary conditions on the fluctuating variables appear somewhat arbitrary, their influence on the solutions seems moderate. The free parameters are chosen to match the observations.

The main results of these calculations concern the temperature fluctuations, which are shown to be controlled mainly by the transfer of radiative energy. They change sign around 110 km above optical depth $\tau = 1$, as the result of penetrative convection, which has an e-folding distance of about 160 km. The center to limb variation of the intensity fluctuations are in satisfactory agreement with the observations.

The same method has also been applied by Nelson (1978b) to an F type star, which has two separate convection zones. His conclusions are similar to those of Latour, Toomre and Zahn mentioned above for an A type star; he too finds that the two convection zones are linked through penetrative motions.

7. WORK IN PROGRESS

Most of the recent results come from calculations that are based on the modal expansion procedure. It should not be forgotten, however, that in all cases the modal expansion was limited to only two, if not one, modes. The consequences of such a drastic truncation are not fully understood yet, and therefore these results should be used with some caution. It is true that most of these calculations have dealt with rather mild convection, involving convective fluxes of only a few percent. The feedback on the mean stratification is therefore very slight, and it is reasonable to assume that a multi-mode solution would not differ much from the superposition of an ensemble of single-mode solutions.

To check this point, it would be very worthwhile to undertake calculations with many more modes. This is now under its way thanks to the efforts of Nordlung, who will present some recent results in this colloquium, and to Marcus (1979). Marcus applies the modal expansions to a spherical geometry, and he just completed a first investigation of Boussinesq convection with as many as 15 modes. This enables him to calculate explicitly the motions on scales that are in the so-called inertial subrange, and which are fed only through the nonlinear interactions discussed earlier.

For this reason, his results should be much less sensitive to the approximation used to close the system.

A less ambitious calculation, involving only three modes in its final form, is undertaken by Latour, Toomre and Zahn to describe the solar convection zone. Inspired by the observed scales separation, the modes are intended to represent respectively the granulation, the supergranulation, and the large cells which seem necessary to render the whole zone nearly adiabatic, even though they are not observed yet with certitude.

The same are also exmaining the penetration of convective motions into an adjacent stable layer, in the Boussinesq limit. Preliminary results have been reported by Zahn (1977): modal calculations with one or two hexagonal planforms predict a penetration which is of the order of the thickness of the unstable layer. Laboratory experiments with water around 0°C lead to the same conclusion.

An important effort is also spent to refine the treatment of radiative transfer in the presence of motions, in the optically thin limit. Legait and Gough (1979) compare various approximations with the exact solutions of the transfer equation, in the case of an anisotropic radiation field; their main goal is to test the validity of the Eddington approximation which is commonly used.

This list is certainly incomplete, but it has the merit to demonstrate that the subject of stellar convection is alive. There is good hope, therefore, that new and perhaps decisive progress will be reported when we meet the next time to discuss turbulence in stellar atmospheres.

REFERENCES

Biermann, L. 1933, *Z. Astrophys.*, 5, 117

Depassier, M.C., Spiegel, E.A. 1979 (in preparation)

Durney, B., Spruit, H. 1979 (preprint)

Gough, D.O. 1969, *J. Atmosph. Sci.*, 26, 448

Gough, D.O., Moore, D.R., Spiegel, E.A., Weiss, N.O., *Astrophys. J.*, 206, 536

Gough, D.O. 1978, *Proc. EPS Workshop on Solar Rotation*, 87 (ed. Belevedere & Páterno, Catania Univ. Press)

Knobloch, E. 1978, *Astrophys. J.*, 225, 1050

Legait, A., Gough, D.O. 1979 (in preparation)

Maeder, A. 1975, *Astron. Astrophys.*, 40, 303

Marcus, Ph. 1979 (in preparation)

Massaguer, J., Zahn, J.-P. 1979, *Astron. Astrophys.*, (in press)

Nelson, G.D., Musman, S. 1977, *Astrophys. J.*, 214, 912

Nelson, G.D. 1978a, *Solar Phys.*, 60, 5

Nelson, G.D. 1978b, *Ph. D. Thesis*, Univ. of Washington

Öpik, E.J. 1950, *M. N. Roy. Astron. Soc.*, 110, 559

Shaviv, G., Salpeter, E.E. 1973, *Astrophys. J.*, 184, 191

Spiegel, E.A. 1965, *Astrophys. J.*, 141, 1068

Spiegel, E.A. 1971, *Ann. Rev. Astron. Astrophys.*, 9, 323

Toomre, J., Zahn, J.-P., Latour, J., Spiegel, E.A. 1976, *Astrophys. J.*, 207, 545

Toomre, J., Gough, D.O., Spiegel, E.A. 1977, *J. Fluid Mech.*, 79, 1

Toomre, J., Latour, J., Zahn, J.-P. 1977, *Bull. Americ. Astron. Soc.*, 9, 337

Unno, W., Spiegel, E.A. 1966, *Pub. Astron. Soc. Japan*, 18, 85

Vitense, E. 1953, *Z. Astrophys.*, 32, 135

Zahn, J.-P. 1977, *Problems of Stellar Convection*, 225 (ed. Spiegel & Zahn, Lecture Notes in Physics, Springer)

INSTABILITIES IN A POLYTROPIC ATMOSPHERE

H.M. Antia and S.M. Chitre
Tata Institute of Fundamental Research
Bombay, India

Abstract

The density in the outer layers of stars varies by several orders of magnitude and it is desirable to include the full effects of compressibility in any study of instabilities arising in stellar convection zones. In an unstable compressible fluid-layer that is thermally conducting both the oscillatory and non-oscillatory motions can simultaneously arise. The conditions under which the oscillatory acoustic modes can be overstabilized in a polytropic atmosphere are examined. It is argued that the linearized perturbation theory breaks down when applied to an inviscid complete polytrope which has vanishing density and temperature at the top for both optically-thin and optically-thick approximations. However, the linearized theory is demonstrated to be self-consistent when viscosity and thermal conductivity are included in the study of complete polytropes.

ON THE DYNAMICS OF THE SOLAR CONVECTION ZONE

Bernard R. Durney and Hendrik C. Spruit
National Center for Atmospheric Research
Boulder, CO 80303 U.S.A.

Abstract

We derive expressions for the turbulent viscosity and turbulent conductivity applicable to convection zones of rotating stars. We assume that the dimensions of the convective cells are known and derive a simple distribution function for the turbulent convective velocities under the influence of rotation. From this distribution function (which includes, in particular, the stabilizing effect of rotation on convection) we calculate in the mixing-length approximation: i) the turbulent Reynolds stress tensor and ii) the expression for the heat flux in terms of the superadiabatic gradient. The contributions of the turbulent convective motions to the mean momentum and energy equation are treated consistently, and assumptions about the turbulent viscosity and heat transport are replaced by assumptions about the turbulent flow itself. The free parameters in our formalism are the relative cell sizes and their dependence on depth and latitude.

THERMAL AND CONTINUUM DRIVEN CONVECTION IN B-STARS

George Driver Nelson
CODE CB, Johnson Space Center
Houston, Texas, 77058

Abstract

Two regions of convective instability are present in the photosphere of a typical B-star (T_{eff} = 30,000 K Log g = 4.0). One is the usual thermal instability caused by the helium ionization. The other is driven by the continuum radiation pressure in a thermally stable layer. Mixing length and anelastic modal representations of these unstable regions show that the rapid radiative cooling of temperature fluctuations limits the velocities to an amplitude of a few meters per second, much too small to account for the observed line broadening and asymmetries.

THE HEIGHT DEPENDENCE OF GRANULAR MOTION

Anastasios Nesis
Kiepenheuer Institut für Sonnenphysik
Freiburg i.Br., W. Germany

Abstract

Spectrograms of Mg I-absorptions lines have been registrated at different positions of the damping wings. The corresponding heights within the atmosphere have been approximated by the Eddington-Barbier approximation. We calculated the coherence between the intensity fluctuations in the continuum and those of the higher layers. We found a rather flat gradient up to a height of 100 km and above this a steep decrease of coherence. From this result we can conclude that above a height of 100 km any fluctuations of intensity are not due to ordinary convective processes. However, up to a height of 400 km we found a coherence between the velocity of granulation and its intensity fluctuations.

NUMERICAL SIMULATIONS OF THE SOLAR GRANULATION

Åke Nordlund
NORDITA, Blegdamsvej 17
Dk-2100, København Ø, Denmark

Abstract

Numerical simulations of the convective granular motions in the solar photosphere are presented. Realistic background physics allows a detailed comparison with observed characteristics of the solar granulation. The numerical methods are based on a bivariate Fourier representation in the horizontal plane, combined with a cubic spline representation in the vertical direction. Using a numerical grid with 16x16x16 grid points, which cover a unit cell of dimension ≃ 3600x3600x1500 km, granular motions have been followed over several turnover times. The simulated motion shows the characteristics of granular motion. The evolution of large granules into bright rings ("Exploding granules") is a consequence of the accumulating excess pressure at the granule center, necessary to support the horizontal velocities required by the continuity equation. The increasing pressure evenutally inhibits further upward motion at the granule center, which cools radiatively and shows up as a dark center in the expanding granule.

DIFFERENTIAL ROTATION IN STARS WITH CONVECTION ZONES

Peter A. Gilman
High Altitude Observatory
National Center for Atmospheric Research
Boulder, Colorado 80307/USA

I. Introduction

The topic I was originally assigned for this colloquium was "Generation of Non Thermal, Non Oscillatory Motions". Being basically a fluid dynamicist, at first I thought this meant I was supposed to talk about the origin of motions which are not thermally driven, i.e., I should not talk about convection. But then I realized all that was meant was that I was to talk about bulk fluid motions, rather than the molecular "thermal" motion of stellar gas that defines its temperature. Obviously the original question was posed by a stellar spectroscopist! Having surmounted that small semantic hurdle, I began to think about all the ways circulatory motions might be generated in a star. All manner of fluid dynamical instabilities come to mind--not only convective instability, but also barotropic or inertial, baroclinic, Kelvin-Helmholz, Rayleigh-Taylor, Goldreich-Shubert, Solberg-Hoiland, etc. The list is large, overlapping, I am sure confusing to an observer (and to many a theoretician). Then there are Eddington-Sweet currents, and several additional motions arising from the presence of magnetic fields--fields which give rise to magnetic buoyancy of flux tubes, and large collection of magnetohydrodynamic instabilities.

I could try to review all of these effects in this talk--but that has been done in several other places, for example in the recent book Rotating Stars by Tassoul, and a soon to be published book on astrophysical magnetohydrodynamics by Parker. The audience is better advised to look there. In addition, I am not convinced that an observer can gain much guidance on how to interpret his stellar spectra from extrapolations that predict the nonlinear consequences of these instabilities (with the exception of convection). With the limited time that I have, I would instead like to address a single question: "What differential rotation should we expect to find in a star that rotates and has a convection zone?"

Why this particular question? Aside from convection itself, differential rotation is probably the largest amplitude motion to be found in a typical stellar photosphere. The star's average rotation is most likely larger still (and easier to measure) but solid rotation in and of itself is fluid dynamically trivial; how it affects other motions to produce differential rotation is, I hope to convince you, very interesting.

Why consider only stars which have convection zones? I personally feel on safer ground for this case, but also I think more definite things can be said. Convection influenced by rotation can come to some sort of statistical equilibrium with the differential rotation it drives, and therefore the differential rotation can be expected to be present over a large fraction of the star's lifetime. By contrast, many of the other instabilities I mention may produce transients in the star, not seen again for a long time because the energy source for the instability is not replenished as fast as for convection (and maybe never). What we are after, then, are statements about differential rotation forced by convection, that might be applicable to a wide variety of stars. All stars rotate, and all stars that convect will have differential rotation. The real question is how much and what kind.

Some further statements are perhaps in order. Differential rotation is from a fluid dynamical point of view a very important quantity to know about a star. It is probably the key to guessing its overall global dynamics. It may tell us what kind of dynamo action the star is experiencing--whether it should be undergoing cycles in its magnetic field. For example, it would be extremely useful to know the rotation and differential rotation of stars for which Olin Wilson sees calcium emission cycles.

Let me acknowledge before I get any further that I know measurements of differential rotation in stars are extremely difficult to achieve, and probably impossible for several classes of stars by present techniques. Measuring rotation itself is bad enough. In the end, it may turn out that we have to be content with predictions of differential rotation for many stars, guided by other information such as rotation, spectral type, abundance, etc.

II. Approach to the Problem

Let me first list some of the physical parameters and processes we might expect to be important in determining the differential rotation at the surface of a convecting star. Then I will review some of the models that have been developed to explain differential rotation in the sun, including some of the problems as well as successes. I have worked on models of this type myself, and I will summarize some of my own results which I think are relevant to the question I have posed. From these results I will then extrapolate to some qualitative statements about other stars--main sequence stars mostly, but red giants will also be considered.

III. Important Parameters and Processes

A. Convection Zone Parameters

Probably the most important single parameter determining what kind of differential rotation to be expected in a stellar convection zone is the ratio of turnover time for the convection, to rotation time for the star. This determines how much rotation affects convection, introducing preference for certain convection patterns, as well as correlations between different velocity components implying nonlinear Reynolds stresses. The larger is this ratio, the more influence rotation has, because as the fluid element moves, coriolis forces have more time to act on it. We will ignore systems with rotation so rapid that the departures from spherical geometry become of order unity.

In addition, certain other parameters are important, for example, convection zone depth, as a fraction of the radius. It is reasonable to expect that stars with shallow convection zones will have different patterns of differential rotation than stars with deep zones. The scale of the dominant convective modes will certainly be larger for deeper zones, and in general, the rotational influence will be stronger, because the turnover time is longer. The particular form the stratification of the zone takes will also be important--how may scale heights it contains as well as how superadiabatic the temperature gradient is. These quantities clearly relate back to the turnover time estimates, as does the total heat flux being carried through the convection zone.

B. Physical Processes Affecting Angular Momentum Distribution

Any process that can redistribute angular momentum in a convection zone is potentially important for determining the differential rotation profile. Since angular momentum can be convected by the fluid itself, any circulation that has a component in either the radial or latitudinal direction, can change the rotation profile in the meridian plane. Conventionally, these circulations are broken into two types, axisymmetric meridional circulation, and global departures therefrom, usually called eddies. The eddies produce net momentum transport when the east-west velocity in them is correlated with either the north-south flow, the radial flow, or both. This correlation is induced by the presence of the Coriolis force and is called a Reynolds stress. Molecular and small scale turbulent diffusion is swamped by the turbulence. The distinction between small scale turbulent diffusion and global eddy transport is an arbitrary one, the former usually representing transport of momentum by those motion scales not resolved in the particular model, the latter being explicitly calculated. Other azimuthal torques, such as produced by electromagnetic body forces, may also be present, but their effect has not generally been taken into account. Buoyancy forces, since they act in the radial direction, do not directly contribute azimuthal torques that can change the differential rotation, but of course they drive motions which transport momentum. Similarly, azimuthal pressure torques average out when integrated around a complete latitude circle.

Figure 1 illustrates angular momentum transport by different circulations. At the top, two schematic meridional circulations are shown, of opposite sense. If either pattern started up in a fluid originally in solid rotation and no other transport processes were acting, high latitudes and deeper layers would tend to speed up, relative to low latitudes and shallow layers, because the circulation would conserve angular momentum. The fluid would move toward a state of constant angular momentum. But now if sufficient diffusion is also present to link different layers and keep the angular velocity more nearly constant, the circulation on the left may produce equatorial acceleration, because fluid moving toward the equator on the outer branch, for example, crossing the dashed line, would contain more angular momentum than fluid moving toward the pole underneath, where the moment arm is shorter.

Competing Mechanisms of Angular Momentum Transport

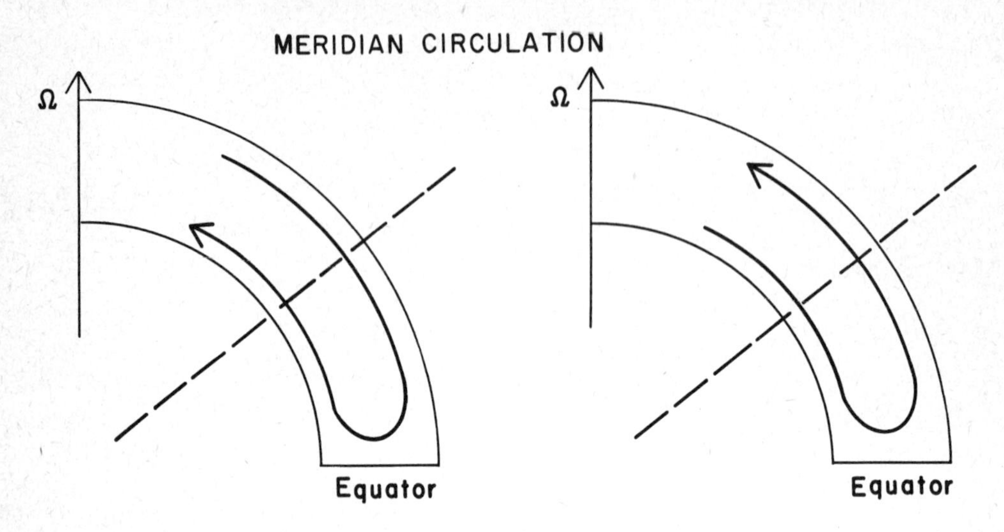

REYNOLDS STRESSES

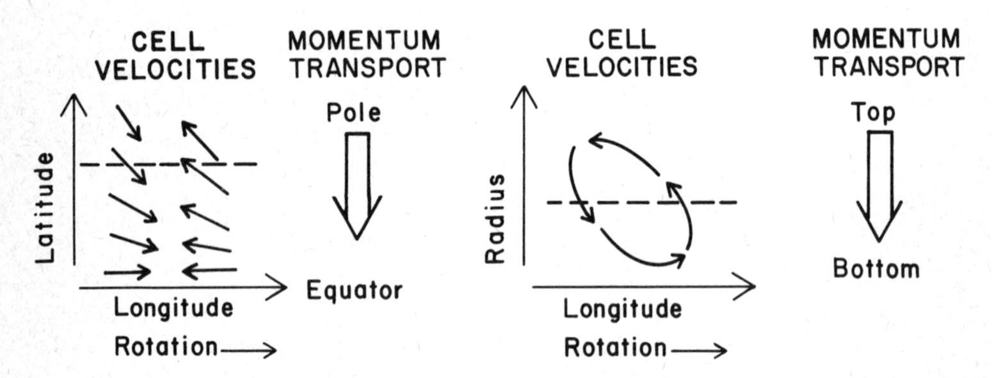

DIFFUSION

In simplest (isotropic) form, tends to equalize angular velocities. May couple with meridian circulation to produce equatorial acceleration.

Figure 1. Angular momentum transport by circulation.

On the other hand, Reynolds stresses will transport momentum toward the equator to drive an equatorial acceleration if the convection patterns adopt a horizontal flow pattern as seen in the left middle schematic. Fluid particles moving eastward in the direction of rotation have a component toward the equator; particles moving west are also moving poleward. If we average along the dashed line, we get a net momentum flux toward the equator, even if there is no net mass flux. In the right middle schematic, we see a pattern of circulation in longitude and radius which leads to a net flux of momentum inwards. The two patterns I have shown are in fact the most common ones seen in actual model calculations.

IV. Models Used in Solar Case

Which angular momentum transport process dominates in producing differential rotation can only be determined by actual nonlinear model calculation. A number of such calculations have been carried out, mostly to explain solar equatorial acceleration.

One group of these models has been axisymmetric, in which all effects of convection are condensed into a few ad hoc parameters. In one such model, an anisotropic turbulent viscosity is assumed, which results in a meridian circulation qualitatively like that at the top left in Figure 1. With suitable choice of the sign and degree of anisotropy, the observed equatorial acceleration can be reproduced. This approach originated with Biermann (1951) and was exploited by Kippenhahn (1963), Cocke (1967) and in greater detail by Kohler (1970). The problem with it is that there is no real way to choose the sign and magnitude of anisotropy, without resort to fitting the correct differential rotation. Also, no account is taken of the influence of rotation upon the parameterization of the convection. Another approach which has been carried further is to assume the convective heat flux is weakly perturbed by rotation, such that it becomes a function of latitude. This was done by Durney and Roxburgh (1971) and later developed further by Belvedere and Paterno (1976, 1977, 1978). Again the meridian circulation and convective parameterization are fitted to the observed differential rotation. The calculations are heavily dependent on the assumption of weak influence of rotation upon convection, which is not likely to be valid in the deep part of the solar convection zone. Also, there is no independent evidence that the heat flux parameterization is valid even under the assumptions made.

As you probably have already guessed, I do not believe these models are the correct ones for the sun, mostly because I suspect the parameterizations upon which they depend are grossly inaccurate. I prefer models in which the dynamics responsible for giving the correct differential rotation are explicitly calculated, with parameterizations of unresolved motions relegated to a less critical role, so their detailed form is less important. Such models have been developed, so far for physics considerably simpler than the real sun, but nevertheless instructive. In particular, models have been developed for nonaxisymmetric convection of a stratified liquid in a rotating spherical shell. Early, mostly linear analyses by Busse (1970, 1973), Durney (1970, 1971) Kato and Yoshimura (1971) and Gilman (1972, 1975) demonstrated the preference for convective modes which transport momentum toward the equator, via the Reynolds stress mechanism described above. More recent, nonlinear calculations by myself (Gilman, 1976, 1977, 1978, 1979) have exploited this fact to determine in detail when equatorial acceleration occurs and with what amplitude relative to the convection which drives it and the basic rotation rate.

I will summarize some results obtained from those calculations. I will give only the most basic, qualitative effects, since the model is incompressible, and many details will change in the compressible case--which we are currently coding. There are other weaknesses of the model, such as larger than observed differentials in heat flux, which I cannot describe in detail, but which we expect to be improved in the compressible case.

V. Summary of Relevant Results from Calculations of Convection in a Rotating Spherical Shell

A. Finite amplitude equatorial acceleration is produced only when the influence of rotation upon convection is strong, i.e., the rotation time is less than the turnover time for convection. Under these circumstances, the angular velocity also decreases inward with depth; when the rotational constraint is very strong, the angular velocity predicted is nearly constant on cylinders concentric with the axis of rotation. If the rotation time is a few orders of magnitude shorter than the turnover time, then the latitudinal profile of differential rotation may be more complicated, but this is an unlikely case for stellar application.

B. When the convection zone is deep, say 1/3 of the radius or more, the equatorial acceleration profile with latitude is broad, with essentially monotonic decrease in angular velocity to the poles.

C. When the convection zone is shallow, say 20 percent or less, and the rotational influence is strong, the angular velocity reaches a minimum in mid latitudes, and then increases again toward the poles. The width of the equatorial acceleration is determined by the depth of the layer.

D. For both deep and shallow convection zones, with weaker rotational influence (increased convective velocities) the profile switches from equatorial acceleration to deceleration. Angular velocity now increases with depth. There is an intermediate stage in which the angular velocity is highest in mid latitudes and lower near the equator and near the pole, while still decreasing with depth.

E. The maximum differential rotation sustainable by the convection is about 40 percent of the average rotation. For larger values, the feedback from the shear on the convection is strong enough to change the dominant patterns and consequently the differential rotation profile, resulting in a new equilibrium with lower amplitude differential rotation.

F. The maximum differential rotation maintainable by the convection has about the same kinetic energy in it as the convection itself. Larger differential rotation may be possible locally in the compressible case.

G. Amplitudes of individual convective modes may change radically with time while the differential rotation amplitude changes hardly at all.

The above statements concerning the profile of differential rotation for deep and shallow convection zones are summarized schematically in Figures 2 and 3 respectively. The extra bumps in the profiles for the shallow layer are intended to indicate that for a shallow layer, the convection, being smaller in horizontal scale itself, should produce more fine structure in the differential rotation profile.

In addition, we can say that first results from linear calculations of compressible convection in a rotating spherical shell indicate that moderate stratification (say a factor of 20 drop in density across the convection zone) will not change any of these results significantly. Larger density variations will produce larger changes, but the qualitative statements we have made with respect to differential rotation with latitude we do not expect to change a great deal. But clearly such conclusions must be tested with a compressible model.

VI. Extrapolation to the Convection Zone of a Star

How do we extrapolate the above results to the convection zone of a star, which, after all, is generally a long ways from being a Boussinesq liquid? In the case of the sun, the theory is generally thought to apply to the deepest layers of the convection zone, which are more nearly Boussinesq, for which the natural scale for convection is nearly global. As Gilman and Foukal (1979) have argued, the small scale but large amplitude granules and super granules are most likely not responsible for the latitude gradient of rotation at all, but perhaps the radial gradients in the shallowest layers. In addition, these small scale motions appear to obscure the global convection or "giant cells" (Simon & Weiss, 1968) as they are sometimes called.

A similar situation probably occurs in convection zones of other stars, as summarized in Figure 4. The little cells at the top are weakly influenced by rotation, and by themselves would produce a weak equatorial deceleration. But they sit on top of slower, larger scale convection below, which is much more strongly influenced by rotation, where the latitude gradient is really produced. This gradient of deep origin is then transmitted to the surface by the small cells.

One conclusion that can be drawn from this argument is that using surface turbulent velocities, to estimate the turnover time for the convection zone would lead to a gross under-estimate, therefore too small a rotational influence and the wrong differential rotation. A better procedure, still very approximate, would be to estimate the turnover time from a mixing length model applied to the whole convection zone. This brings us to a prescription for estimating the differential rotation to be expected in certain main-sequence stars.

What we have done is to estimate convection zone depths and turnover times near the bottom of the zone, using a stellar envelope code, and then compare these with observed or estimated rotation times calculated from the well known summary of stellar rotation, by Kraft (1967). Part of this procedure is similar to one employed by Durney and Latour (1978) for another purpose, which calculation I discovered as I was beginning these calculations. In fact, I used Latour's envelope code, which in turn is similar to an earlier one applied to estimate convection zone depths by Baker (unpublished). I chose essentially Baker's values for a sequence of stars from A to early K, the luminosity, radius, and effective temperature for which are shown in Figure 5. We then used these values, together with a composition of X = .70, Y = .27, Z = .03 in Latour's code to estimate convection zone depth for these stars, shown in Figure 6, given as a function of α, the ratio of assumed mixing length to pressure scale height. The results are quite similar to those of Baker, though not identical. As expected, convection zone depth shrinks as a fraction of stellar radius as the mass and luminosity increase. Depending on the α chosen, the convection zone effectively shrinks to zero somewhere between early F and late A. Larger α implies deeper convection zone.

Then, following Durney and Latour (1978) we find a convective velocity in meters/second one pressure scale height up from the bottom, shown in Figure 7, which we use to compute a turnover time = scale height/velocity, shown in Figure 8. Note that it goes from 1-2 x 10^6 sec. for G and late F stars, down quickly to 10^4 sec. and less by F0.

But early stars generally rotate much faster too, so the rotation time drops rapidly with increased mass, as seen in Figure 9, derived from Kraft (1967). (Here, for stars later than the sun, we assume an upper limit of the same angular velocity as the sun.) So how does the ratio change? This is plotted in Figure 10. We see that for α = 1.0, the sun is already in the steep part of the curve, with convection in slightly later stars much more strongly influenced by rotation, convection in slightly earlier stars much less so. For larger α, earlier stars out to F0-F2 are as strongly influenced, or more so, by rotation than the sun is. Which α is most reasonable (if any)? For the sun, α = 1 yields by my earlier arguments too shallow a convection zone (< 20%) to sustain an equatorial acceleration with angular velocity decreasing all the way to the pole, so I favor a larger α (at least 2). Suppose we take α = 2. Then from all our results we should expect broad equatorial acceleration on stars later than the sun, (unless they have much lower rotation rates than the sun) and also for slightly earlier stars, with an increasingly narrow equatorial acceleration out to log $M/M_0 \approx .12$. Maximum amplitude in this equatorial acceleration would be about 40% of

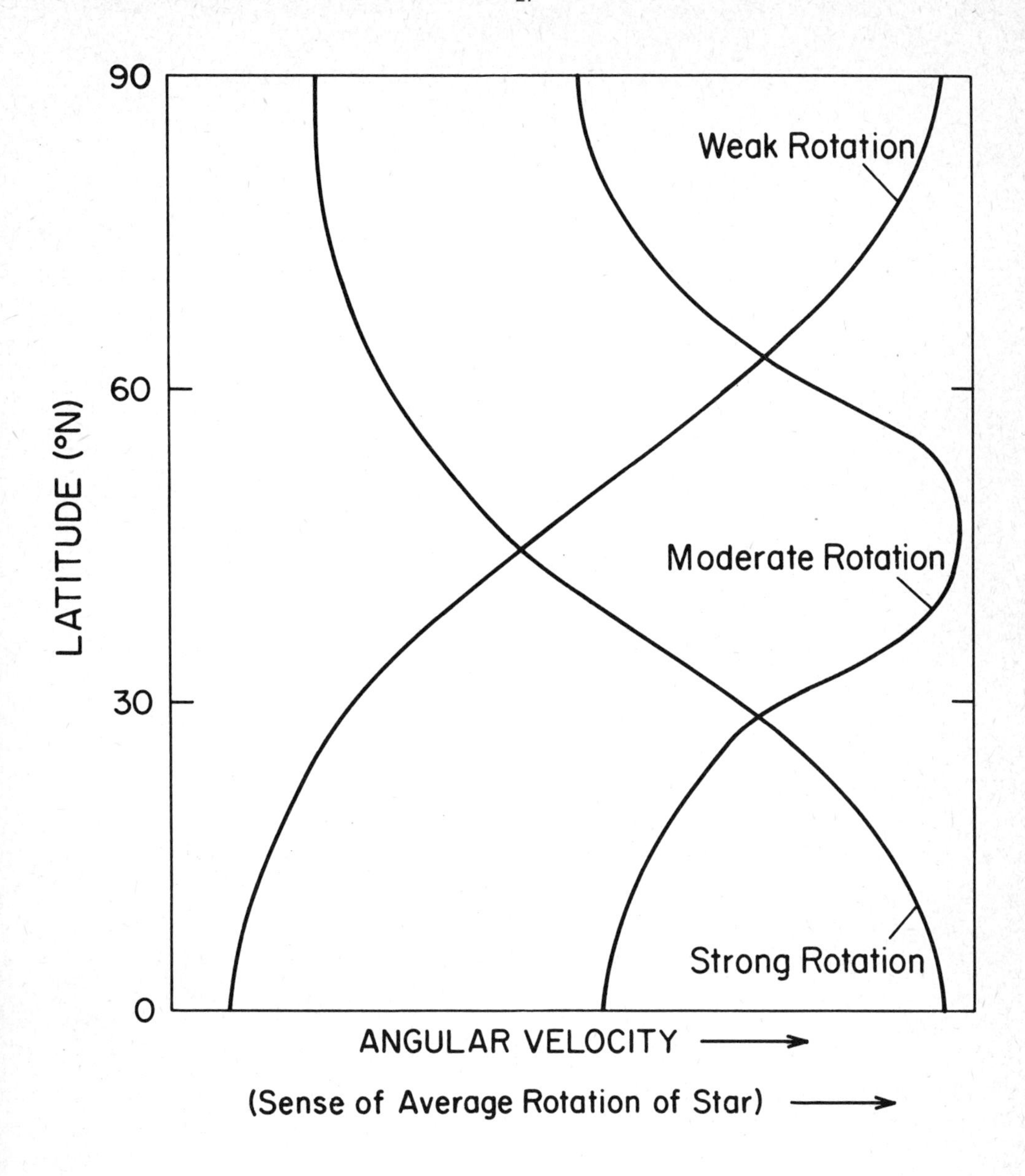

Figure 2. Qualitative schematic of differential rotation regimes for a deep convection zone.

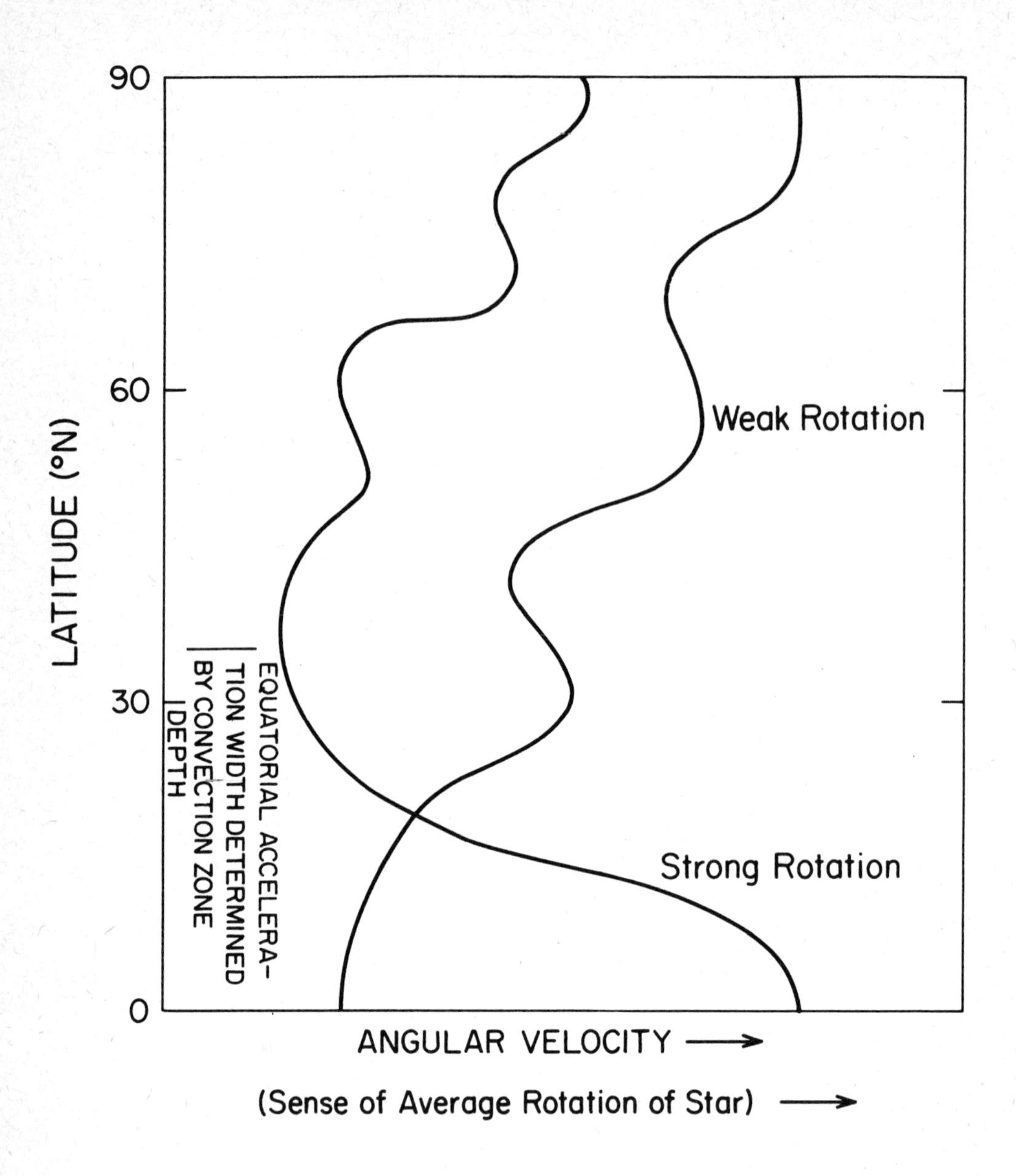

Figure 3. Qualitative schematic of differential rotation regimes for a shallow convection zone.

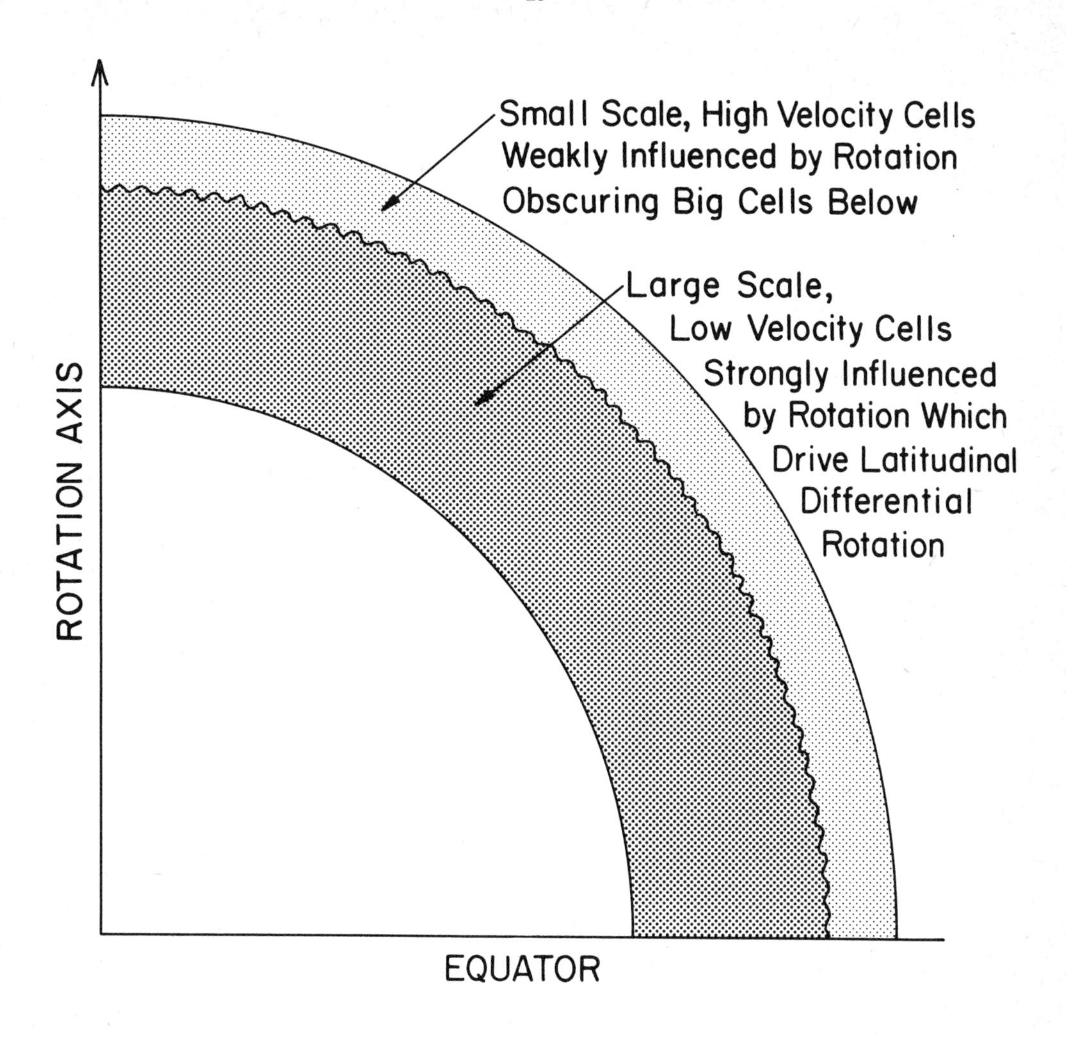

Figure 4. Schematic of a stellar convection zone.

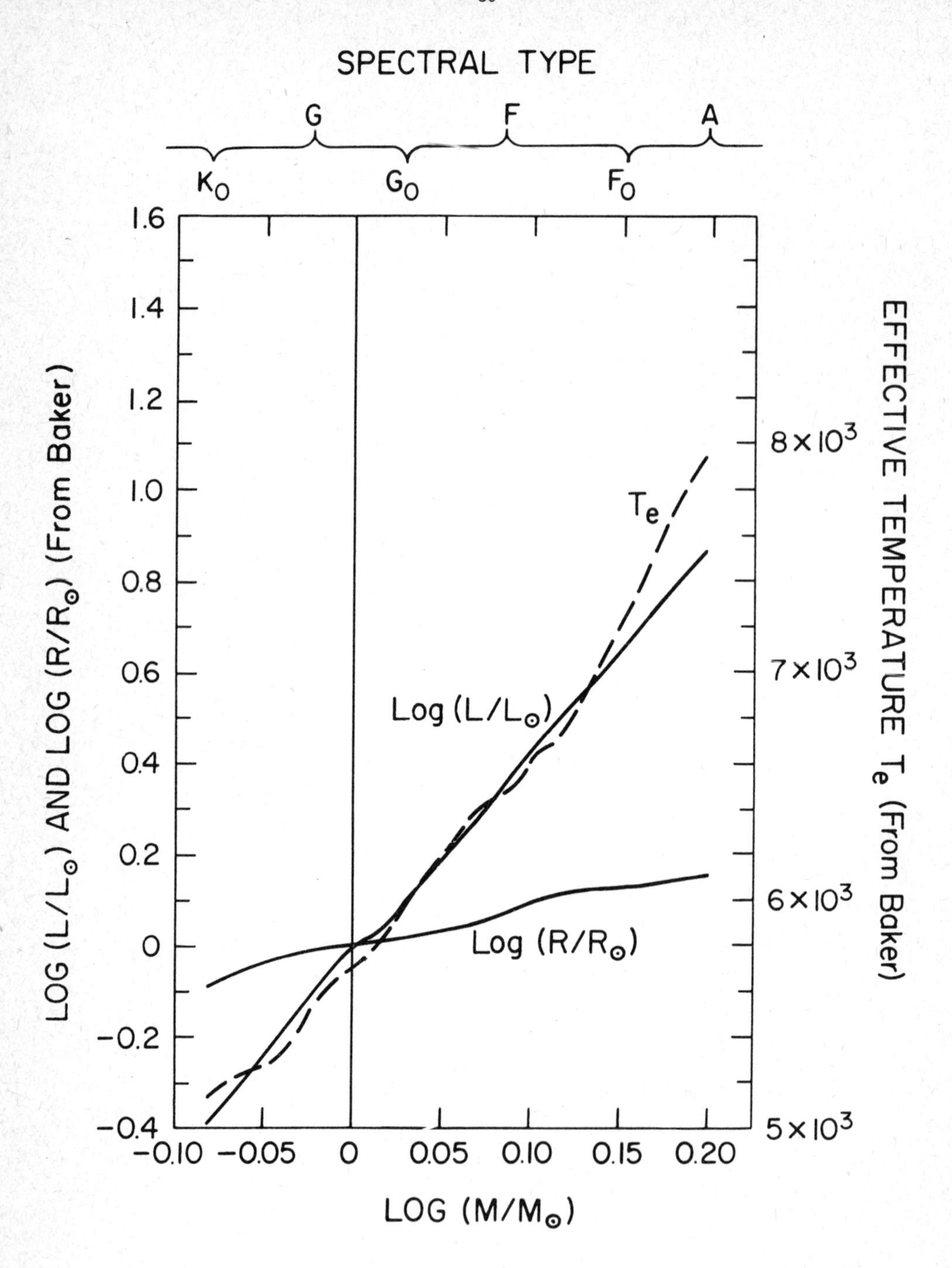

Figure 5. Relations between luminosity, radius, and effective temperature.

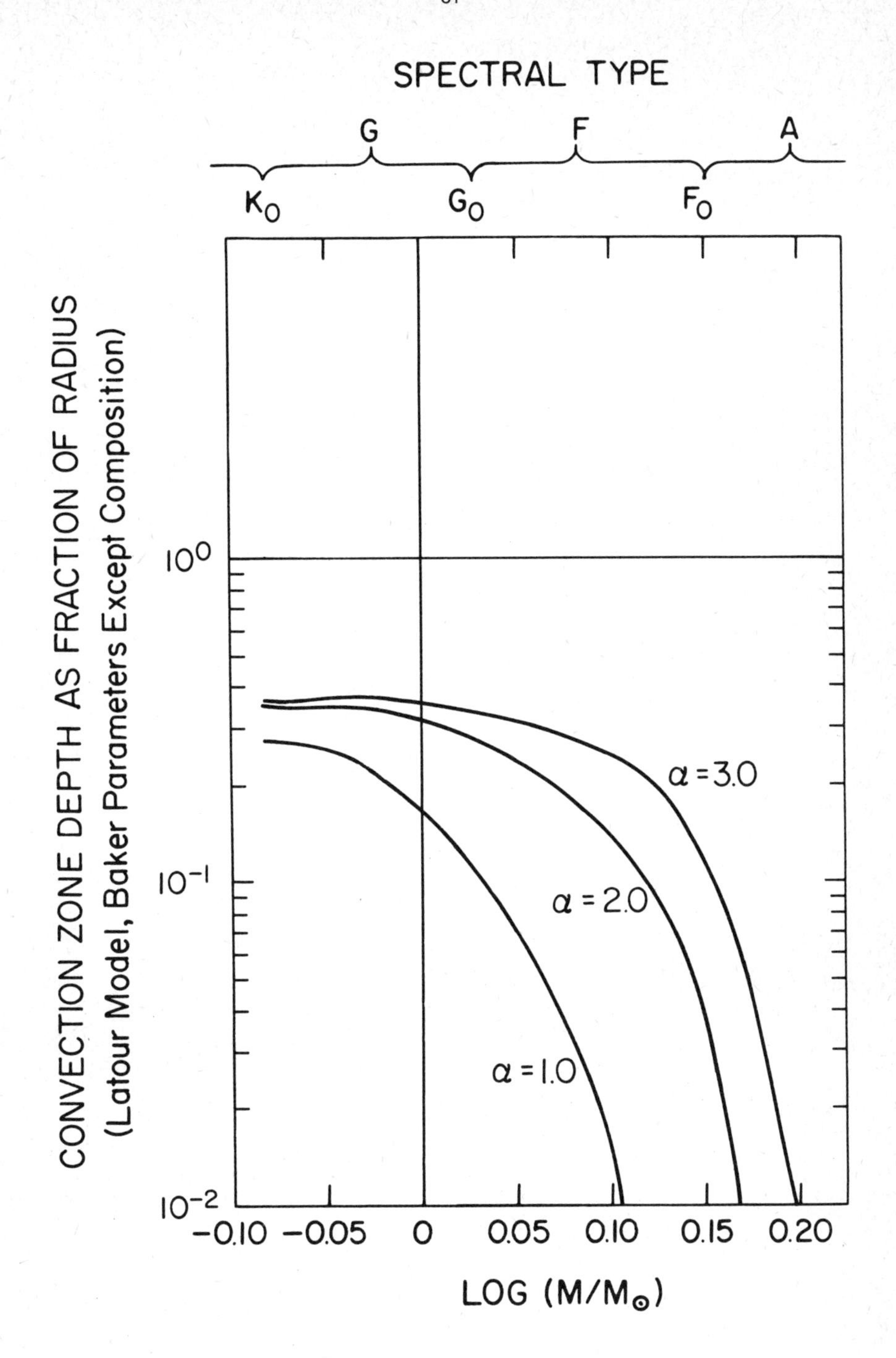

Figure 6. Convection zone depths.

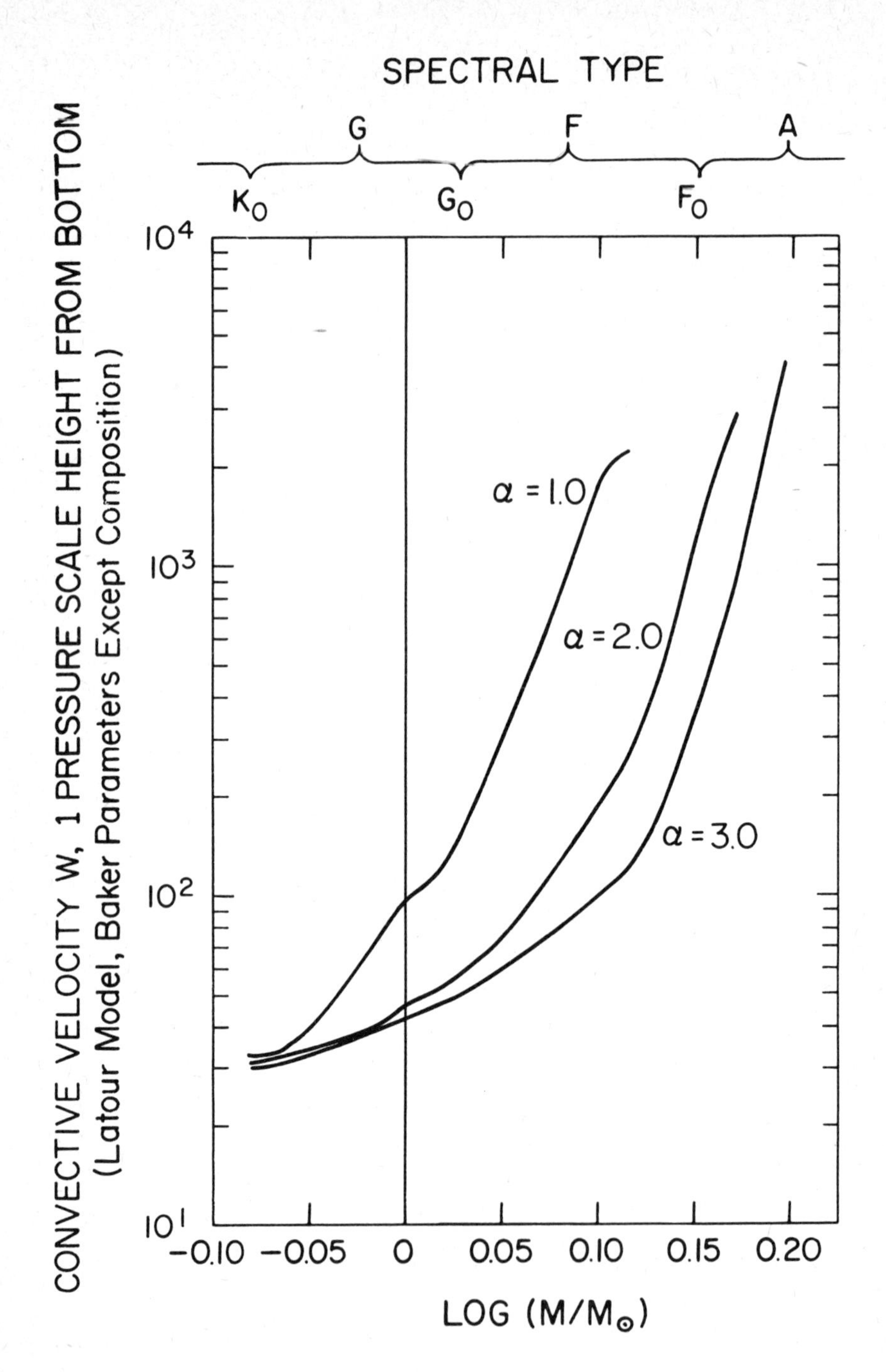

Figure 7. Convective velocities.

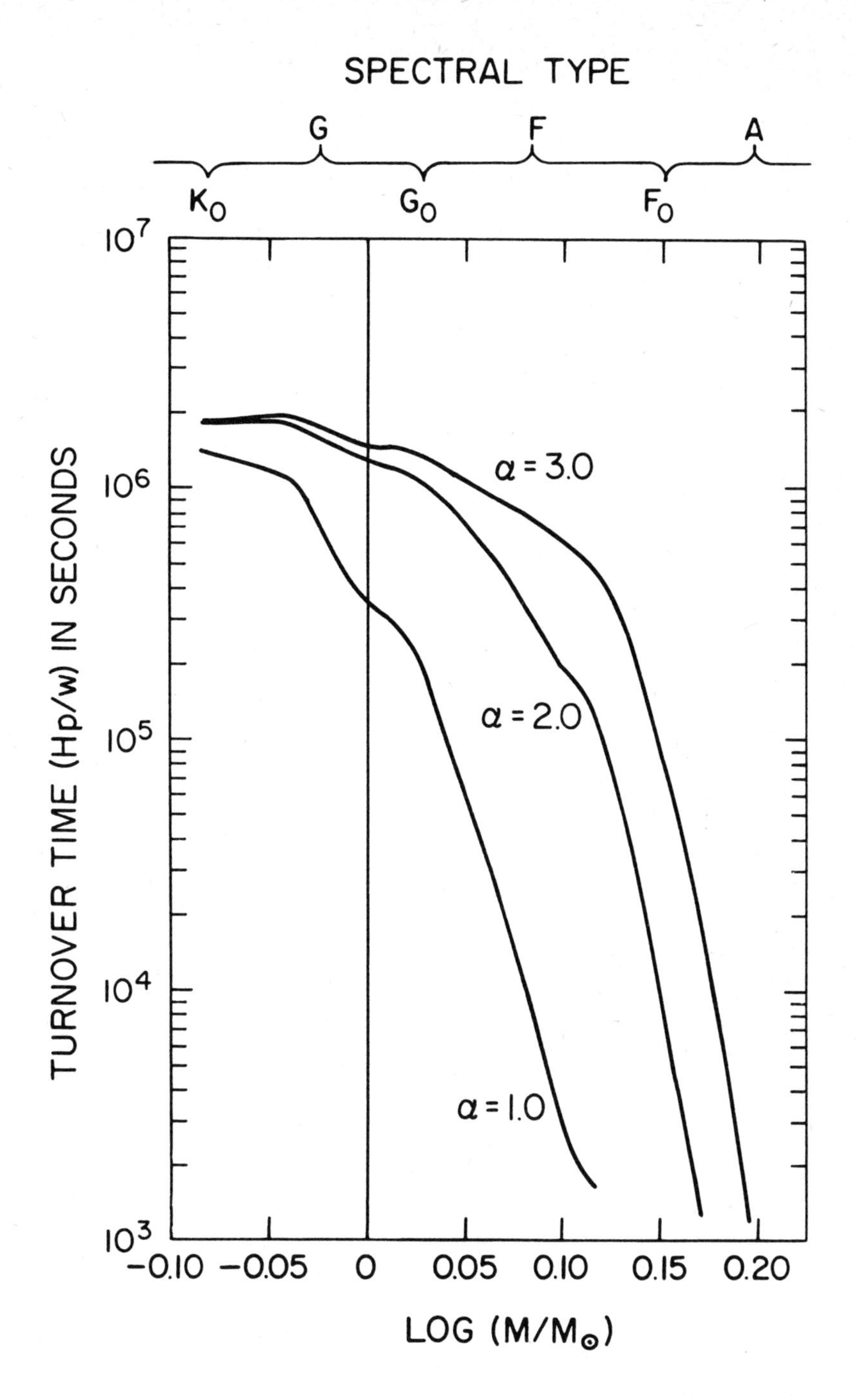

Figure 8. Convective turnover times.

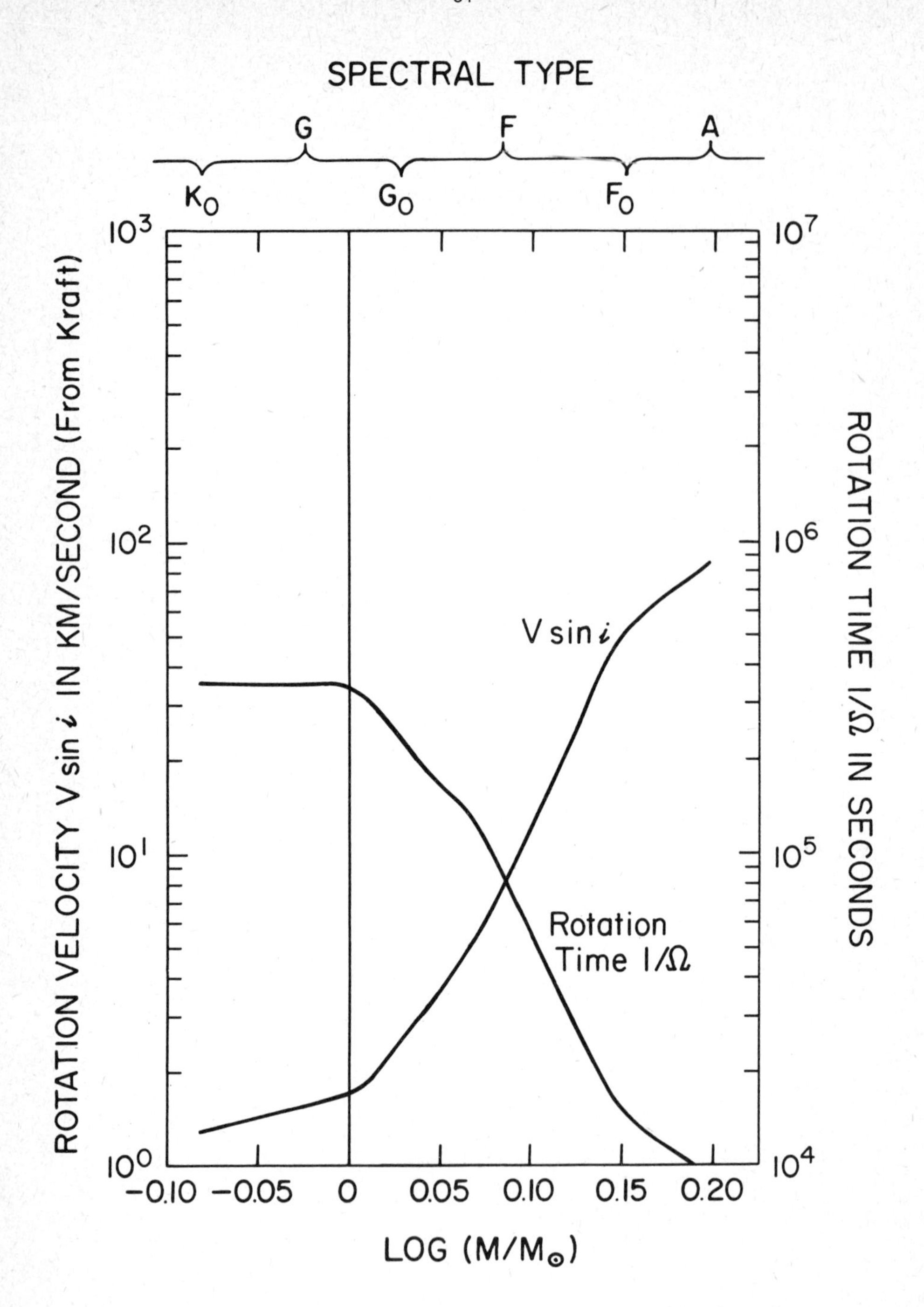

Figure 9. Rotation velocity and rotation time as a function of mass.

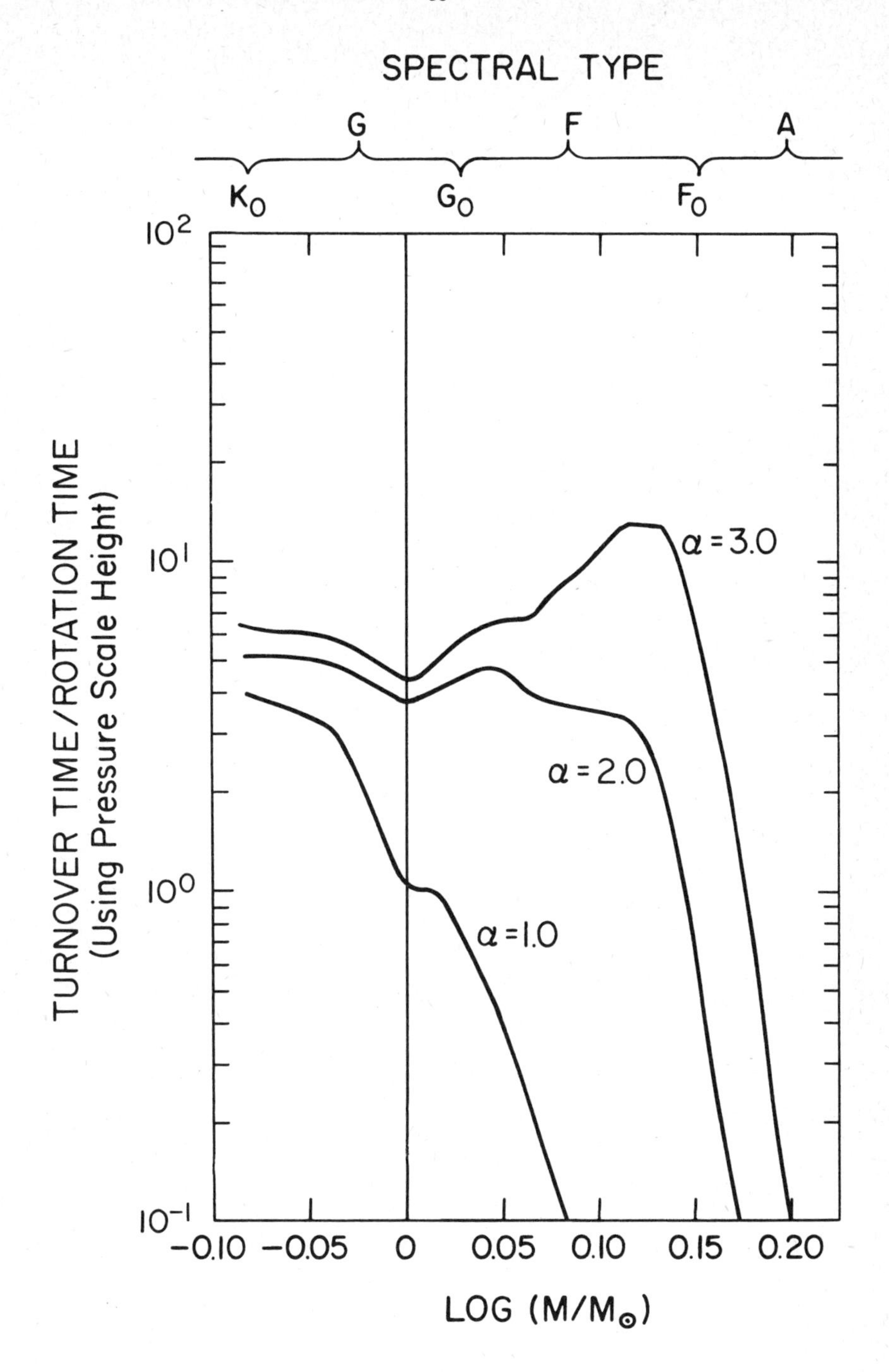

Figure 10. The ratio of turnover time to rotation time as a function of mass.

the average rotation rate. Beyond that, there would be a rapid transition to equatorial deceleration, with a lot of structure in the profile, due to the dominance of small scale modes in the shallow zones there. The conclusion for $\alpha = 3$ would be similar.

It would seem that rather analogous arguments could be made for red giants. Those with deep convection zones should have either broad equatorial accelerations or decelerations, depending on their rotation rate. Perhaps the calculations from which these inferences are made apply better here than for main sequence stars, since in most red giants the scale height is a large fraction of the convection zone depth so the convection is more nearly like that in a stratified liquid.

In closing, let me remark again that if one knows the rotation rate of the star, and a few other basic properties, one may be able to estimate its differential rotation even if one cannot observe it. For stars with surface emission features, from which an independent estimate of rotation rate, and perhaps even differential rotation, can be made, we may be able to test our conclusions in detail.

I close with a question for the observers: Can you tell the difference by spectral measurements between a star in solid rotation, and one with nearly the same surface average angular velocity that in fact is rotating differentially?

Acknowledgments. I thank Jean Latour for making available to me his stellar envelope code, and my student, Gary Glatzmaier, for running a number of calculations with this code for me. Dimitri Mihalas was kind enough to read the manuscript for me. Previous discussions with Douglas Gough on this topic have been useful.

The National Center for Atmospheric Research is sponsored by the National Science Foundation.

References

Belvedere, G. and Paterno, L., 1976: Solar Phys. 47, 525.

Belvedere, G. and Paterno, L., 1977: Solar Phys. 54, 289.

Belvedere, G. and Paterno, L., 1978: Solar Phys. 60, 203.

Biermann, L., 1951: Z. Astrophys. 28, 304.

Busse, F., 1970: Astrophys. J. 159, 629.

Busse, F., 1973: Astron. Astrophys. 28, 27.

Cocke, W.J., 1967: Astrophys. J. 150, 1041.

Durney, B.R., 1970: Astrophys. J. 161, 1115.

Durney, B.R., 1971: Astrophys. J. 163, 353.

Durney, B.R. and Latour, J., 1978: Geophys. Astrophys. Fluid Dyn. 9, 241.

Durney, B.R. and Roxburgh, I.W., 1971: Solar Phys. 16, 3.

Gilman, P.A., 1972: Solar Phys. 27, 3.

Gilman, P.A., 1975: J. Atmos. Sci. 32, 1331.

Gilman, P.A., 1976: in IAU Symposium #71, Basic Mechanisms of Solar Activity, ed. V. Bumba and J. Kleczek (Dordrecht: Reidel), p. 207.

Gilman, P.A., 1977: Geophys. Astrophys. Fluid Dyn. 8, 93.

Gilman, P.A., 1978: Geophys. Astrophys. Fluid Dyn. 11, 157.

Gilman, P.A., 1979: Astrophys. J. (in press).

Gilman, P.A. and Foukal, P., 1979: Astrophys. J. 229, 1179.

Kippenhahn, R., 1963: Astrophys. J. 137, 664.

Kohler, H., 1970: Solar Phys. 13, 3.

Kraft, R.P., 1967: Astrophys. J. 150, 551.

Simon, G.W. and Weiss, N.O., 1968: Z. Astrophys. 69, 435.

Yoshimura, H. and Kato, S., 1971: Publ. Astron. Soc. Japan 23, 57.

GENERATION OF OSCILLATORY MOTIONS
IN THE STELLAR ATMOSPHERE

Yoji Osaki
Department of Astronomy,
University of Tokyo,
Tokyo, Japan

Abstract

Thermal overstability of non-radial eigenmodes of stars is discussed as one of possible causes for generating non-thermal motions in the stellar atmosphere. The nature of oscillatory motions in stars is first considered both in the local and the global stand points. Then, the excitation of eigen-oscillations is discussed and results of numerical studies so far made are reviewed for the vibrational stability of various stellar models against non-radial oscillations. It is found that many of non-radial p-modes of high tesseral harmonics are likely excited in various stars of the HR diagram and that they possibly manifest themselves as non-thermal velocity fields in the stellar atmosphere.

1. Introduction

Two kinds of phenomena are known to indicate the existence of non-thermal motions in stellar atmospheres. One of them is the spectral line broadening, asymmetry, and shift indicating the existence of velocity fields, and the other is the mechanical source of energy that is required to heat the stellar chromosphere and corona. One of the most important questions to be addressed in this colloquium will be, what kinds of motions are involved in these phenomena? What mechanisms (or instabilities) can generate them? It is generally thought that thermal convection is one of the most likely sources for generating these motions in stellar atmospheres. However, it will be evident that all of non-thermal motions are not necessarily generated by thermal convection because various activities such as evidenced by micro- and macro-turbulences and by X-ray emissions due to the hot corona are observed also in early-type stars where any appreciable surface convection zones are not expected. Thus, some other instabilities that can generate motions in the stellar atmosphere have to be investigated as well. In this paper, I will discuss the thermal overstability of trapped waves in stars as one of possible mechanisms. This mechanism can generate wave motions of some finite amplitude from perturbations of infinitesimally small thermal fluctuation.

The thermal overstability is the basic mechanism for generating pulsation motions in variable stars such as Cepheid and RR Lyrae. However, modes of oscillations with which we are concerned here are different from ordinary radial pulsation, but they are those of non-radial type. The existence of non-radial oscillations as stellar eigenmodes have been known theoretically for a long while, but their importance in relation with observational phenomena in stars seems to have been recognized only recently. The five minute oscillation in the solar atmopshere is now understood as global non-radial p-modes of the sun (Deubner 1975, 1977). Although velocity fields due to thermal convection, which is manifested as the granulation in the solar atmosphere, are dominant at the photospheric level, the five minute oscillation as the velocity fields becomes increasingly important with height, and the latter dominates in the upper photosphere and the chromosphere. Non-radial oscillations have been inferred as the possible modes of observed motions in stars such as line-profile variable stars (Smith 1977) and early type supergiants (Lucy 1976a, b). Lucy (1976) has shown that the semiregular variability in radial velocity observed in the A-type supergiant star α Cygni may be due to the simultaneous excitation of many discrete pulsation modes of non-radial type and the "macroturbulence" required for large line-broadening in this star may be explained by the superposition of these oscillation modes.

One might think that two pictures of "turbulence" and "eigenmodes" are seemingly contradicting since turbulence is essentially a stochastic phenomenon with energy cascading from larger eddies to small eddies while eigenmodes are discrete in frequency and wavenumber and are coherent in time and space. However, they do not necessarily contradict each other since "turbulence" used in stellar spectroscopy is not turbulence as understood in aerodynamics and it simply means non-thermal motions responsible for spectral line broadening. On the other hand, non-radial eigenmodes are rich in physical properties and have a very dense spectrum although they are still discrete. Thus, if extremely large numbers of eigenmodes are excited in a star and are superposed, they show up spectroscopically as unresolved velocity fields. One may visualize the relation between turbulent convection and eigenmodes of the star in the power spectrum of velocity fields against the wavenumber k. The turbulent convection occupies a high wavenumber domain with $k>1/H$, while the non-radial eigenmode oscillation occupies a low wavenumber domain with $1/H \lesssim k \lesssim 1/R$, where H and R denote the scale height of the atmosphere and the star's radius, respectively. Thus, observationally, turbulent convection is more related to "microturbulence" due to its small scale and eigenmode oscillations are to "macroturbulence".

In this paper, I will argue that the non-radial p-mode oscillations of high tesseral harmonics are very likely to be excited in various stars in the HR diagram and they might possibly be responsible for the non-thermal motions in some of these stars. As far as I am aware, Christy (1962) was the first to propose in 1962 that non-spherical pulsation-type motions are present in late-type stars and they are important for the hydrodynamics in the atmospheres of these stars. However, this suggestion seems not to have been taken seriously for a long while, and it is only in recent years that numerical stability analyses for non-radial oscillations have been carried out in a number of stars. We shall discuss the nature of oscillations in stars from the local and the global points of view in sections 2 and 3. Excitation mechanisms of oscillations and results of numerical calculations will be reviewed in section 4.

2. Waves in the Stellar Atmosphere

In this section, we discuss wave motions in the stellar atmosphere from the local point of view. Wave motions in gravitationally stratified fluids have extensively been studied in geophysics (see, e.g., Eckart 1960). It is well known that two kinds of forces act on fluid elements: pressure force due to compressibility and buoyancy force due to gravitational stratification. If the medium is stably stratified (i.e., convectively stable), both of them are restoring forces and they give rise to oscillation of a fluid element if it is displaced from static state. Corresponding to this, two kinds of waves, i.e., acoustic waves and gravity waves, occur in the stellar atmosphere.

We now consider the wave propagation in a plane-parallel isothermal atmosphere under a constant gravitational field, in which the pressure p_0 and density ρ_0 of static state are known to vary with height z as

$$p_0,\ \rho_0 \propto \exp\,(-z/H) \tag{1}$$

where $H=p_0/(\rho_0 g)=\mathcal{R}T_0/(\mu g)=\mathrm{const}$ is the scale height, and g the gravitational acceleration, T_0 the temperature, μ the mean molecular weight, $\mathcal{R}$ the gas constant. In this case, the linearized system of equations in hydrodynamics for adiabatic perturbation allows a simple solution of plane waves for velocity $\underset{\sim}{v}$, the pressure variation p', and the density variation ρ' of the form

$$\underset{\sim}{v},\ \frac{p'}{p_0},\ \frac{\rho'}{\rho_0} \propto \exp\left(\frac{z}{2H}\right)\ \exp[i(-\omega t+k_x x+k_y y+k_z z)] \quad . \tag{2}$$

Here the exponentially growing factor with height z in equation (2) arises so as to conserve wave energy $\rho_0 v^2$ in the vertical direction since the density ρ_0 in the atmosphere decreases exponentially with height.

Substitution of these expressions into equations of hydrodynamics yields a dispersion relation between frequency ω and wavenumber k, which is written as

$$k_z^2 = f(\omega, k_h)$$

$$= \frac{1}{c^2}(\omega^2 - \omega_{ac}^2) + k_h^2 \left(\frac{N^2}{\omega^2} - 1\right) \quad , \tag{3}$$

where $k_h = \sqrt{k_x^2 + k_y^2}$ is the horizontal wave number. Here quantities ω_{ac} and N, having a dimension of frequency, are called the acoustic cut-off frequency and the Brunt-Väisälä frequency, respectively, and they are given by

$$\omega_{ac} = \frac{c}{2H} = \frac{\gamma g}{2c} \quad , \tag{4}$$

and

$$N = \frac{g}{c}\sqrt{\gamma - 1} \quad , \tag{5}$$

where c is the sound velocity, and γ is the ratio of the specific heats of gas.

If ω and k_h are given, equation (3) determine k_z^2. If $k_z^2 > 0$, waves can propagate vertically. On the other hand, if $k_z^2 < 0$, no waves can propagate, and the energy density of perturbations decreases exponentially with height (evanescent waves), if perturbations are coming from below. This situation is most conveniently shown in the diagnostic (k_h, ω)-diagram, which is illustrated schematically in Figure 1.

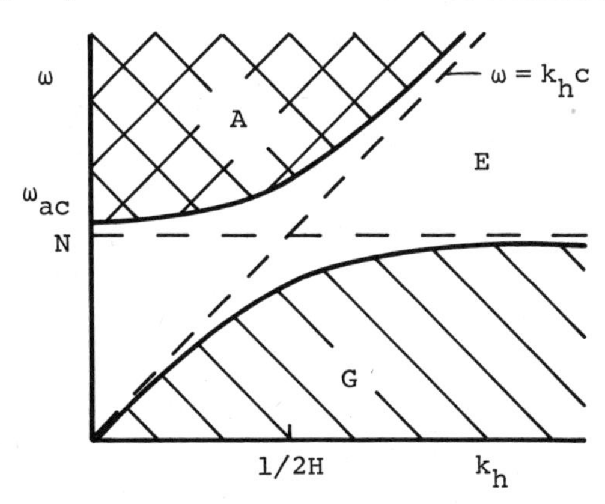

Figure 1. Diagnostic diagram

The diagnostic diagram is divided into three regions: (1) region A with $k_z^2 > 0$ where modified acoustic waves can propagate, which is given approximately by the conditions $\omega > \omega_{ac}$ and $\omega > k_h c$. (2) region G with $k_z^2 > 0$ where the modified gravity waves can propagate, which is approximately given by $\omega < N$ and $\omega < k_h c$. (3) region E with $k_z^2 < 0$ where waves are evanescent.

Oscillations of our interest are trapped waves, but the isothermal atmosphere has not such resonance property. Thus, in order to have a proper waveguide character of the stellar atmosphere and envelope, we have to take into account the variation in temperature with height, and this will be considered in the next section. For global eigenmodes of the star, the atmosphere act as an outer reflecting boundary. Waves are evanescent in the atmosphere for such oscillations. It is important to note here that the velocity amplitude of evanescent waves does not necessarily decrease with height, but rather it usually increases slightly within the atmosphere. This is because the energy density of perturbations decreases with height less rapidly than the density ρ_0 itself and thus the factor of $\exp(z/2H)$ in equation (2) plays an essential role.

3. Eigenmodes of Stars

So far we have considered wave motions in the stellar atmosphere from the local stand point. We now turn to discuss stellar oscillations from the global point of view. Any persistent oscillations of a star may be considered as a superposition of normal modes of the star. There are two kinds of normal modes in stars: the radial oscillations and the non-radial oscillations. However, radial oscillations may be regarded as one of special cases of non-radial oscillations with the spherical harmonic index $\ell=0$. If we assume that the unperturbed state of the star is in spherically-symmetric time-independent equilibrium, the eigenfunction of a non-radial mode for perturbations of physical variables (e.g., density pergurbation ρ') can be expressed in the spherical polar coordinates (r, θ, ϕ) by

$$\rho' = \rho'_{k,\ell}(r)\, Y_\ell^m(\theta, \phi)\, e^{-i\omega t} , \tag{6}$$

where $Y_\ell^m(\theta, \phi)$ is the spherical harmonics. The quantity ω is the eigenfrequency of the nonradial oscillation which is specified by three integrals (k, ℓ, m). For a given ℓ, the quantum number m takes integer values from $-\ell$ to ℓ, and eigenfrequencies of these $(2\ell+1)$-modes are degenerate in a spherically symmetric (non-rotating, non-magnetic) star.

The existence of two extra indices (ℓ, m), describing the horizontal dependence of eigenfunctions, makes non-radial oscillations a complicating appearance and it also gives them an extra richness in the eigenvalue spectrum. Besides that, as noted in the previous section, two different kinds of restoring forces (i.e., the pressure force and the buoyancy force) operate in non-radial oscillations, and there exist therefore two different kinds of modes: pressure (acoustic) modes and gravity modes.

The pressure modes (p-modes) form a sequence of increasing eigenfrequency with the order of modes specified by the number of nodes, i.e.,

$$\omega_{p_1} < \omega_{p_2} < \omega_{p_3} < \cdots \to \infty \quad ,$$

while the gravity modes (g-modes) form a sequence of decreasing frequency, i.e.,

$$\omega_{g_1} > \omega_{g_2} > \omega_{g_3} > \cdots \to 0 \quad .$$

A system of equations, which describes the linear adiabatic non-radial oscillations, forms a boundary-value problem of fourth-order ordinary differential equations in the radial coordinates r, and for a given stellar equilibrium model eigenvalues and eigenfunctions are to be calculated numerically. Although these equations look complicated, they can be reduced under a certain assumption to a form analogous to the Schrödinger equation of quantum mechanics. We can then make some qualitative discussion with the help of the so-called propagation diagram (see, Unno et al. 1979). Under a given stellar model and for a fixed spherical harmonic index ℓ, we exhibit in the propagation diagram the spatial variations of the Brunt-Väisälä frequency, N, and the Lamb frequency, L_ℓ, as functions of the radial coordinates r in the stellar interior. Here two frequencies N and L_ℓ are defined by

$$N^2 = g\left(\frac{1}{\Gamma_1}\frac{d\ln p}{dr} - \frac{d\ln\rho}{dr}\right) \quad ,$$

and

$$L_\ell^{\,2} = \frac{\ell(\ell+1)c^2}{r^2} \quad ,$$

and they represent the local buoyancy frequency and the frequency (the reciprocal time) of horizontal acoustic propagation with a given horizontal wavenumber ℓ.

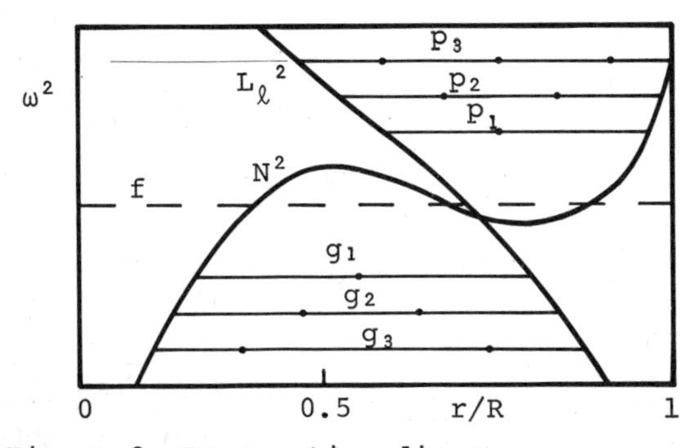

Figure 2. Propagation diagram

Figure 2 illustrates a schematical propagation diagram for a star near the main-sequence with a low order ℓ. In this diagram, the acoustic waves can propagate locally in the region with $\omega^2>N^2$ and $\omega^2>L_\ell^{\,2}$ (P-zone) and the gravity waves can propagate in the region with $\omega^2<N^2$ and

$\omega^2<L_\ell^2$ (G-zone) and waves cannot propagate (i.e., evanescent) in other regions with $L_\ell^2<\omega^2<N^2$ or $N^2<\omega^2<L_\ell^2$ (E-zone). This diagram shows therefore in what part of the stellar interior a wave with a given frequency has locally a propagating or non-propagating character. The propagation diagram may be comparable to the potential energy function of the one-dimensional Schrödinger equation of quantum mechanics in such a way that a propagation zone corresponds to a potential well and an evanescent zone to a potential wall. Since eigenmodes are standing waves, they occur in a region of a potential well that is enclosed by potential walls at both ends. The most important difference between the problem of non-radial oscillations and that of quantum mechanics lies in the fact that in the case of non-radial oscillations there are two different kinds of potential wells: one is that opened upward (p-wave zone) and the other is that opened downward (g-wave zone). The main characteristics of a star as an oscillator can be essentially represented by the propagation diagram. The complicated behavior of non-radial oscillations in evolved stars are mainly caused by complicated spatial variations in the Brunt-Väisälä frequency as a function of the position in the stellar interior.

Generally speaking, non-radial p-modes are oscillations trapped near the surface and g-modes are those trapped in the deep interior. Since we are interested in those oscillations that may produce observable motions in the stellar atmosphere, the most important modes are those of non-radial p-mode oscillations. Furthermore, oscillations of non-radial p-modes are more strongly concentrated near the surface as the spherical harmonic index ℓ is increased. This can be seen in the propagation diagram as the Lamb frequency L_ℓ moves upward with the increase of ℓ. The horizontal wavenumber of oscillations in the stellar surface becomes almost continuous for high value of ℓ and it is given by $k_h=\sqrt{\ell(\ell+1)}/R \sim\ell/R$. In the diagnostic (k_h, ω)-diagram, non-radial p-modes then appear as several discrete ridges corresponding to each radial-order mode (i.e., p_k-mode), which has been observed beautifully in the case of the solar five minute oscillations (Deubner 1977, Deubner et al. 1979).

4. Overstability of Eigenmodes

Since the ordinary dissipation gives damping of oscillations, some special mechanisms of excitation must operate inside the star in order for an eigenmode to be excited and to be observed. To study this, we must explicitly consider non-adiabatic effects of oscillations, i.e., the energy exchange of fluid elements with the surroundings during a cycle of oscillation. The thermodynamic excitation of oscillations have extensively been studied in connection with the driving of pulsation in Cepheid and

RR Lyrae stars. Two kinds of excitation mechanisms are known which may operate in the outer layers of stars. The first one is the κ-mechanism (opacity-mechanism) of the hydrogen and helium ionization zones in the stellar envelope, and this is the mechanism that drives pulsations in Cepheids. The second one is the so-called Cowling-Spiegel mechanism which operates in the zone with the super-adiabatic temperature stratification. The latter mechanism operates only for non-radial oscillations and not for radial oscillations.

To get some idea how these mechanisms work, we describe briefly the operation of the κ-mechanism below. The opacity κ of partial ionization zones of the hydrogen and the helium increases at the phase of compression in a cycle of oscillation and the radiative flux flowing from the center to the surface is blocked there at that phase. The heat thus accumulated will give stronger repulsion at the next phase of expansion, and the amplitude of oscillation will tend to grow with time if the effect of excitation of this κ-mechanism is stronger than that of damping in other regions. As for the Cowling-Spiegel mechanism in a super-adiabatic temperature stratification, if some other restoring forces such as the stabilizing chemical composition gradient, magnetic fields and rotation, exist, this mechanism is known to be effective to give overstability of some non-radial modes (Moore and Spiegel 1966). However, if there exists no other restoring force, it is then not clear whether this mechanism acting as a sole agent of excitation can destabilize oscillatory modes. In this respect, Unno (1977) has pointed out that the Cowling-Spiegel mechanism associated with the variation of convective flux is more important than that of radiative flux. Then the variation of convective flux has to be taken into account in order for the problem to be settled.

Stars have been thought in the past to be less susceptible to non-radial oscillations than to radial oscillations because of the increased dissipation of the oscillations by the lateral radiative heat exchange. However, recent investigations show that much variety in the geometrical and physical properties of non-radial oscillations should give some of non-radial modes a better chance to get excited. In fact, those stars which are unstable against radial pulsations are likely to be unstable against some of non-radial modes as well. Furthermore, stars which are unstable against some of non-radial modes will probably occupy a wider range in the HR diagram than radially pulsating stars. The most important reason why some of non-radial modes are easy to get excited is that they can be "trapped" locally in the place where some excitation mechanism works effectively so that the dissipation of oscillation in other regions of the star is kept minimal. If one of non-radial modes is unstable in a star, several of other modes are very likely to be excited because their physical characters are very similar. Thus, if non-radial oscillations are ever excited in a star, it is most probable that many of non-radial modes are excited simultaneously.

The question of overstability of an eigenmode is a problem of delicate balance between excitation and damping and it can only be answered adequately by the numerical analysis of stability using a realistic stellar model. Thus, the full equations of linear non-adiabatic non-radial oscillations have to be solved numerically. The radiative heat exchange both in the sub-photospheric layers and in the atmosphere is important in the case of our interest, and, exactly speaking, the problem of time-dependent three dimensional radiative transfer has to be treated. It looks so much difficult that it is the usual practice at present to treat this by using the Eddington approximation developed by Unno and Spiegel (1966). The basic equations are then reduced to the fourth order linear differential equations with complex coefficients (under the Cowling approximation in which the Eulerian perturbation of the gravitational potential is neglected), and they form together with adequate boundary conditions an eigenvalue problem with a complex eigenvalue and complex eigenfunctions. The real part ω_R of the complex eigenfrequency ($\omega=\omega_R+i\omega_I$) gives the frequency of oscillation and its imaginary part ω_I determines the growth (or damping) rate of oscillations. This system of equations for linear non-adiabatic oscillations have been solved in a few cases of our interest and they will briefly be reviewed below.

(1) The solar non-radial p-modes.

Ando and Osaki (1975) performed extensive calculations of eigenfrequencies of non-radial p-modes of the sun in relation to the solar five minute oscillation. They solved equations of linear non-adiabatic non-radial oscillations for a realistic solar envelope model and obtained complex eigenfrequencies of non-radial p-modes covering a wide range in the spherical harmonics ℓ (ranging from $\ell=10$ to 1500) and in several overtones of the radial order. It was found that many of non-radial p-modes were overstable and that most unstable modes occupy a region in the diagnostic (k_h, ω)-diagram centered with a period of 300 sec and with a wide range of horizontal wavenumber.

It has been seen from eigenfunctions that the vertical velocity of oscillations has an increasing amplitude with height in the atmosphere, even though the kinetic energy of oscillations per unit volume $\rho|\underset{\sim}{v}|^2$ is decreasing within the atmosphere. Many of non-radial p-modes are found to be unstable and they are excited by the κ-mechanism of at the hydrogen ionization zone. The Cowling-Spiegel mechanism seems to work at the convective-radiative transition zone, but quantitatively the κ-mechanism is found much more effective than the latter. The radiative dissipation in the atmosphere ($\tau<1$) contributes appreciably to the damping of oscillations and its importance increases with the increase in frequency of modes, and higher p-modes become stabilized ultimately.

The biggest uncertainty in this analysis is the problem of the interaction between convection and oscillation, which has been neglected in the work of Ando and Osaki (1975) because of lack of the definitive theory. The damping due to turbulent viscosity of convection is thought to have a stabilizing tendency for otherwise unstable p-modes. Goldreich and Keeley (1977) have made a rough estimate of its effect and found that it is as important as the thermodynamic excitation of oscillations. However, they have failed to reach the definitive conclusion about the overstability of p-modes of the sun because of uncertainty in convection theory.

(2) Cepheids.
Cepheids are pulsationally unstable against radial modes due to the κ-mechanism in the hydrogen- and helium-ionization zones. Since this mechanism works also for non-radial p-modes, it will be natural to ask whether or not Cepheids are vibrationally unstable against non-radial modes.

There is, however, an important difference between radial pulsations and non-radial p-mode oscillations. Dziembowski (1971) first noticed that, in the case of non-radial oscillations of giant stars, even high-frequency envelope p-modes behave as internal gravity waves of extremely short wavelength in the deep interior, and he once concluded that the excitation of non-radial oscillations would be prevented by strong radiative dissipation in the core of these evolved stars. However, there exists another important character of non-radial oscillations, that is the wave-trapping phenomenon, and this effect was not taken into account in Dziembowski's (1971) discussion. Shibahashi and Osaki (1976) have shown that non-radial modes with high-order spherical harmonics can be very clearly divided into two types in evolved stars, one type being a gravity mode trapped in the core and the other a mode trapped in the envelope. There exist, therefore, envelope p-modes with high ℓ that are almost completely free from the influence of the core.

By taking into account both the effect of dissipation in the core and the non-adiabatic effect in the envelope, Osaki (1977) has examined vibrational stability against non-radial modes in a Cepheid model. It is found that non-radial modes with lower-order spherical harmonics (i.e., $\ell \lesssim 5$)

are stable because of heavy leakage of wave energy from the envelope to the core, but that those of higher ℓ (i.e., $\ell \gtrsim 6$) are trapped well within the envelope and some of them (i.e., f- and p_1-modes) are unstable due to the negative dissipation in the hydrogen- and helium-ionization zones. The growth rates of unstable non-radial modes are found to be of the same order as those of radial modes. A similar result was obtained independently by Dziembowski (1977) who found that the overstability of non-radial modes with very high-order spherical harmonics extends far beyond the boundary of the classical Cepheid instability strip.

(3) Other stars.
The vibrational stability of non-radial p-mode oscillations has been investigated for various stellar envelope models in the wide range of the HR diagram by Ando (1976) and by Dziembowski (1977).

Ando (1976) examined the stability of stellar envelopes in late-type dwarfs, giants, and super-giants against non-radial p-modes with high-order ℓ which are well trapped near the surface. He found many of non-radial p-modes are overstable due to the κ-mechanism of the hydrogen ionization zone in stars that lie to the right of the Cepheid instability strip in the HR diagram. He then suggested that these overstable acoustic modes might be responsible for the formation of the chromosphere and the corona and for the Wilson-Bappu effect in late type stars. However, the biggest uncertainty of this result is again the problem of coupling between convection and oscillation, whose effect has been ignored in this investigation.

Dziembowski (1977) made a similar study for stars lying within and to the left of the Cepheid instability strip. He found that δ Scuti stars are unstable both for radial and for low ℓ non-radial modes, but that the maximum of instability occurs for non-radial modes with very high ℓ (i.e., $\ell \simeq 500$). He also studied two models of an early type supergiant corresponding to α Cyg (A2Ib) and of a medium type supergiant ($T_e \sim 5000$°K). The overstability was found in the case of the early-type supergiant only for non-radial f-modes with very high ℓ-values (i.e., $\ell = 32 \sim 63$). On the other hand, the overstability of low-order non-radial modes ($\ell \simeq 3$-5) was found in the case of the medium-type supergiant.

5. Summary and Discussion

We have considered the nature of oscillations in stars from the local and the global standpoints. Overstability of non-radial oscillations has then been discussed. Results of numerical analysis show that many of non-radial p-modes are very likely unstable in various stars of the HR diagram. Observationally, non-radial oscillations of high tesseral harmonics (i.e., excepting the low-ℓ modes) do not give rise to neither the stellar light variability nor the stellar radial velocity variability, but they show up as unresolved velocity fields responsible to spectral line broadening and the mechanical source of energy to heat the upper atmosphere (i.e., the non-thermal velocity fields in the stellar atmospheres).

We have so far discussed the linear stability of non-radial modes, but we have not mentioned non-linear effects which are thought to be responsible for limiting amplitudes of overstable oscillations. Non-linear effects that are important in limiting amplitudes in pulsating variables are;

1. The saturation of the excitation mechanism.

The saturation of the κ-mechanism in the second helium ionization zone is thought to be the most important in determining the final amplitude of Cepheids and RR Lyrae stars.

2. The enhancement of dissipation by shock waves.

When the amplitude of oscillations becomes sufficiently large, shock waves are generated in the atmosphere, which greatly enhances the dissipation. Besides these,

3. Non-linear mode coupling, will be important, in the case of non-radial oscillations. This will redistribute kinetic energy of oscillations between various non-radial modes.

Even in the linear stability analysis, there remain several unresolved problems. The problem of coupling between convection and oscillation was mentioned previously. We have not yet succeeded in finding the instability mechanism of pulsations in β Cephei stars (early-type pulsating variables). This means that our stability analysis is not still accurate enough. Possible insufficiencies in our knowledge are suspected to exist with respect to the opacity and the effect of radiation pressure. Much theoretical investigation is thus needed.

References

Ando, H. 1976, Publ. Astron. Soc. Japan, **28**, 517.

Ando, H. and Osaki, Y. 1975, Publ. Astron. Soc. Japan, **27**, 581.

Christy, R. F. 1962, Astrophys. J., **136**, 887.

Deubner, F.-L. 1975, Astron. Astrophys., **44**, 371.

Deubner, F.-L. 1977, in *The Energy Balance and Hydrodynamics of the Solar Chromosphere and Corona*, ed. R.-M. Bonnet and Ph. Delache (G. de Bussac, Clermont-Ferrand), p.45.

Deubner, F.-L., Ulrich, R. K., and Rhodes, E. J., Jr. 1979, Astron. Astrophys., **72**, 177.

Dziembowski, W. 1971, Acta Astron., **21**, 289.

Dziembowski, W. 1977, Acta Astron., **27**, 95.

Eckart, C. 1960, *Hydrodynamics of Oceans and Atmospheres* (Pergamon Press, London).

Goldreich, P. and Keeley, D. A. 1977, Astrophys. J., **211**, 934.
Lucy, L. B. 1976a, Astrophys. J., **206**, 499.
Lucy, L. B. 1976b, Paper read at IAU Colloq. No.29 (Budapest, Sept. 1975).
Moore, D. W. and Spiegel, E. A. 1966, Astrophys. J., **143**, 871.
Osaki, Y. 1977, Publ. Astron. Soc. Japan, **29**, 235.
Shibahashi, H. and Osaki, Y. 1976, Publ. Astron. Soc. Japan, **28**, 199.
Smith, M. A. 1977, Astrophys. J., **215**, 574.
Unno, W. 1977, in Problems of Stellar Convection, ed. E. A. Spiegel and J.-P. Zahn (Springer-Verlag, Berlin), p.315.
Unno, W., Osaki, Y., Ando, H., and Shibahashi, H. 1979, Nonradial Oscillations of Stars (University of Tokyo Press, Tokyo).
Unno, W. and Spiegel, E. A. 1966, Publ. Astron. Soc. Japan, **18**, 85.

THE EVOLUTION OF AN AVERAGE SOLAR GRANULE

R.C. Altrock
Air Force Geophysics Laboratory
Sunspot NM, U.S.A.

Abstract

High-resolution photographic spectra of the center of the solar disk have been obtained with the Vacuum Tower Telescope at Sacramento Peak Observatory. Two weak iron lines and the neighboring continuum were recorded with 40 sec time resolutions and better than 1" spatial resolution over a period of 40 min. Intensity and velocity fluctuations were obtained in the two lines and continuum as a function of time and space, and 300 sec oscillations were filtered out. The resulting fluctuations, due solely to granulation, were assembled into an ensemble average of the center of a granule and the center of an intergranular lane, as a function of time. The intensity-fluctuation data have been analyzed through calculation of model line profiles to yield temperature fluctuations in a granule as functions of time and height. We find that the line parameters are distinctly out of phase with continuum brightness, so that, for example, maximum brightness at line center occurs approximately 100 sec prior to maximum continuum brightness. A series of one-dimensional model atmospheres representing the granule at various stages of its lifetime is presented.

OBSERVED SOLAR SPECTRAL LINE ASYMMETRIES AND WAVELENGTH SHIFTS DUE TO CONVECTION

Dainis Dravins
Lund Observatory
S-222 24 Lund, Sweden

Abstract

Convective motions are manifest in the solar spectrum as slight spectral line asymmetries and wavelength shifts. These have been studied for 311 Fe I lines. Most lines are blueshifted because the larger contribution of blueshifted photons from bright and rising granules statistically dominates over the contribution from dark and sinking intergranular lanes. Fainter lines (formed deeper) show larger blueshifts than strong lines; high-excitation lines usually are more blueshifted (preferentially formed in the hotter granules) and short-wavelength lines are more blueshifted because of increased granulation contrast there. Detailed studies of line bisector behavior as function of line parameters permit the construction of model atmospheres incorporating convective hydrodynamics. The method does not require spatially resolved observations and can be extended to studies of stellar convection.

DIFFERENTIAL LINE SHIFTS IN LATE TYPE STARS

A. Stawikowski
N. Copernicus Astronomical Center, Astrophysics Laboratory
Torun, Poland

Abstract

The differential line shifts method for studying convective type motions has been described in the review of Glebocki and Stawikowski at this colloquium. Here the results of differential line shifts for twenty late type stars are presented. Line shifts were determined either from published wavelength measurements (α CMi, β Peg, α Boo, α Ser, α UMa, α Per, α Sco, γ Cyg, α Car and HD 19445) or from my own wavelength measurements on 8.1 Å/mm dispersion spectra (56 Ori, η Peg, η Psc, 61 Cyg A, ε Eri, δ Eri, o Tau, ε Cyg, α Aqr and ι Ari). Because of low accuracy of the wavelength determinations, mean shifts (VR) were calculated for narrow ranges of lower excitation potentials (LEP). These shifts were plotted against excitation potential and VR-LEP relations were obtained for all analysed stars. Because of the scatter of points in these diagrams linear approximation was assumed and the slopes (denoted A and expressed in Km/s/eV) of the VR-LEP relations were calculated. The stellar parameters and the derived slopes for the whole sample of investigated stars were analysed, and correlations with microturbulence, luminosity and gravity were found. These correlations demonstrate that the slope of VR-LEP relation describes some component of velocity field in late type stars.

The detailed description of the procedure and results will be published in Acta Astronomica.

TEMPORAL AND SPATIAL FLUCTUATIONS IN WIDTHS OF SOLAR EUV LINES

R.G. Athay and O.R. White
High Altitude Observatory
Boulder, CO 80302, U.S.A.

Abstract

Analyses of some 300 hours of time sequences of solar EUV line profiles obtained with OSO-8 show large fluctuations in line widths. At a given location on the sun, line widths fluctuate temporally on time scales ranging from less than a minute to over an hour. At any given time, line widths fluctuate spatially on a variety of scales ranging from active region size to arc second size. Temporal and spatial fluctuations are of approximately the same amplitude. Thus, the sun can be characterized by an aggregate of small cells in each of which line widths are fluctuating in time and which have random phases with respect to each other.

Spatial fluctuations in line width are correlated with large scale spatial fluctuations in brightness for some lines but not for others. Temporal fluctuations in width are sometimes correlated with either Doppler shifts or intensity fluctuations, but more often such correlations are absent.

For a given line, the line width varies through an extreme range of about a factor of two. Nonthermal components of line width vary from approximately the local sound speed to a small fraction of the sound speed.

FORMATION OF THE PROFILES OF ABSORPTION LINES IN THE INHOMOGENEOUS MEDIUM

R.I. Kostik
Main Astronomical Observatory of the Ukrainian Academy of Sciences
Kiev-127, U.S.S.R.

Abstract

The shapes of weak fraunhofer lines, including their asymmetry, are explained by the influence of acoustic waves and granular motions. The following pattern of granula is adopted: the ascending (granular) and descending (intergranular) velocities V_1 and V_2 as well as the relevant temperatures T_1 and T_2 are maximum (minimum) at the center of granula (intergranula) and decrease (increase) sinusoidally outwards. There is also tangential outflow of the matter from the center of granula with the velocity V_3 = const. and temperature T_3 equal to the temperature of the photospheric model T_0. The areas occupied by ascending, descending, and tangential motions are S_1, S_2, S_3 respectively.

The inhomogeneities caused by granulation are perturbed by the acoustic waves with the velocity amplitude $V_0 = 0.4$ km/s and with the period T = 300 sec. The formula for the absorption line coefficient was derived, and the center-to-limb profiles of 11 weak fraunhofer lines were calculated. The comparison of calculated and observed line profiles have resulted in determining "best fit" values of V_1, V_2, V_3, S_1, S_2, S_3 and $\Delta T = T_1 - T_2$. Tangential motions occupy the larger part of the solar surface $S_3 = 0.7$, $V_3 \approx 2.7$ km/s, $V \approx 1.0$ km/s, ΔT ranges between 270° - 400°.

THE DETERMINATION OF STELLAR TURBULENCE BY LOW RESOLUTION TECHNIQUES

R. Głębocki
Institute of Physics, Gdańsk University, 80-952 Gdańsk, Poland

A. Stawikowski
N. Copernicus Astronomical Center, Astrophysics Laboratory,
87-100 Toruń, Poland

INTRODUCTION

In this review the problem of measuring velocity field in stellar atmospheres from the observer's point of view is presented. Our purpose here is to discuss observational methods in which detailed analysis of line profiles is not necessary, i.e. methods based on measurements of total absorption of lines (curve of growth and narrow-band photometry), methods based on measurements of widths of lines (Goldberg-Unno method and curve of line width correlation) and measurements of differential line shifts. Therefore, our review will be limited to the discussion of basic assumptions of each method, to the analysis of their advantages and disadvantages, to a specification of a quantity which can be derived from a given method and finally to a brief presentation of the results. Usually, the low resolution methods provide an information about a particular component of the stellar velocity field. But the problem is more complicated as we do not know if particular components can be isolated from the stellar velocity field, thus, we do not know the real physical meaning of the measurable parameters. In this paper we shall adopt classical concept of micro- and macro-turbulence and convective type velocity. This simplified picture was criticized since 15 years on every colloquium devoted to hydrodynamic phenomena in the stellar atmospheres, but until now the theoreticians have not succeded in developing a theory which would satisfactorily interpret the observations.

Theoreticians would be satisfied if an observational method could be found allowing to distinguish between different types of oscillations and distributions of the sizes and velocities of turbulence. The reality is however more pedestrian, because the observables are averages over the angles and depth in whole stellar atmosphere. Therefore, even the most sophisticated theory of velocity field in stars can not be verified with observations unless it predicts some average parameters suitable for observations in the integrated light of stars. Even for the Sun, where observations of high resolution in space and time are available, the microturbulence derived from integrated light is not uniquely attributed to any local event, although, the theory of some local phenomena is quite satisfactory.

As already mentioned, our review paper is devoted to classical methods of observations. It does not mean that there is no progress in observational techniques (see the review of Gray), but it merely reflects the fact that the huge observational material accumulated over a half of a century is still open to theoretical interpretation.

THE CURVE OF GROWTH METHOD

The method of determination of the microturbulent velocity from the curve of growth is the most commonly used, though subjected to serious uncertainties. This method has been applied for determination of the ξ - parameter for about 700 stars in a wide range of T_e and M_v.

The microturbulence is determined by comparison of the observed COG (i.e. $\log(W/\lambda)$ versus $\log(gf\lambda) - \chi\Theta_{exc}$) with the theoretical one (i.e. $\log(W/\lambda)(c/v)$ versus $\log A$). The displacement of the empirical COG vertically and horizontally to fit the theoretical curve establishes c/v and N. Various types of theoretical curves of growth are available depending on the definition of log A: Wrubel's COG based on S-S and M-E approximations, Unsold's COG based on Minnaert's empirical line-profile formula, weighting function method and precise calculation of theoretical COG for a given stellar atmosphere model. It should be noticed however that the choice of the theoretical curve has a minor influence on the ξ determination. The last of the mentioned above methods is similar to the fine analysis with a one, but important difference. In the fine analysis theoretical and observational profiles are compared, while in the COG method the equivalent widths are calculated from the theoretical profiles and compared with the observational values of W. Profiles of weak and medium-strong lines can be determined only from high resolution spectrograms. They are subjected to instrumental and rotational broadening, both being unimportant for equivalent width determination. Thus, the most important advantage of the COG method lies in its applicability to fainter stars, for which high resolution spectrograms are not available.

The COG method assumes that the geometrical scale of turbulent elements is so small that the non-thermal velocity distribution can be convoluted with the thermal velocity distribution. The vertical shift of the observed COG relative to the theoretical one yields c/v, where $v = (v_{th}^2 + \xi^2)^{1/2}$. The assumption of Gaussian distribution of velocity of turbulent elements is of course a simplification which may not be fully justified.

In most investigations the microturbulence was derived as a by-product of the abundance analysis. Only few studies have been entirely devoted to the determination of the ξ, usually for narrow ranges of T_e and M_v (Chaffee, 1970; Andersen, 1973; Głębocki, 1972; Foy, 1976). These investigations are of great importance because of the homogenuity of the observational material. But a comparison of ξ derived for the same star by different authors usually displays a considerable discrepancies.

Statistical analysis of the behaviour of ξ on the HR diagram based on the published data has been made by different authors (for references see Gray, 1978). The results of these studies are very important for our understanding of the physics of velocity fields in stellar atmospheres. Unfortunately, these results should be regarded with caution. During the last twenty years not only the method of the COG evolved, as mentioned above, but what is more important the numerous new determinations of the oscillator strengths with increased accuracy, allow for investigation of second order effects in the COG method. These second order effects can noticeably change the derived values of microturbulence. There is even a believe that many uncertainties in the COG method preclude a coherent discussion of the distribution of microturbulence on the HR diagram when based on the data published by different authors. These uncertainties are well known and can be listed in the following order: errors in equivalent widths, errors in oscillator strength values, lack of data and/or correct formula for damping constant, magnetic intensification, hyperfine structure, non-LTE, errors in temperature and pressure gradients, non-gaussian distribution of turbulent velocity. The first two factors are the most responsible for the disagreement in the ξ determinations for the same star by different authors.

Errors in equivalent widths. The stellar curves of growth are based on spectrograms with dispersion from about 1.5 Å/mm to 30 Å/mm. The observers know that the accuracy of W determination in mÅ is of the order of the dispersion of spectrogram given in Å/mm. But, it should be kept in mind, that systematic errors in W are more essential than the internal accuracy of measurements. Wright (1966) and Smith (1973) compared several W scales. The accuracy of W determination is of the order of 5% for lines stronger than 100 mÅ, but may be as low as 50% for weak lines (25 mÅ). The tendency of equivalent widths to increase with decreasing resolution is a well known effect though its origin is not quite well understood. E.g. a comparison of W from 8.5 Å/mm and 2.7 Å/mm spectrograms showed that lower dispersion equivalent width scale is 30% larger than the higher, and is roughly independent of equivalent width (Smith, 1973). The equivalent widths determined from spectrograms of 16 Å/mm differ by a factor of two when compared to the data taken from 2 Å/mm spectrograms. A factor of two in equivalent width scale would mimic increase of microturbulence by about 3 km/s for γ Equ (Smith, 1973). The only way of reducing the errors in ξ due to systematic differences in W scale is to compare the equivalent widths of the program stars with those of the standard star, for which the W scale has been well established. Unfortunately only few observers take the spectra of their program stars together with a standard star in order to compare their W scale with that derived from the high dispersion spectrograms. Besides, there is still a lack of commonly accepted standard stars for equivalent widths calibration. Thus, it is often impossible evaluate the error of the individual ξ determination caused by uncertainty in W.

Uncertainties in oscillator strengths. In the last years the improvement in the determination of the oscillator strengths, especially for iron, has been significant. The number of neutral iron lines with very accurate gf values (0.1 dex, e.g. Blackwell and Shallis, 1979) has considerably increased. The situation is worse for other elements and especially for ions. The new scale of gf values differs from the old one not only by a constant factor but, what is essential by a factor including upper-level excitation potential. The new scale may change considerably the determination of microturbulence and explain the spurious dependence of ξ on excitation potential and height in the stellar atmosphere found by e.g. Bell (1951), Wright (1951), Warner (1964), Zeinalov (1970), Osmer (1972). For example the χ - dependent turbulent

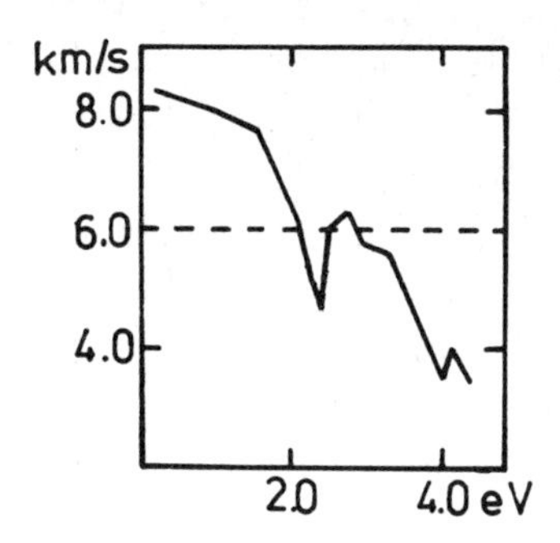

Fig. 1. Variation of the turbulent velocities with LEP derived by Zeinalov (1970) - full line, and constant velocity derived from revision by Hasegawa (1978) - dashed line.

velocity derived by Zeinalov (1970) for γ Cyg is caused by systematic errors in gf and W. Zeinalov used Corliss and Warner (1964) gf values. Hasegawa's reanalysis of that star demonstrated that adopting the new scale of gf χ- independent value of ξ is derived as shown in fig. 1 (Hasegawa, 1978). The investigation of Andersen (1973) demonstrated that ξ is reduced by about 30% for main sequence stars when using gf values of Garz and Kock (1969) instead that of Corliss and Warner (1964). For other stars this reduction of microturbulence may be even larger (Elste and Ionson, 1971). We think, that the well known discrepancies in ξ for the same stars analysed by different authors are in great extent due to differences in gf values. Almost every set of new gf determinations is applied to the solar curve of growth yielding a revised value of ξ . There is however a general lack of revision of microturbulence determination for standard stars (εVir, αPer, α CMi), what in principle unables the use of older ξ determinations for statistical investigation.

Uncertainties in the damping constant. The now available high accuracy of gf values especially for iron allows to investigate the influence of the second order effects (e.g. splitting of the COG due to collisional damping) which previously has been completely lost in the scatter of points in the COG. Some authors claim that value of damping constant is connected with the parity of multiplets. This effect has not been fully confirmed, but Foy (1972) found a splitting of the flat part of the solar COG. This splitting when translated into microturbulence is equivalent to 0.5 km/s. The approximations in the theoretical formulae for damping constant as well as poor experimental data make a detailed discussions of damping constant problem suspicious. The theoretical values of Γ are usually smaller than the observed ones. Many authors use an artificial enhancement factor in order to fit theoretical and observed profiles. The calculation by Smith (1973) showed that a change of Γ by a factor of 7.5 would mimic a change of ξ by 1.25 km/s. Similar results were obtained by Evans and Elste(1971) who showed that for the flat part of the COG the effect of an increased damping by a factor of 2 could be compensated by a decrease of microturbulence by an amount of 0.5 km/s. The influence of an error in Γ becomes negligible for lines near to the turn off point in the COG.

Magnetic intensification of lines. Spectral lines having rich Zeeman structure may increase their equivalent widths in the presence of strong magnetic field. The magnetic intensification is pronounced for medium-strong lines for which every π and σ component is saturated. For very strong lines collisional damping acts more effectively than Zeeman broadening. Zeeman intensification will rise the flat part of the COG analogously to the increase of microturbulence. Evans claims that 1 kGauss field mimics an increase of microturbulence of 1 km/s, and that this effect roughly scales (Smith, 1973), what is confirmed by the investigation of Hensberge and De Loore (1974). This conclusion is important when microturbulence of Ap stars is derived from horizontal part of the COG. The calculations of Havnes and Moe (1975) show that the Zeeman intensification is negligibly small (10% for 5000 Gauss) for equivalent widths near to the turn off point in the COG.

Hyperfine structure. Lines of some elements, like Co, Cu, Mn, Eu, are sensitive to their hyperfine structure. Curves of growth including HFS were calculated by Bely (1966) and Landi Degl'Innocenti (1975). Van Paradijs (1973) found that the intensification of Mn and Cu lines due to HFS mimics the value of microturbulence of 3.5 km/s in comparision to the adopted value of 1.2 km/s. Iron lines are insensitive to HFS intensification. The lines of Cu, Mn and some rare earths should be avoided in determination of ξ, especially in Ap and Am stars, where a joint effect of magnetic and HFS intensification may significantly change the derived value of microturbulence (Landi Degl'Innocenti,1975).

Non-LTE and uncertainties in models of stellar atmosphere. Non-LTE may be important for strong resonance lines in the early type stars. The profiles of these lines can be significantly changed when non-LTE procedure is applied, what translated to microturbulence can introduce corrections as high as 3 km/s (Mihalas, 1973). But for most faint and

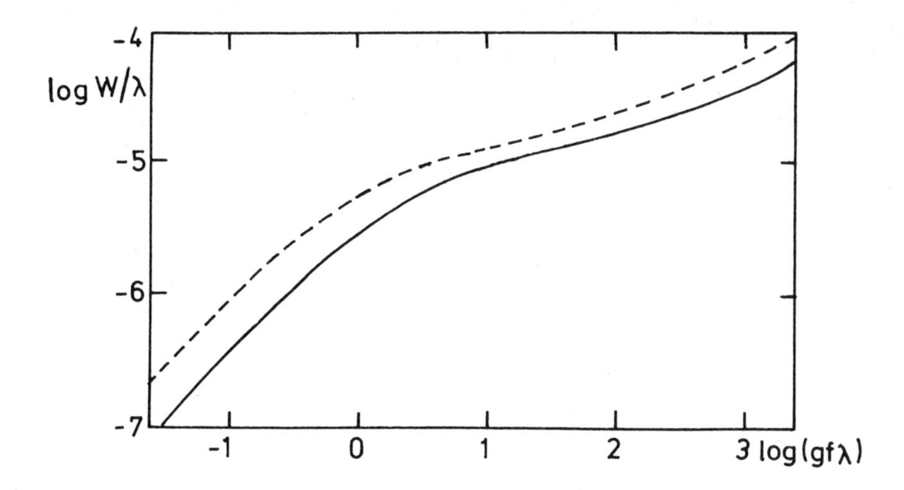

Fig. 2. Solar curves of growth with non-LTE - dashed line, and LTE - full line. Adopted from Dumont et al. (1975).

medium-strong lines used in the COG method the influence of non-LTE is negligible. Dumont et al. (1975) calculated the influence of non-LTE for a hypothetical 4-5 level atom plus continuum. Fig. 2 shows that the non-LTE calculations does not significantly change the COG. Application of non-LTE calculations for iron lines in Arcturus by Smith (1974) shows that the effect is on a level of detectability. Ignorance of non-LTE leads to an overestimate in ξ of 12%, i.e. 0.2 km/s.

Models of stellar atmospheres are still uncertain in that sense that they do not include the effects of stellar velocity fields. The heating by mechanical flux and blanketing effect can considerably change the temperature gradient in the uppermost layers of the atmosphere. The influence of a change in dT/dh on the slope of the COG was analysed by several authors. This influence can not be significant because weak and medium-strong lines used in the COG analysis originate in deeper photospheric layers. The results of Böhm-Vitense (1972) and Bonnell and Branch (1979) confirm the negligible effect of changes in dT/dh on the ξ determination from the COG, but it should be included in the abundance analysis. This conclusion is even more valid when microturbulence is determined from the lines near the turn off point in the COG.

The determination of ξ by the COG method is a valuable extension of the study of velocity field in the stellar atmospheres. No matter what is the real physical meaning of ξ we all agree that its quantitative change on HR diagram is important for our understanding of physical processes underlying the origin of turbulence and its connection with rotation, the depth of convective zone, mass losses and stellar winds. In this aspect the statistical investigations of the behaviour of ξ on HR diagram are essential. The COG method supplies most of the data for that kind of analysis. Unfortunately, as it follows from our review, the COG method is subjected to many uncertainties. Moreover, ξ is often derived as a by-product or fitting parameter, thus less methodological attention is paid to its determination. The following conclusions can be drawn from our discussion.

- Neutral iron lines should be used to the determination of ξ by the COG method. Iron lines are the most numerous in a wide range of effective temperatures. Their oscillator strengths have the greatest accuracy of 0.1 dex, also for faint lines.
- Spectrophotometric standard stars should be chosen in the whole range of T_e and M_V. These standards should have a very good determination of equivalent widths based on the highest resolution spectrograms and a carefully determined parameters of stellar atmospheres and velocity field. This would allow for a careful reanalysis of ξ for a huge number of stars.
- The microturbulence should be determined from the turn off point rather than from the flat part of the COG in order to avoid the influence of errors in damping constants, magnetic and HFS intensifications. These second order effects act effectively on strong and medium-strong lines rising the horizontal part of the COG. The high accuracy of gf values for faint lines permits now for a more precise evaluation of the turn off point with a partial elimination of personal factor which is also a source of discrepancies in the ξ determination by the COG method.
- Our experience shows that some discrepancies in ξ determinations for the same star by different authors can be removed if careful reanalysis is applied. The situation is worse for early type stars. The precision of oscillator strength for ions (Fe II, Cr II, Ti II) is still unsatisfactory for reliable determination of microturbulence.

Fast computer facilities supersede recently the COG method. Quasi-fine analysis like WIDTH program, where ξ is a fitting parameter are often used when analysing high dispersion spectrograms. However the above mentioned conclusions remain valid for this modification of the COG method.

NARROW-BAND PHOTOMETRY

The effect of line saturation in the COG can be utilized for detection of microturbulence by narrow-band photometry. Spectral regions with a bulk of lines from different parts of the COG can be found in stellar spectra. In the region where lines from the flat part are dominating blanketing effect is mainly a function of microturbulence, while in the region crowed with weak lines blanketing effect is sensitive to the metal abundance. In the standard narrow-band photometry systems colour indices are influenced by both factors. The sensitivity of colour indices on different atmospheric parameters was discussed by Gustafsson and Bell (1979) in their analysis of synthetic spectra. It turned out that some of the commonly used narrow-band indices like m_1, c_1, Uppsala systems, DDO systems are mildly dependent on ξ value. For illustration see figures in Gustafsson and Bell (1979). Analysis of the blanketing effect lead these authors to the suggestion that it is possible to invent a narrow-band photometric system suitable for the evaluation of microturbulence. For this purpose the following regions in the stellar spectra should be found: one with many weak lines and another with numerous medium-strong lines plus two reference regions free of any lines. The reference regions should not be too distant from their line regions. Index CI defined as

$$CI = 2.5 \log \frac{\text{flux of line region}}{\text{flux of reference region}}$$

would measure either metal abundance or microturbulence. Nissen and Gustafsson (1978) have proposed such a system based on measurements at the bands near to 4800 Å and 4980 Å for F type dwarfs. They observed 52 stars with $5800° < T_e < 7200°$ and $3.9 < \log g < 4.4$ and found that in this region of HR diagram, ξ is only weakly dependent on T_e changing from 1.2 km/s for $T_e \sim 5800°$ to 1.9 km/s for $T_e \sim 7200°$. Their result is in qualitative accordance with determinations by the COG method (Głębocki, 1973), see fig. 3. Systematic difference is caused by the adopted values of ξ for the Sun (Nissen and Gustafsson - 0.8 km/s, Głębocki - 1.3 km/s). Another narrow-band determination of ξ with the use of a very similar photometric system made by Gustafsson et al. (1974) for G8-K3 type giants. They observed 48 stars and concluded that microturbulence was constant for varying gravity and metal content. The mean value for this sample is equal to 1.7 ± 0.4 km/s. Their results confirm the conclusion of Głębocki (1973) based on analysis of the COG

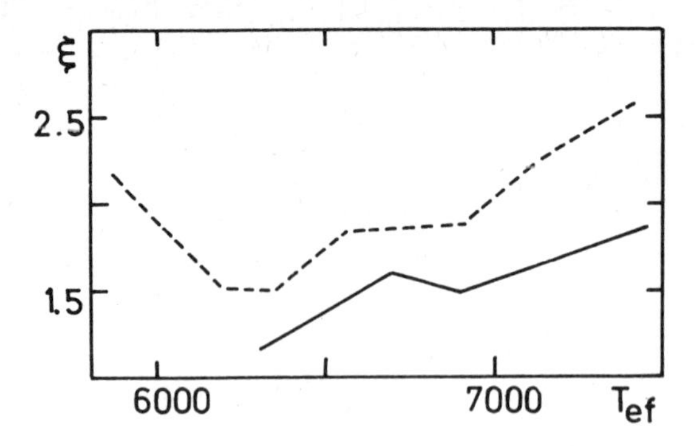

Fig. 3. Variation of microturbulence with T_e for F type dwarfs. Dashed line - narrow-band photometry data (Nissen and Gustafsson, 1978), full line - COG data (Głębocki, 1973).

data that the microturbulence is determined by the position of the star on HR diagram (see however Foy (1978) analysis indicating an age dependence of turbulence).

The narrow-band photometry method of ξ determination is based on the same unclear physical assumptions about the stellar turbulence as the COG method although it is methodologically quite different. It has an advantage of differential analysis avoiding errors in atomic data and errors in abundance of a particular element because of averaging the fluxes over the whole spectral region containing lines of different elements and different excitation potentials. Besides, photoelectric measurements of colour indices are more economical and accurate than the spectrophotometric determinations of absorptions of numerous lines. Troubles with establishing continuum in line crowded areas typical for spectrophotometric analysis are unessential in the narrow-band method. Besides, photoelectric measurements are much less time consuming especially in the elaboration of observations and they are more favourable in reaching fainter stars in comparison to high dispersion spectrophotometry.

There are however important drawbacks of the narrow-band photometry method. Its fundamental weakness lies in the calibration of the zero points. Even, when using sophisticated models of Bell and Gustafsson (1978) uncertainty in the calibration of zero point with real stars can be as large as about 1 km/s caused by errors in damping constant ($\pm$ 0.8 km/s), errors in the assumed value of ξ for the Sun ($\pm$0.3 km/s), errors in log g ($\pm$ 0.3 km/s) and because of non-LTE effects ($\pm$0.4 km/s) (Nissen and Gustafsson, 1978). These errors can change by an approximately constant value all the results for a given photometric system. Internal consistency of measurements is much better and a typical error does not exceed ± 0.15 km/s.

There are two obstacles which prevent a widespread use of narrow-band photometry for the determination of microturbulence in whole range of T_e and M_V. The most suitable photometric system for ξ determination consists of narrow spectral bands centered on regions with medium-strong lines, weak lines and reference regions free of lines. These regions can be established for stars in rather narrow ranges of temtemperature and luminosity. For stars with considerably different T_e and M_V medium-strong lines can change into strong or weak lines and vice versa, and spectral regions free of lines may become overcrowded with lines. Therefore, the best narrow-band system should be established for rather narrow boxes on HR diagram. But in that case, for each box and each system good spectrophotometric standards with well known microturbulence parameter are necessary in order to establish the zero point of the ξ scale. It should be stressed once more that such standards are necessary for narrow-band photometry method, for COG method and for other methods as well.

A comparison of photoelectric narrow-band determinations of ξ with those from the COG method is shown in fig. 4. In this diagram the data are taken from different sources and because of the lack of standards

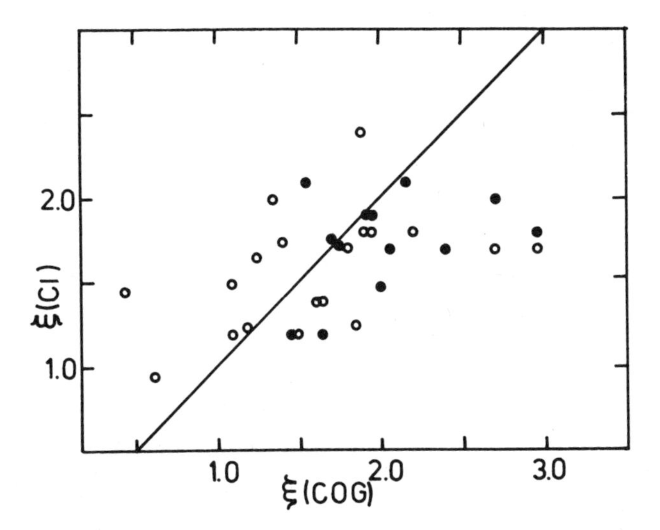

Fig. 4. Correlation between narrow-band and COG determinations of microturbulence. Circles denote stars with single COG determinations, points denote mean values from several determinations for the same star.

they are not reduced to one system. In spite of the large scatter of points caused by errors in both methods the results are in rather satisfactory agreement. But, this figure explicitely reveals that any differences in microturbulence of the order of 0.5 km/s when found from inhomogeneous data can not be taken seriously.

GOLDBERG - UNNO METHOD

The method suggested by Goldberg (1958) and extensively applied to the solar lines by Unno (1959 a, b) rests on the comparison of two line profiles from the same multiplet to determine the Doppler width and hence ξ.

The basic assumptions are as follows:

a) the source functions of the two lines of the same multiplet are equal at the same geometrical depth;

b) the source functions do not depend on wavelength;

c) the profile is Gaussian, $\phi(\Delta\lambda) \sim \exp -(\Delta\lambda/\Delta\lambda_D)^2$;

d) the Doppler width, $\Delta\lambda_D$, is constant in the region of line core formation;

e) the continuous opacity is equal for the both lines.

When this assumptions are fulfilled line absorption coefficients are equal for the two lines at points of equal emergent intensity. This leads to a simple formula for the Doppler width

$$\Delta\lambda_D = \frac{0.4343\,(\Delta\lambda_A{}^2 - \Delta\lambda_B{}^2)}{\log\,(gf)_A - \log\,(gf)_B}$$

where $\Delta\lambda_A$ and $\Delta\lambda_B$ are widths of A and B lines at the same emergent intensity. In order to avoid errors connected with the pressure broadening, only the lines with Voigt parameter $a < 0.01$ should be used and measurements must be limited to $\Delta\lambda < \Delta\lambda_D$. Source functions of the both lines should be the same with the tolerance of only about 2%. In spite of these severe limitations it is not difficult to find at least few pairs of lines useful for the Goldberg-Unno method. The best results are obtained when the two lines are of comparable (but not equal) intensity, their lower excitation potentials are the same and they are not very distant ($\lesssim$ 50 Å) in the wavelength scale. The influence of errors in determination of the $\Delta\lambda$ is reduced when the measurements are made at the level of intensity equal to the center of the weaker line ($\Delta\lambda_A$=0). Unfortunately, because of the discussed above assumptions and restrictions the Goldberg-Unno method can be used only if the spectrograms with dispersion better than about 5 Å/mm are available.

Up to date nine stars were studied with the use of this method α CMi and α Boo (Sikorski, 1976) α Boo, ε Dra, δ Dra, ε Cyg, η Cep, γ Cep and τ Cep (Stenholm, 1977) and τ Uma (Zaremba, 1979). The resulting average ξ values for these stars are in good agreement with those derived from the COG method. An important advantage of the Goldberg-Unno method is the possibility of evaluating the changes of ξ with height in the stellar atmosphere. Using pairs of lines of different central depths, ξ values can be found for different emergent intensities, i.e. for different layers in the stellar atmosphere. The most serious problem is the estimation of a layer of formation of a given point in the line profile. How troublesome and misleading could be such a determination has been demonstrated by Athay (1972). The problem is even more complicated when the calculated contribution function differs considerably from the response function. Besides, the layer of formation of a given point in the profile extends sometimes up to 3/4 of the thickness of the photosphere. Therefore, the attribution of the ξ- value to a given depth in the atmosphere is subjected to serious errors. With all the caution given to these uncertainties, results for the Sun and the stars suggest an increase of microturbulence in the upper photospheric layers ($\log\tau_5 \lesssim -3$) and in some cases a slight increase in deep layers ($\log\tau_5 > -1$). The depth dependence of turbulence was studied in three K2 giants and the Sun using widths of weak lines formed in the wings of the H and K lines. Ayers (1977) found that non-thermal velocity component in the Sun is constant at 1.6 km/s between $\log\tau_5$=0 and -3, while Stencel (1977) found that the velocities increase outward in K2 giants and the higher the luminosity, the steeper the gradient.

It should be remembered that the Goldberg-Unno method can not be applied when broadening by convection, macroturbulence and rotation is not negligible. But for late type stars the results based on weak and medium-strong lines are undoubtedly reliable. In spite of the discussed above limitations the Goldberg-Unno method should be recommended mainly because of its simplicity in practical application.

THE CURVE OF LINE-WIDTH CORRELATION METHOD

In order to utilize measurements of the line shapes for the studies of stellar turbulence a method called curve of line-width correlation (COLWC) was proposed by Huang and Struve a quarter of century ago. It is well known that microturbulence causes only broadening of line without changing its total absorption. If the velocity field in the stellar atmosphere is assumed to be depth-independent and can be represented by one value of microturbulence and one value of macroturbulence, then a unique relation log W versus log $D_{1/2}$ should exist, where $D_{1/2}$ is the half width. This relation is called curve of line-width correlation. Theoretical COLWC has been calculated by different authors for given model of stellar atmosphere with parameter b defined as the ratio of macro- to micro-turbulent velocity, $b=v_{macro}/\xi$. In order to establish ξ and v_{macro} for a given star the theoretical curves are slid both vertically and horizontally until a curve for a particular b gives the best possible fit to the observations. Because of the difficulties with the unique fit usually, the microturbulent velocity is independently derived from the COG method, while the obtained value of b permits to derive the macroturbulent velocity.

The COLWC method though simple in principle is burdened with serious drawbacks. The scatter of observational points is usually too

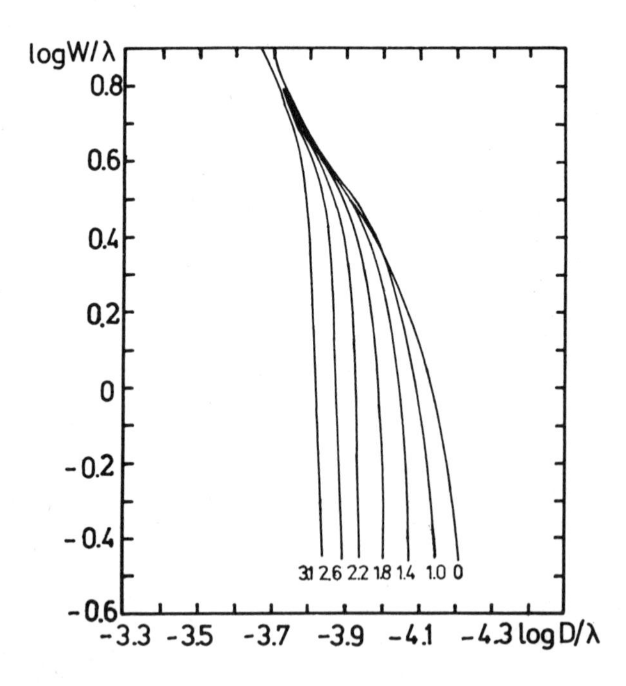

Fig. 5. Theoretical curves of line-width correlation for different ratios of macro- to micro-turbulence velocities.

large for a unique determination of the b parameter. Some improvement of the procedure is possible when mean values rather than individual observational points are plotted on diagram (Bell and Rodgers, 1964, 1965). But even then the observations can not be usually fitted by a single curve. That is why the COLWC method has not been widely applied. The macroturbulence has been derived by this method only for α Cyg, δ CMa and the Sun (van den Heuvely, 1963), β Dor (Bell and Rodgers, 1964), δ CMa (Bell and Rodgers, 1965) and α CMi (Evans et al., 1975). Some modifications were introduced to the COLWC method by Evans et al. (1975). They measured widths of lines at 1/2 and 3/4 of central line depth and constructed plots of log W versus $\log(D_{3/4}/D_{1/2})$. This modification did not however removed the disadvantages of the classical COLWC. The modified COLWC was tried unsuccessfully in an analysis of early type stars by Slettebak (1956).

From the practical point of view the developments of this method seem to be purposeless. Only high resolution spectrograms could reduce errors in the measurements of line-widths and diminish the scatter of points on the COLWC diagrams. But for high dispersion spectra fine and/or Fourier analysis provide much more reliable evaluation of the parameters describing the velocity fields in the stellar atmospheres.

DIFFERENTIAL LINE SHIFTS METHOD

Differential shifts of stellar lines can occur when the convective type velocity field in a stellar atmosphere is coupled with the differences in the depth of line formation. If the stellar disc is covered by small convective type cells, similar to solar granulation, then a relation between differential shifts of lines, $\lambda^{*} - \lambda_{lab}$, and their low excitation potentials, χ, should be expected. High-excitation spectral lines are preferentially formed in the hot, rising and thus blueshifted elements, while low-excitation lines are preferentially formed in the cooler sinking and redshifted elements. In the solar atmosphere a correlation between continuum brightness and upward velocity is most pronounced for weak lines formed in the deeper layers and decreases for lines in the higher layers (Canfield and Mehltretter, 1973). Thus a correlation between differential line shifts and the strength of lines should be expected. This effect known as Burn's effect has been observationally confirmed by Głębocki and Stawikowski (1969, 1971) and Beckers and Nelson (1978). The relation between line shifts and low excitation potential (VR-LEP) have been found by Lambert and Mallia (1968) and Głębocki and Stawikowski (1969, 1971).

The measurements of differential line shifts require very high dispersion spectrograms. It is possible to measure stellar line positions in high quality spectrograms with an accuracy of 2 microns what for a dispersion of 2 Å/mm corresponds to 0.1 km/s. Because of uncertainties in laboratory wavelengths, measurements of stellar line shifts should be made either relative to very accurate low-pressure laboratory wavelengths or relative to the Sun, whose wavelengths are known with an accuracy of 1 mÅ or better. Otherwise pressure shifts or errors in laboratory λ scales may introduce spurious effects (Głębocki and Stawikowski, 1971). The differential line shifts are expressed in velocity units, VR = $(\Delta\lambda/\lambda)c$. The solar line shifts change their sign at the limb. This might be important when differential shifts are analysed in stars, where integrated light is observed. The contribution of the limb to the integrated light is however small, especially when limb darkening is taken into account.

The results of the differential line shifts measurements are usually presented by the gradient in the VR-LEP relation $A = \Delta VR/\Delta\chi$. Fig. 6 shows the VR-LEP relation for α Boo obtained by Dravins (1974). The points represent individual shifts for neutral metal lines derived from Griffin's spectrograms of 1.5 Å/mm (Griffin and Griffin, 1973). The gradient A amounts -0.45 km/s/eV. Stawikowski (1976) obtained the A

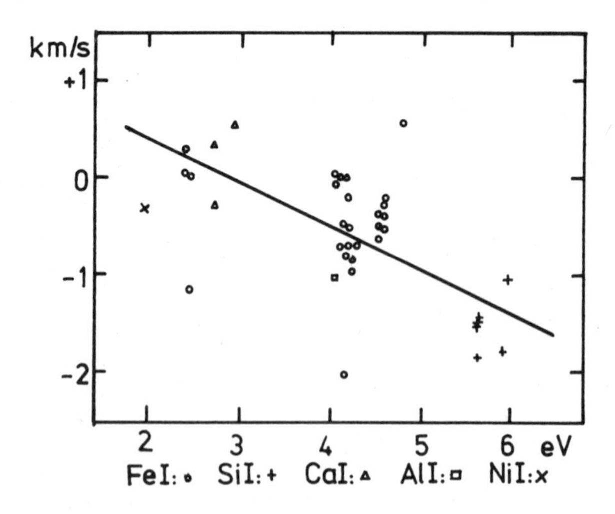

Fig. 6. VR-LEP relation for α Boo according to Dravins (1974). Symbols represent measurements of individual line shifts.

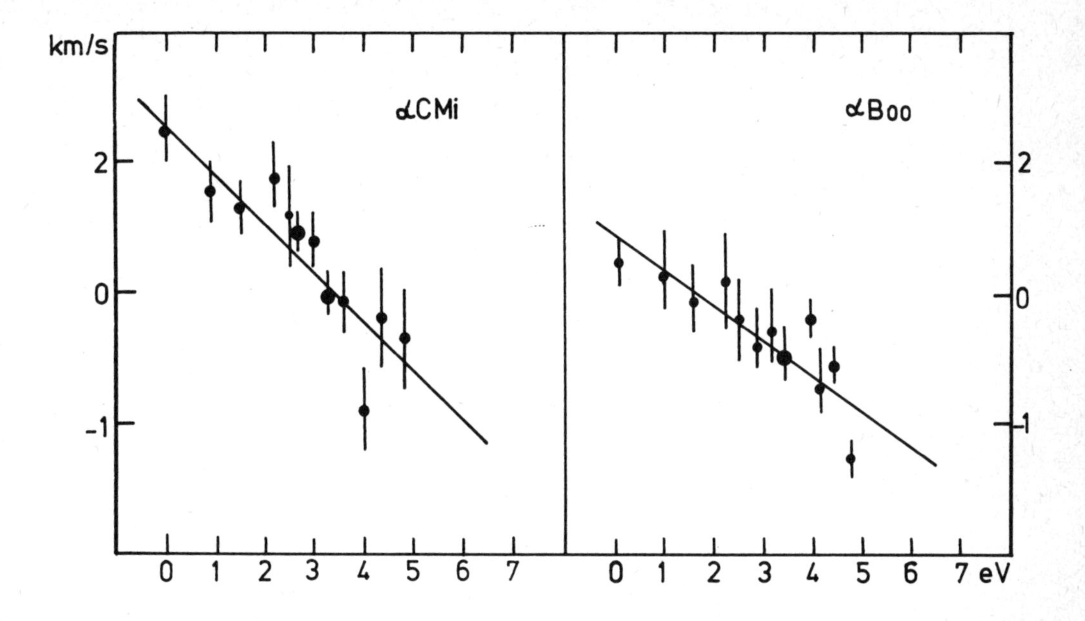

Fig. 7. VR-LEP relations according to Stawikowski (1976). Each point represents mean value of line shifts for several dozens of Fe I lines. Gradients (full lines) has been fitted to the points by the least square method.

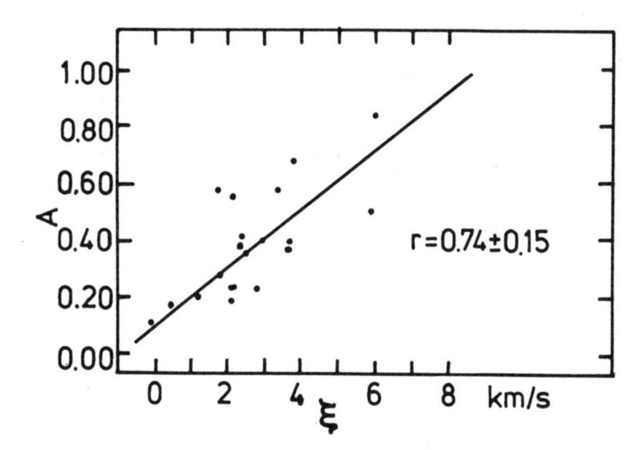

Fig. 8. Correlation between the value of gradient A and COG microturbulent velocities. Correlation coefficient amounts 0.74.

values for 20 F, G and K type stars. His results were based on inhomogeneous observational material with the dispersion of 2 to 8 Å/mm. Examples of VR-LEP relation are shown in fig. 7. Each point on these diagrams represents an arithmetic mean of several doznes VR measurements for lines with a given value of χ. A correlation between the A values and microturbulence is shown in fig. 8.

The VR-LEP method suffers from two serious shortcomings. First, it requires very high dispersion spectrograms; second and the essential, the results of line shifts when presented as a gradient A can not be explicitely converted to the convective type velocity in the stellar atmosphere. Even when a simple two-column model of atmosphere is adopted with a very crude mathematical treatment, the observed slope in the VR-LEP relation depends not only on convective velocity, but on the mean excitation temperature, on the temperature difference $\Delta T = T_{hot} - T_{cold}$ and on the percentage of disc area covered by the hot columns, S. Even for the Sun these quantities are poorly known. For stars ΔT and S have to be treated as free parameters in a function relating the observed gradient A with the convective velocity.

Schatzman and Magnan (1975) presented a slightly different interpretation of the observed VR-LEP relation. They assumed that the velocity of the ascending and descending convective bubbles can be represented by a Gaussian distribution. Additionally, they made an assumption that a unique relation exists between the temperature of the bubble and its velocity. Owing to the differences in lower level populations between hot and cold bubbles coupled with the upward and downward velocities a theoretical relation between differential shifts and lower excitation potential and the line strength has been derived. However, the number of free parameters in this approach has not been reduced in comparison with the column model. The authors stressed that the principal weakness of their interpretation lies in the ignorance of the correlation length of the turbulence. They have considered only the microturbulent situation, i.e. the mean free path of the photon being smaller than the size of convective element. The opposite situation corresponds to the theory of column model.

In spite of the difficulties with the interpretation of VR-LEP relation this method is a valuable complement of the stellar velocity field studies. It presents the interesting possibility of separating stellar turbulence into its physical components by comparing differential line shift results with other measurements of turbulence in the same star.

REFERENCES

Andersen,P.H. 1973, Publ. Astron. Soc. Pacific, 85, 666.
Athay,R.G. 1972, Radiation Transport in Spectral Lines, D.Reidel Publishing Comp., Dordrecht
Ayers,T.R. 1977, Ap.J., 214, 905
Beckers,J.M. and Nelson,G.D. 1978, Solar Phys., 58, 243
Bell,B. 1951, Harvard Univ. Special Rep., No.35
Bell,R.A. and Gustafsson,B. 1978, Astron. & Astroph.Suppl.in press
Bell,R.A. and Rodgers,R.W. 1964, M.N.R.A.S., 128, 365
Bell,R.A. and Rodgers,R.W. 1965, M.N.R.A.S., 129, 127
Bely,F. 1966, in Abundance Determination in Stellar Spectra, IAU Symp. No.26, ed.H.Hubenet, Academic Press, p.254
Blackwell,D.E. and Shallis,M.J. 1979, M.N.R.A.S., 186, 673
Böhm-Vitense,E. 1972, Astron. & Astroph., 16, 81
Bonnell,J. and Branch,D. 1979, Ap.J., 229, 175.
Canfield,R.C. and Mehltretter,J.P. 1973, Solar Phys., 33, 33
Chaffee,Jr.,F.H. 1970, Astron. & Astroph., 4, 291
Corliss,C.H. and Warner,B. 1964, Ap.J.Suppl., 8, 395
Dravins,D. 1974, Astron. & Astroph., 36, 143
Dumont,S.,Heidmann,N.,Jefferies,J.T. and Pecker,J.-C. 1975, Astron. & Astroph., 40, 127
Elste,G.H.E. and Ionson,J. 1971, Bull. Am. Astr. Soc., 3, 380
Evans,J.C. and Elste,G.H.E. 1971, Astron. & Astroph., 12, 428
Evans,J.C.,Ramsey,L.W. and Testerman,L. 1975,Astron.& Astroph.,42,237
Foy,R. 1972, Astron. & Astroph., 18, 26
Foy,R. 1978, Astron. & Astroph., 67, 311
Garz,T. and Kock,M. 1969, Astron. & Astroph., 2, 274
Głębocki,R. 1972, Acta Astron., 22, 141
Głębocki,R. 1973, Acta Astron., 23, 135
Głębocki,R. and Stawikowski,A. 1969, Acta Astron., 19, 87
Głębocki,R. and Stawikowski,A. 1971, Acta Astron., 21, 185
Goldberg,L. 1958, Ap.J., 127, 308
Gray,D.F. 1978, Solar Phys., 59, 193
Griffin,R. and R. 1973, M.N.R.A.S., 162, 255
Gustafsson,B. and Bell,R.A. 1979 Astron. & Astroph., in press
Gustafsson,B.,Kjaegaard,P. and Andersen,S. 1974,Astron.& Astroph.,34,99
Hasegawa,T. 1978, Journal of Hokkaido Univ. of Education, 28, 61
Havnes,O. and Moe,O.K. 1975, Astron. & Astroph., 42, 269
Hensberge,H.and De Loore,C. 1974, Astron. & Astroph., 37, 367
Heuvel,van den,E.R.J. 1963, Bull.Astr.Inst.Neth., 17, 148
Lambert,D.L. and Mallia,E.A. 1968, Solar Phys., 3, 499
Landi Degl'Innocenti,E. 1975, Astron. & Astroph., 45, 269
Mihalas,D. 1973, Ap.J., 179, 209
Nissen,P.E. and Gustafsson,B. 1978, in Astronomical Papers dedicated to Bengt Stromgren, eds.A.Reiz and T.Andersen,Copenhagen Univ.Obs., p.43
Osmer,P.S. 1972, Ap.J.Suppl., 24, 255
Schatzman,E. and Magnan,C. 1975, Astron. & Astroph., 38, 373
Sikorski,J. 1976, Acta Astron., 26, 1
Slettebak,A. 1956, Ap.J., 124, 173
Smith,M.A. 1973, Ap.J., 182, 159
Smith,M.A. 1974, Ap.J., 192, 623
Stawikowski,A. 1976, thesis, Toruń Univ.
Stencel,R.E. 1977, Ap.J., 215, 176
Stenholm,L.G. 1977, Astron.& Astroph., 61, 155
Unno,W. 1959a, Ap.J., 129, 375
Unno,W. 1959b, Ap.J., 129, 388
Van Paradijs,J. 1973, Astron. & Astroph., 23, 369
Warner,B. 1964, M.N.R.A.S., 127, 413

Wright,K.O. 1951, Publ. Dom. Astroph. Obs., 8, 1
Wright,K.O. 1966, in Abundance Determinations in Stellar Spectra, IAU Symp. No.26, ed.H.Hubenet, Academic Press, p.15
Zaremba, D. 1979, Acta Astron., in press
Zeinalov,S.K. 1970, Izv. Crimean Astroph. Obs., 41, 298

ANALYSIS OF HIGH RESOLUTION STELLAR LINE PROFILES

David F. Gray
Department of Astronomy
University of Western Ontario
London, Ontario, Canada

1. Introduction

High resolution implies that we obtain some information on spectral line shapes. In late-type stars, we need to measure velocities of a few km/sec to accomplish this. Increasing the spectral resolution and the signal to noise ratio allows us to progress step by step toward deeper physical understanding. The steps we take often lead to good debate and "stimulate" our lives. I am sometimes amused at the urgency we feel to press on to the next step. We rarely seem to pause and enjoy the completion of previous steps. Perhaps this is because we always see shortcomings in completed work. Quite typically one will "discover" the importance of some physical phenomenon (It makes little difference how many others already know about it.), and in his eyes everything done previously becomes wrong because this phenomenon was not included. We used to hear how Milne-Eddington or Schuster-Schwarzschild model atmospheres were inadequate - we had to use instead properly computed depth dependence. We used to hear how LTE models were no good - we had to use more detailed physics. Now we talk about line analyses being inadequate because it has not included velocity fields. The curious thing is that we believe that including our pet phenomenon gives the correct models. We ignore all those other phenomena as yet unseen! (Is this a mechanism for maintaining sanity?) I think it really amounts to a statement of what we are able to measure, compute, or understand.

Observationally, I view the situation in steps of "toughness", i.e. how difficult it is to obtain certain types of information from spectral line shapes. Roughly these can be grouped into

1) line widths - typically a half width is measured but we get no information about what causes the broadening.
2) line broadening - where shapes as well as widths are measured and we begin to discern the characteristics of broadening resulting from rotation or macroturbulence.
3) the details - we see line asymmetries or other structure; we measure true central line depths. Questions concerning the physical mechanisms for turbulence can be tackled, $T(\tau)$ derivations become possible, and non-LTE effects can be disentangled.

Generally, each degree of toughness involves higher spectral resolution and better signal to noise. The detrimental effects of line blends also become more bothersome as we progress to a tougher category.

I will concentrate on the second and third categories of toughness and largely omit the first one. Part of line analysis is knowing how good the data is. We now consider modern capabilities in stellar spectroscopy.

2. Observational Considerations

The number of stars available for high resolution studies is restricted in practice by photon rate limitations. Typically 20 mÅ/pixel is appropriate for work on late-type stars for a resolution of ~50 mÅ (~3 km/sec). A 20 mÅ slice is pretty tiny when it comes to building up good photon statistics. Everyone knows that signal to noise goes as $N^{1/2}$ but also recall that the time required to collect a given number of photons varies as the cube of the resolution (or if multielement detectors are used and the size of the field is not of interest, then as the square of the resolution). In the resolution regime I am considering here, even 3-4 magnitude stars begin to look faint. Large telescopes are important for bright star work.

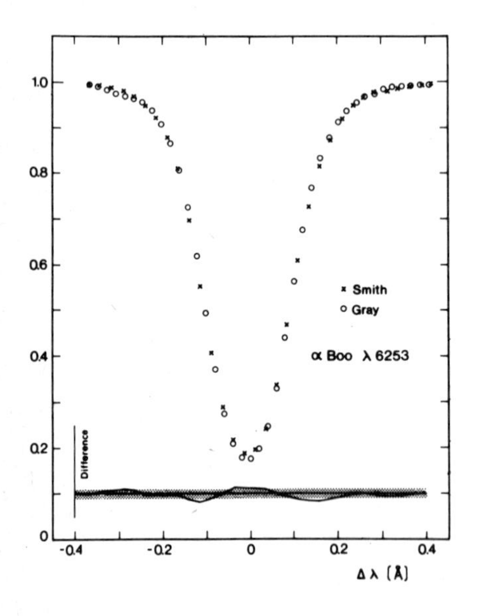

Fig. 1. Two sets of measurements of Fe I λ6253 for Arcturus are compared. The differences are plotted on the same scale below the profile. The dotted strip is ±1% in height.

I have recently compared line measurements made with four totally different sets of equipment in order to see what errors exist with modern measurements. As you know, past comparisons have shown errors of 10-20% to be common (e.g. Wright et al. 1964,

Conti and Frost 1977). Fig. 1 shows Fe I λ6253 in Arcturus as measured by M. Smith using a Reticon on the McDonald 2.71 m telescope coude and by myself using a Digicon on the U.W.O. 1.22 m telescope coude. After allowances for differences in scattered light, the average deviation over ±0.4 Å is 0.7%. Fig. 2 shows a similar comparison

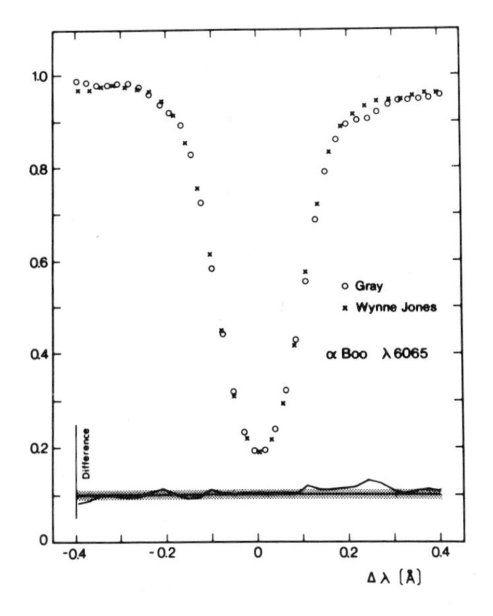

Fig. 2. Measurements of Fe I λ6065 for Arcturus are compared.

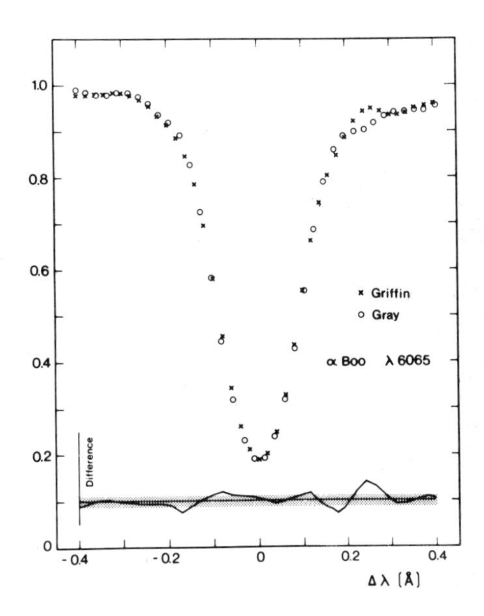

Fig. 3. Measurements of Fe I λ6065 for Arcturus are compared.

for Fe I λ6065 in Arcturus. The measurements of Wynne-Jones were done using the 2.50 m Herstmonceux telescope to feed a Michelson interferometer. The average deviation here is 0.8%. Fig. 3 shows a comparison of photographic measurements made by Griffin with mine. The average deviation is 1.0%. A more detailed discussion of this material is being published in the P.A.S.P.

I conclude that line shapes can currently be measured with errors ≲1%. This is still not as good as in solar work, but it is good enough to start doing some physics.

Caution though. We still have a range of up to 5% in zero-level (scattered light?) differences. This is a crucial point for T(τ) derivations because there we need line center residual fluxes. Relatively large uncertainties in zero level is one reason why I place the measurements of central line depths in toughness category 3.

3. Analysis in the Fourier Domain

a) Basic Concepts

The general discussion of line analysis in the Fourier domain is available in several places (e.g. Gray 1976, 1977, 1978; Smith and Gray 1976) and I will assume that you are somewhat familiar with the process.

First, I wish to show you fig. 4. It is a reconstructed profile - or rather two versions of a reconstructed profile - with slightly different Fourier noise filters. The noise level in the original data is about 2% (not bad by stellar standards). Both filters are reasonable and possible choices. I think you will agree that differences in these reconstructions are uncomfortably large. In principle it is possible to get "better" data, i.e. data with lower noise and higher resolution. But that is costly in observing time and may be impossible in practice. Rather, one can avoid the noise filter problem by performing the analysis in the Fourier domain.

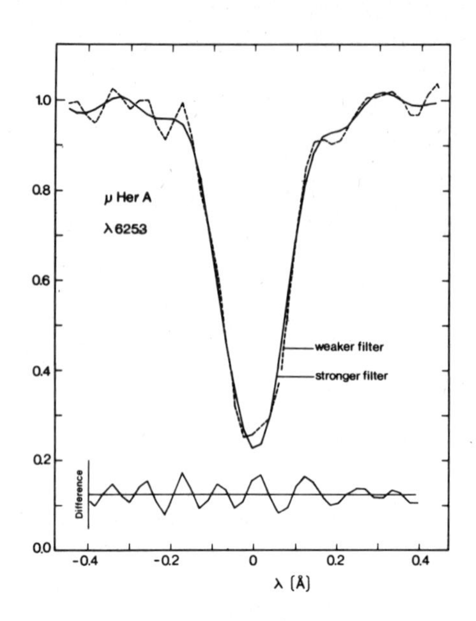

Fig. 4. Two reconstructions of the same set of data showing the effect of the noise filter. The difference is plotted below the profiles.

A basic point of what happens is illustrated in fig. 5. Here several common transforms are compared. Clearly, in order to distinguish between them, it is necessary to measure high enough in frequency and with low enough noise to resolve the differences where they occur.

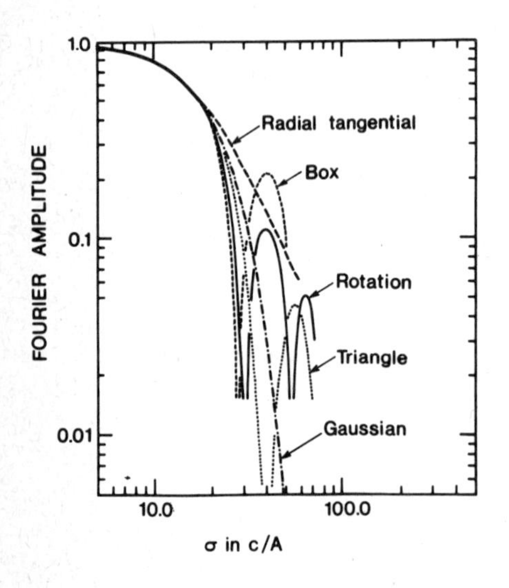

Fig. 5. Transforms of several commonly encountered functions are compared. Differences become larger toward high frequencies.

Fig. 6 emphasizes the signal/noise behavior. For a noise level n_1, only the top of the main lobe can be seen. As higher photon counts are reached, the high frequency, low amplitude structure is revealed - the details we need to do the science. Notice that the signal/noise is a strong function of σ and that the signal is concentrated to limited σ bands, e.g. the first sidelobe.

It was with these tools that I attacked the classical problem of separating macroturbulence from rotation. If we think way back, we recall that most people used to assume isotropic Gaussian macroturbulence, and it was repeatedly stated that one could not distinguish macroturbulent from rotational broadening except in a statistical sense. Fig. 5 already showed why this was the case and what needed to be done to solve the problem. At low spectral resolution, the two broadeners look the same. It was necessary to push the observations to the region where they are unique. That has been done. Further, under certain assumptions, it is possible to allocate a certain fraction of the line's macrobroadening to macroturbulence and the remainder to rotation. Obviously this process has meaning only within the context of its basic assumptions - no different than any other operation in science.

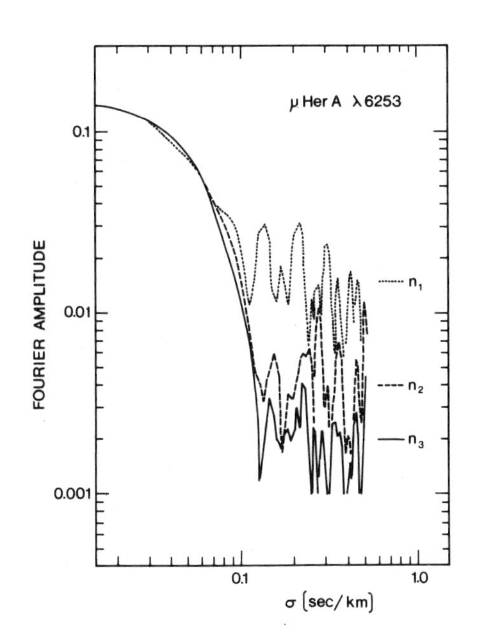

Fig. 6. The noise level can often obscure the information contained in the higher frequencies.

b) Convolutions and Disc Integrations

Initially I used the convolution approximation to combine the effects of macroturbulence and rotation. This is a valid approximation for some types of macroturbulence and the classical treatment of rotation. But as more involved models come into use, it was necessary to do integrations over the disc (rather costly). The convolution approximation can be used if the functions are independent of position on the disc (θ). Otherwise integration over the disc cannot be avoided. Still, as a thinking tool, the convolution approximation can be very useful.

Numerical experiments also show that in many applications the full convolution approximation is good to a few per cent. If, for example, we replace the $G(\lambda)*M_{RT}(\lambda)$ convolution, where $G(\lambda)$ is the rotation profile and $M_{RT}(\lambda)$ is the radial-tangential macroturbulence profile, with a disc integration, we gain the inclusion of limb darkening and a more rigorous treatment of the macroturbulence broadening which is θ dependent. Assuming equal weighting of radial and tangential components, the disc integrations turn out to give 6 to 8% larger values for ζ_{RT} than does the convolution formulation. This is almost exclusively a result of limb darkening reducing the broadening of the tangential component.

In addition, we should include the center to limb variation in the I_ν profile for a complete and proper disc integration. For the solar case, where we can measure this change, it is small enough so that neglecting it still gives reasonable results for ζ_{RT} and v sin i (Gray 1977). For other stars, we do not know the θ dependence so we cannot put it into the disc integrations even if we want to. We may, of course, postulate a model which gives the θ dependence and then the complete integration can indeed be done. The postulated model will have to produce a substantial θ dependence to have much of an effect on the analysis.

Another basic uncertainty exists in the $I_\nu(\theta)$ profile. That is the depth dependence of modeled broadening parameters, e.g. microturbulence dispersion, ξ. It is possible to postulate such large gradients in ξ that virtually any shape can be manufactured for $I_\nu(\theta)$ profiles. It quickly becomes a numerical exercise reminiscent of epicycles. Fortunately in many cases the $I_\nu(\theta)$ profile is significantly narrower than the macrobroadening.

c) Precision in Fourier Analysis

Now let me be somewhat more specific in discussing the precision involved once a model has been adopted. We expect quite generally that the greatest precision will be obtained for the dominant component of broadening. If we apply the micro-macroturbulence plus rotation model to K giant analyses, then macroturbulence dominates, rotation comes next, and microturbulence is smallest. In fact it is even surprising that any information can be obtained on microturbulence in the presence of the two larger broadeners. But we are fortunate in this case because the first zero and the side-lobe structure of the transform are sensitive to the saturation of the line.

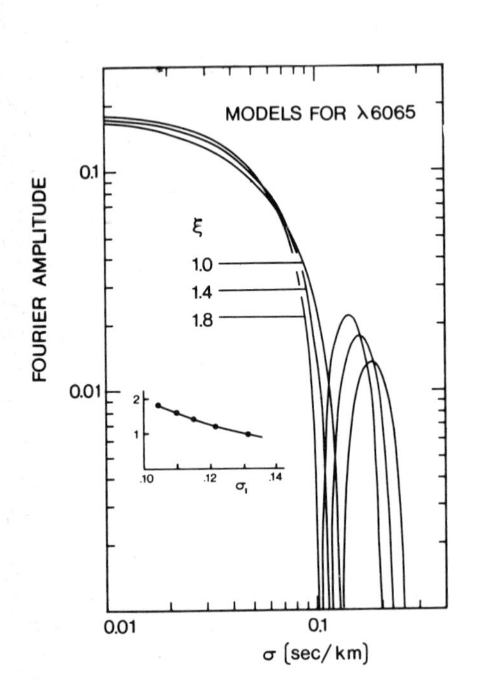

Fig. 7. The position of the first zero in the transform of a saturated line is very sensitive to the degree of saturation. The insert graph shows ξ (km/sec) as a function of the position of the first zero (σ_1).

If classical depth independent microturbulence of dispersion ξ is introduced into the calculation, it is possible to fix ξ to within 0.1 km/sec (fig. 7). Larger errors appear in the observed position of the first zero making the errors in ξ more typically $\pm$0.1 km/sec.

Systematic errors can also come in here because the derived value of ξ depends on $T(\tau)$ and non-LTE effects. In a recent comparison (Gray and Martin 1979) of SMR K giants with normal K giants, where temperature differences of up to 180 °K occur near log

$\tau_{5000} = -3$, the derived values of ξ change by 0.2 to 0.3 km/sec which is about 20% of the value of ξ.

The toughest part of the K giant line broadening analysis though is choosing the ratio v sin i/ζ_{RT}(fig. 8). Errors in this ratio lead to modest errors in ζ_{RT}, typically ±0.3 km/sec (about 7% of ζ_{RT}) but significantly greater errors in v sin i, typically ±0.5 km/sec (≳20% of v sin i). The leverage is good for macroturbulence but poor for rotation. The leverage will be reversed in early type stars where rotational broadening dominates.

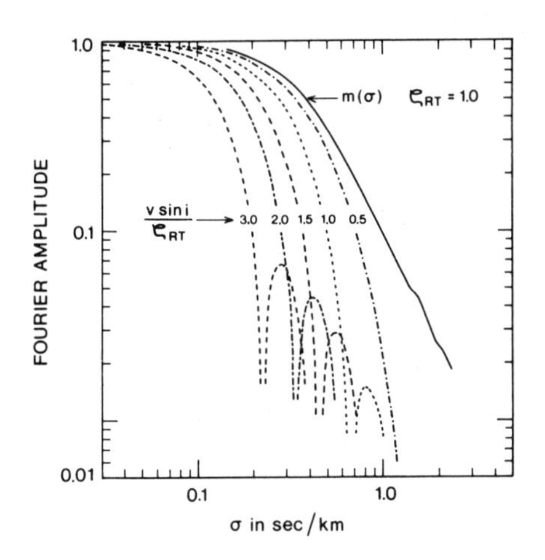

Fig. 8. The separation of macroturbulence from rotation depends on choosing one of these curves from the ensemble on the basis of its shape (from Gray 1977).

Some idea of external consistency can be gained by a comparison of results between workers. For four stars measured by both M. Smith and by me, the average deviation from the mean in ζ_{RT} was 6%. Rotation values were only consistent to ~35%.

d) Versatility of Transform Analysis

As you can see, the ideas of Fourier transforms are very useful in conceptualizing data collection and analysis. Curiously enough, I received a manuscript not long ago from a competent theoretician in which he spoke of "overcoming" the most serious "flaw" that exists in present Fourier-analysis techniques. By this he apparently meant that he preferred his own models over those used in the past. From my point of view this is an asset not a flaw. We can incorporate new modeling into the Fourier analysis at will. The classical micro-macroturbulence-rotation model is only one of several possibilities and the value of the Fourier domain is in no way restricted to this model.

The Fourier analysis is particularly suited for discerning small systematic differences in profile curvature where measurements over the whole profile can be brought to bear. (This is exactly the case for macroturbulence vs. rotational broadening.) In some other instances it will be more suitable to use the profile itself.

4. Analysis in the Wavelength Domain

It is pretty hard to see the central depth of a line in the Fourier domain. More generally, if we seek source function information, we will want to use the profile itself. I am thinking of the Eddington-Barbier relations and their generalization in terms of contribution functions. We are then forced to live with the uncertainty of the noise filter and the observations become tougher.

A typical wavelength domain analysis consists of synthesizing the theoretical line profile and comparing it to the observations. If significant Fourier noise filtering has been made on the data, then the same filter should be applied to the theoretical profile.

But rather than dissipate my remaining time discussing various innuendoes of profile fitting, I will concentrate on one interesting case where profiles tell more than their transforms, namely the measurement of line asymmetries in photospheric lines of late-type stars. One might expect that the imaginary component of the Fourier transform would be the appropriate tool since it is a measure of asymmetry in functions. But it turns out that the imaginary component is too sensitive to small blends and noise in the continuum. Furthermore, the asymmetries in these lines involve only a fraction of the whole profile (unlike UV lines in early type stars) which effectively removes one of the leverages of the Fourier analysis. Using the profile against its

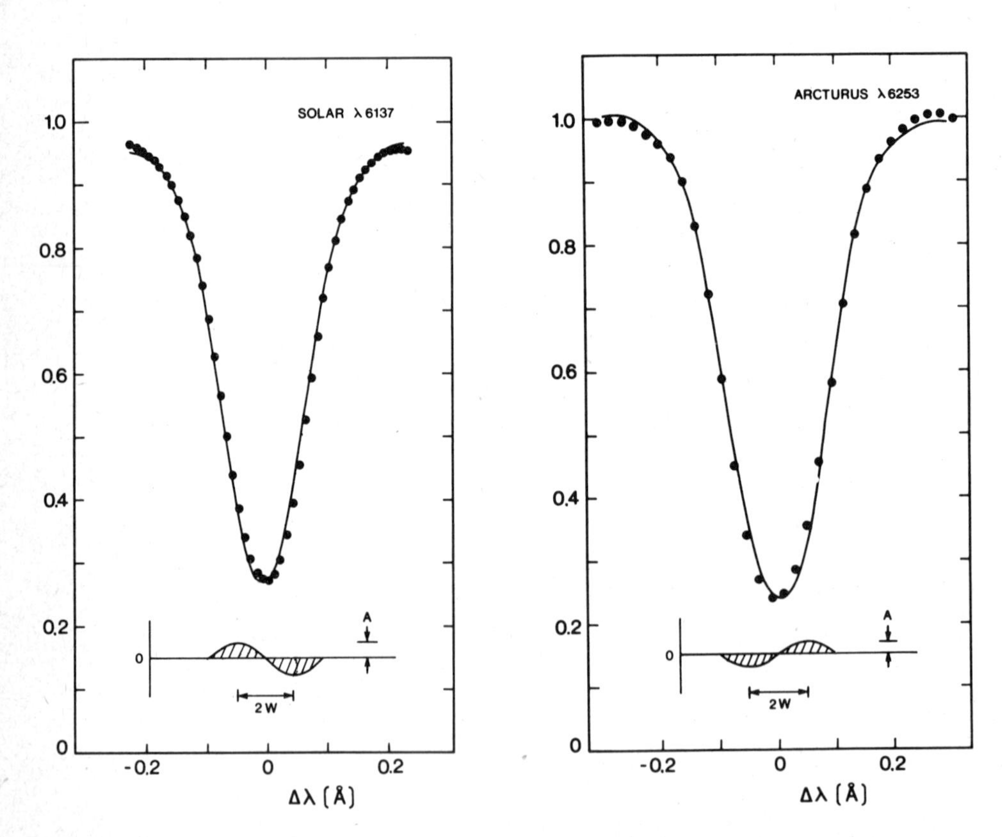

Fig. 9. The asymmetry in solar flux lines shows a red-shifted core.

Fig. 10. The asymmetry in lines of the Arcturus spectrum show blueshifted cores.

mirror image produces a result like the one in fig. 9. These are solar flux measurements taken from the atlas of Beckers et al. (1976). The process shows very well the familiar red shifted core but neglects the small red shift of the far wings which can be seen only with difficulty.

The traditional measure of asymmetry is the line bisector. It has been used frequently in solar work but is less appropriate in the stellar case. To use it, one needs oversampling of the data and excellent signal to noise. The bisector allows us to compare small sections of the profile with each other. The profile reflection scheme of fig. 9 reduces the subdivision of the profile to the point where we are comparing the core to the wings. The amplitude and width of the difference curve is used to parameterize the asymmetry.

Another example is given in fig. 10 where we see the asymmetry in a profile of Arcturus. Here the core is blue shifted.

Notice that the accuracy of the spectrophotometry must be $\sqrt{2}$ times better than the amplitude of the difference curve just to detect the effect. Difference curve amplitudes are 2-3%. (Details are being published in the Ap.J.) In addition, one has to worry about spectrograph focus errors and aberrations which can make the instrumental profile asymmetric. These are reasons for placing line asymmetry measurements in toughness category 3.

5. Summary and Comments

High resolution observations (~3 km/sec) with photometric errors of $\lesssim 1\%$ can be made and are capable of giving physical information about stellar turbulence. There is no <u>one</u> technique for analyzing such data. We can use the advantages of the wavelength domain or the Fourier domain or other transformations yet to be invented according to which of them proves to be the most suitable for the job at hand.

Details of Fourier analysis philosophy and application illustrate the versatility and power in separating line shapes. Uncertainties in modeling remain the biggest obstacle to complete analysis. Adoption of the microturbulence-macroturbulence-rotation model of the broadening allows precision of a few per cent in determining the size of the largest broadener (macroturbulence for late-type stars).

Line asymmetries in late-type stars are more readily measured in the wavelength domain than in the Fourier domain. Examples of asymmetries as seen in Arcturus and the solar flux spectrum are given.

We are in a reasonable position for identifying physical mechanisms responsible for stellar line broadening. It is important that theoretical calculations based on various models be explicit in tagging features which are distinct for that model. Then if these features can be observed, they will serve as discriminatns between the physical mechanisms now lumped together under the name turbulence. I am optimistic as I look toward future possibilities.

References

Conti, P.S. and Frost, S.A. 1977 Ap.J. 212, 728.

Beckers, J.M., Bridges, C.A., and Gillman, L.B. 1976 Sacramento Peak Observatory Project 7649, Vols. 1, 2.

Gray, D.F. 1976 The Observation and Analysis of Stellar Photospheres (New York: Wiley).

——— 1977 Ap.J. 218, 530.

——— 1978 Solar Physics 59, 193.

Gray, D.F. and Martin, B.E. 1979 Ap.J. 231, 139.

Smith, M.A., and Gray, D.F. 1976 P.A.S.P. 88, 809.

Wright, K.O., Lee, E.K., Jacobsen, T.V., Greenstein, J.L. 1964 Publ. Dom. Astrophys. Obs. Victoria 12, 173.

EXAMPLES OF NON-THERMAL MOTIONS AS SEEN ON THE SUN

Jacques M. Beckers*

Sacramento Peak Observatory†
Sunspot, NM 88349/USA

SUMMARY

On the sun we can identify many of the motions derived from stellar spectral analysis. A summary is given of the observed solar velocity phenomena. Many of these (e.g. meridional flow, giant cells, solar differential rotation, supergranulation) are of great interest in astrophysics especially for interior structure and chromospheric and coronal structuring but contribute virtually nothing to the velocities derived from a solar irradiance spectrum analysis. Others (granulation, very small scale motions and to a lesser extent, oscillations) do contribute substantially to the integrated sun velocity analysis. Some of the properties of these motion fields are described.

1. INTRODUCTION

The word "turbulence" in astrophysics is used generally to decribe motions which cause line broadening and changes in line saturation (curve of growth effects) but which are otherwise intangible as a specific kind of motion because of the absence of additional observational information like spatial resolution, characteristic line profiles, line shifts, etc. Turbulence in astrophysics may therefore have nothing or little to do with hydrodynamic turbulence except for the fact that statistical techniques are used also in astrophysics to define the magnitude of the associated non-thermal motions because of the lack of sufficient observations. Astrophysical "turbulence" can include convection, waves, stellar winds, large scale flow patterns and even stellar rotation if the spectral resolution is insufficient to identify the characteristics of any of these non-thermal motions. Even "microturbulence," or the quasi-thermal motions derived from line saturation changes, can result from e.g. systematic velocity gradients along the line of sight say in stellar winds, in convection, or in acoustic waves propagating along the line of sight. Observations on the sun, where the abundance of precise observations allows the identification of many, although not all, of these "turbulent" motions, substantiate the

---

*Now at the Multiple Mirror Telescope Observatory (MMTO), University of Arizona, Tucson, AZ 85721. The MMTO is jointly operated by the University of Arizona and the Smithsonian Astrophysical Observatory.

†Operated by the Association of Universities for Research in Astronomy, Inc. under contract AST 78-17292 with the National Science Foundation.

above. Rather than talking about "turbulence" I therefore will use the term "non-thermal motions" to refer to the kind of motions discussed in this colloquium and the terms "micro-" and "macro-velocities" as introduced by Canfield and Beckers (1976) for what is often referred to as "micro-" and "macroturbulence." As a statistical measure of the non-thermal motions I will use rms velocities which are $\sqrt{2}$ x less than the usual turbulent velocities in the case of a gaussian velocity distribution.

In this review I will first summarize the non-thermal motions derived from observations of the sun as a star. Then I will describe the resolved and unresolved solar velocity fields as known today and compare these with the sun as a star results. Solar observations thus serve to identify the astrophysical mechanisms responsible for the stellar non-thermal motion observations. I refer to other recent reviews for a more detailed description (Beckers and Canfield, 1976; Canfield and Beckers, 1976; Deubner, 1977; Zirker, 1979; Beckers, 1980).

2. RESULTS OF SOLAR IRRADIANCE SPECTROSCOPY

High resolution, good photometric precision spectra of the integrated sun radiation obtained by Beckers _et al._ (1976) were analyzed by Gray (1977) and Stenholm (1977). Figure 1 shows part of this spectrum atlas. Stenholm (1977) derives a total

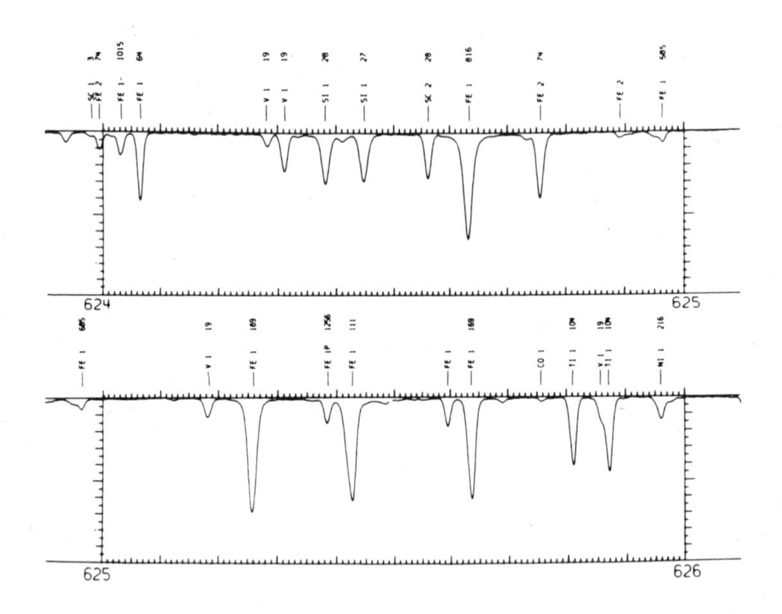

Figure 1. Portion of the spectrum atlas of integrated sunlight by Beckers _et al._ (1976).

non-thermal rms velocity field of 1.0, 0.6 and 0.3 km/sec for continuum optical depths of 10^{-1}, 10^{-2} and 10^{-3} respectively using Goldberg's (1958) multiplet method. These values are impossibly small considering that the solar rotation alone should result in rms velocities in excess of 1 km/sec.

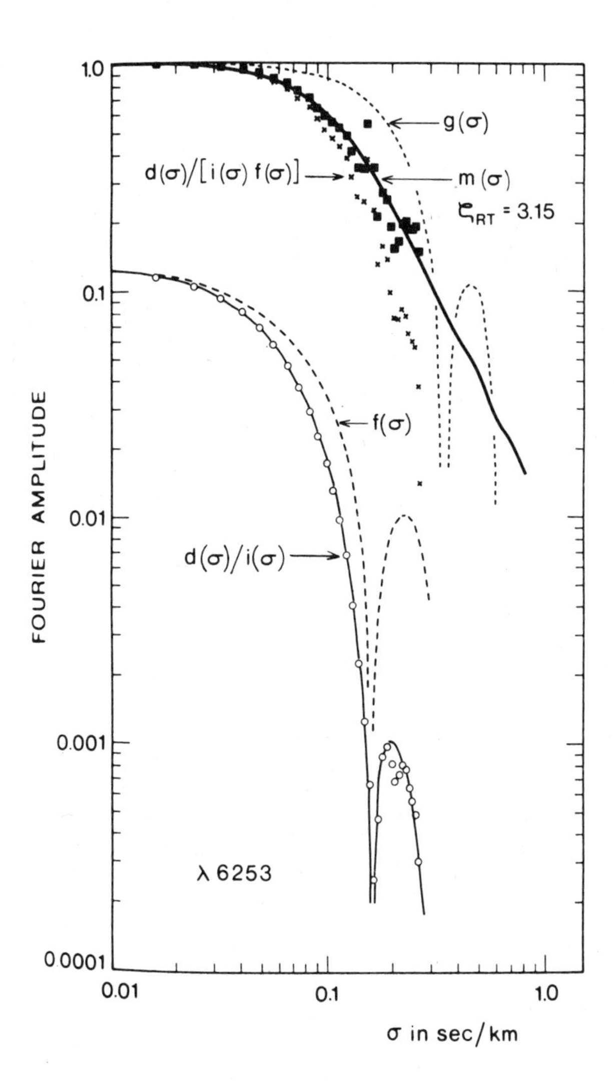

Figure 2. Analysis by Gray (1977) of the integrated sun λ6252.6 profile in the Fourier transform domain. σ = frequency in spectral domain (in cycles/Å); d(σ) = line profile; i(σ) = instrument profile; g(σ) = solar rotation; m(σ) = macrovelocities (from Gray, 1977).

Gray's (1977) analysis of solar line profiles in the Fourier transform domain gives results which look more reasonable. Figure 2 is from his publication and shows the separation (in the Fourier transform domain) of the Fe I λ6252.6 solar profile, after correction for instrumental smoothing $(d(\sigma)/i(\sigma))$, in solar rotation $g(\sigma)$, macrovelocities $m(\sigma)$ and the inherent line profile (including microvelocities), $f(\sigma)$. The solar rotation velocity derived from g equals 1.92 km/sec which compares well with the solar synodic equatorial rotation rate of 1.80 km/sec. The rms microvelocity derived from the zero(s) in $f(\sigma)$ (0.35 km/sec $\pm$ 0.1 km/sec) is somewhat less than the 0.5 and 0.7 km/sec vertical and horizontal microvelocities determined from "resolved" disk observations at $\tau \approx 0.01$ (Canfield and Beckers, 1976). The rms macrovelocities range from 2.1 to 1.6 km/sec when derived from weak to medium strong ($\approx$150 mÅ) lines. These exceed the vertical (horizontal) macrovelocities derived from "resolved" disk observations (insufficiently resolved however to resolve the macrovelocity contributors) of 1.2 (1.6) km/sec and 0.7 (1.3) km/sec for τ_0 = 0.1 and 0.01 respectively by almost a factor of two. An analysis of integrated sun spectra by Smith (1978) leads to similarly high values for the solar macrovelocities. The differences are probably the result of the different methods used in analyzing line profiles. It would be of interest to analyze the same set of solar spectral lines for macro- and microvelocities in both integrated sunlight and "resolved" disk spectra using the same analysis technique The result of such an analysis should remove the integrated sun - "resolved" solar disk discrepancies thus establishing a firmer link between solar and stellar velocity observations.

3. RESOLVED SOLAR MOTIONS

Figure 3 and Table 1 summarize the kinds of motions which have been resolved in the solar atmosphere. The important contributors to the total non-thermal velocity field are the solar rotation, the 5-minute oscillations and the solar granulation. The interesting solar motions associated with the supergranulation, non-radial pulsations and meridional flows would entirely escape detection in line profile analysis of the sun as a star but may show up in other ways. For example, the reversals in the integrated sun Ca^+ H and K profiles result to a large extent from the brightening in the chromospheric network which in turn originates in the supergranulation. I will restrict myself here to a short discussion of the three main known contributors to the macrovelocity field.

3.1 SOLAR ROTATION

There are many different ways of measuring solar rotation using both spectroscopic techniques and the proper motions of tracers such as sunspots, coronal and chromospheric structures, standing 5-minute acoustic waves, etc. Recent reviews of methods and results are given by Howard (1978) and Paternó (1978). Of special interest are:

a. A systematic difference between the rotation rates of the photospheric

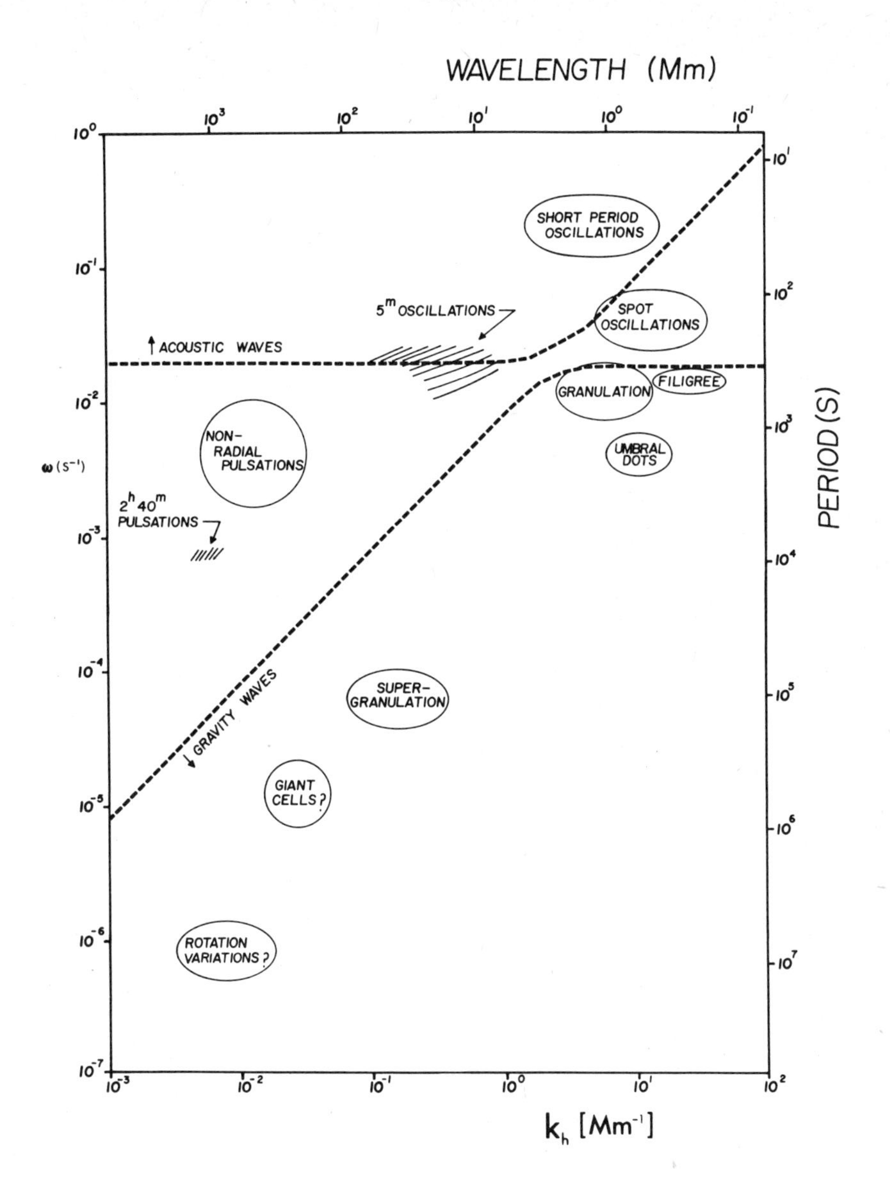

Figure 3. Location on the (k_h,ω) diagram of different types of solar velocity phenomena.

plasma as determined from Doppler shifts and that of longer-lived magnetic phenomena like sunspots, and other active region phenomena (Figure 4). The latter rotate $\sim$ 4% faster than the plasma or the short-lived magnetic phenomena. This difference is generally interpreted as the result of an increase of

TABLE 1

Summary of Resolved Velocity Structures on the Quiet Sun

Structure	Size (km)	Lifetime	Velocity (rms*) (τ_0 = 0.1)		Comments
			Vertical	Horizontal	
Rotation	700000	10^{10} yr	0	1.8 km/sec	- 1.8 km/sec is the synodic rotation velocity - Varies with latitude (differential rotation) - 4% faster for "magnetic phenomena" (spots, faculae, etc.) - Differential rotation absent for coronal holes and large scale magnetic patterns - (Apparent) temporal variation of $\sim$ 2%
Meridional Flows	700000	?	---	30 m/s	- Towards poles
$20-60^m$ Non-radial Pulsations	10^5?	?	.5 m/s	---	- Detected as solar diameter variations and maybe in Doppler shifts
160^m Pulsations	700000	?	.5 m/s	---	- Detected by Doppler shifts
Large Scale Flows	200000?	?	---	30 m/s?	- Probably responsible for apparent rotation variations - May be identical to so-called giant cells
Supergranulation	32000	1 day	30 m/s?	150 m/s	- Packed pattern of large convection cells - Perhaps driven by He ionization zone - Causes a structuring of the photospheric magnetic field which in turn causes the chromospheric network
5^m Oscillations	20000	$\geq 100^m$	400 m/s	---	- Standing acoustic waves in upper convection zone

TABLE 1 (continued)

Structure	Size (km)	Lifetime	Velocity (rms*) (τ_0 = 0.1)		Comments
			Vertical	Horizontal	
5^m Oscillations (continued)					- Show interference patterns both in temporal and spatial domain
Granulation	1500	500^s	460 m/s	890 m/s	- Associated with large intensity fluctuations (11% at 5000 Å) - Is convective overshoot phenomenon - Larger velocity amplitudes have been claimed
Short Period Oscillations	500?	?	250 m/s?	200 m/s?	- Periods < 100^s - Presence and magnitude under debate
Filigree	150km	500^s?	(500 m/s)	---	- Small scale magnetic regions of ∿ 1500 gs strength - Cover only very small fraction of sun and therefore do not contribute to total velocity field
TOTAL RESOLVED MOTIONS EXCLUDING ROTATION:			660 m/s	920 m/s	

*rms except for solar rotation and meridional flows for which actual velocities are given.

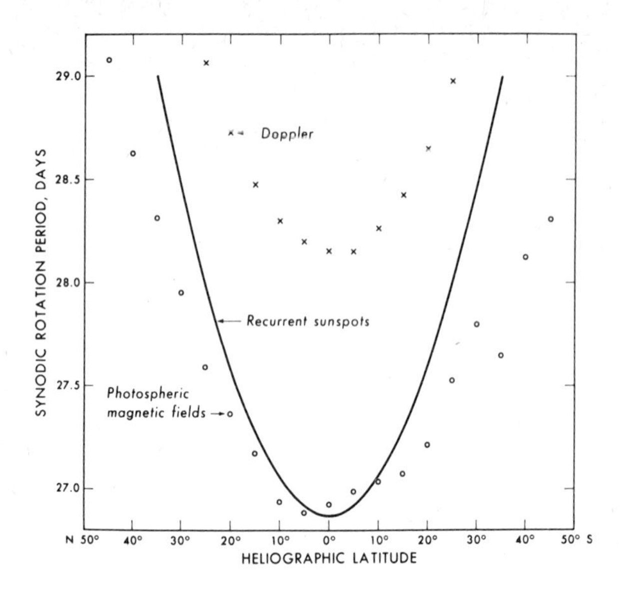

Figure 4. Solar rotation as a function of heliographic latitude for magnetic phenomena (sunspots) and photospheric plasma (Doppler) (from Wilcox and Howard, 1970).

rotation rates inwards into the sun in maybe the first $\sim$ 25000 km, this increase being reflected by deep-seated tracers like sunspots. This view has recently received additional support by the observation of the increase of solar rotation rates with depth using the standard 5^m period acoustic waves (Deubner et al., 1979). For stellar observations this may mean that rotation rates determined from line profiles and from e.g. periodic fluctuations of Ca^+ H and K profiles need not be the same.

b. A major variation of the differential rotation with latitude depending on the type of tracers used. Some, especially the long-lived coronal holes and general quiet sun magnetic field patterns show almost no differential rotation. This is believed to be due to a more rigid rotation of the solar interior probably in the deeper convection zone ($\sim$ 100000 km?).

c. Temporal variations of the solar rotation of the order of a few percent. Short term changes (1-1000 days) are probably due to the passage of large scale velocity cells on the solar surface. Longer term changes like solar cycle-related changes of $\sim$ 3% (Howard, 1976; Livingston and Duvall, 1979) and secular changes ($\sim$ 100 years) of $\sim$ 2% (Schröter and Wöhl, 1978) cannot be due to solar angular momentum changes on these time scales but must be the result of a redistribution of this momentum in the convection zone.

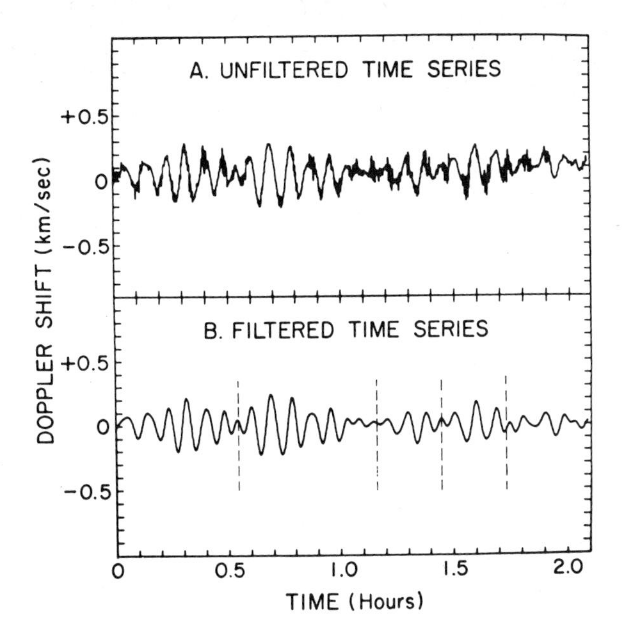

Figure 5. Typical velocity curve for individual oscillation (averaged over 10 x 10 arc sec) (from White and Cha, 1973).

3.2 FIVE MINUTE OSCILLATIONS

Figure 5 shows the time variation of the Doppler shift at one point on the solar disk. The velocity oscillations with 5 minute periods increase in amplitude with height. The rms velocity amplitude is given by Canfield (1976) as

$$V_{rms} = 0.35 \exp(h/1100) \text{ km/sec} \tag{1}$$

where h equals the height above τ_{5000} (= τ_0) = 1 in km. The 5 minute oscillations in the photosphere are evanescent waves but just below the solar surface they can be identified with propagating acoustic waves (or p-mode waves) which propagate inward into the sun to some depth at which they are reflected upward again to be reflected again near the solar surface. The trapping of the waves results in an interference pattern which shows up at the solar surface as a ridge structure in the (k_h,ω) diagram as predicted by Ulrich (1970) and Ando and Osaki (1975) and as observed by Deubner (1975, 1978) and Rhodes *et al*. (1977, see Figure 6). The 5 minute oscillations are therefore an important tool for the study of the solar interior and have so far led to somewhat improved models of the solar convection zone (Rhodes *et al*., 1977), an estimate of the mixing length parameter ℓ/H of 2-3 (Rhodes *et al*., 1977)

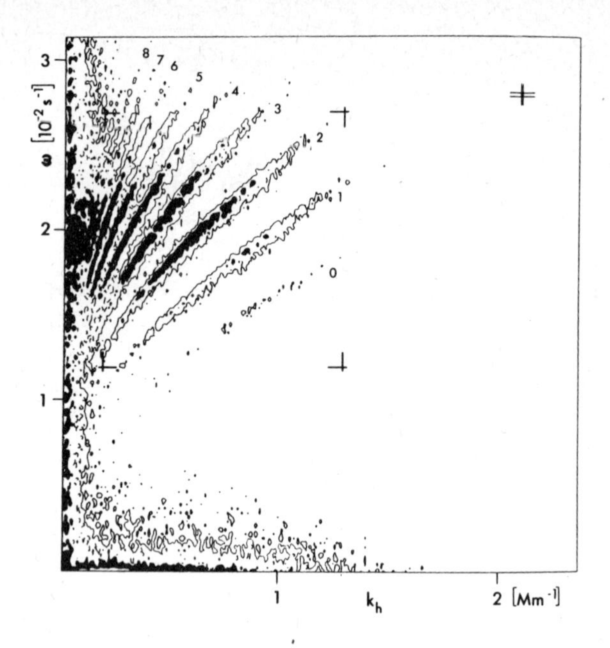

Figure 6. (k_h,ω) diagram of solar 5^m oscillation showing the resolved interference patterns (from Deubner, 1978).

and a measure of the depth variation of the solar rotation rate (Deubner et al., 1979).

3.3 SOLAR GRANULATION

The solar granulation is the result of the convective overshoot of the motions in the superadiabatic convection zone below the visible solar surface into the sub-adiabatic photosphere. Although in deeper layers convection carries most of the energy flux, at the layers we observe the granulation, only a small percentage of the energy is carried as convective energy flux. The solar granulation is close enough in size to the resolution limits of solar telescopes and of the atmosphere that corrections for finite resolution are major. This is especially true for velocity observations where longer exposure times, and more complex instruments, are needed. There is therefore disagreement among investigators as to the velocity fluctuations associated with the granulation. The values used in Table 1 were taken from Canfield (1976) who gives the rapidly decreasing (with height) vertical rms velocity associated with the solar granulation as

$$V_{rms} = 1.27 \exp(-h/150) \text{ km/sec} \qquad (2)$$

a result which has been substantially confirmed by Keil (1979) by a better separation of granular and oscillatory velocities.

Durrant et al. (1979) however strongly disagree with the Canfield and Keil results. They derive a much slower height variation of

$$V_{rms} = 0.98 \exp(-h/1700) \text{ km/sec} \tag{3}$$

which leads to much larger velocities at the heights where lines are observed. This implies much more convective overshoot and lends some support to granule models by Nordlund (1976, 1977, 1978, 1979). These models have much larger velocities than those listed in Table 1, so large in fact that the almost entire macrovelocity field is contained in solar granular motions.

The observations by Keil and Canfield (1978) are the basis for the theoretical model of the solar granulation by Nelson (1979) shown in Figure 7. This model shows

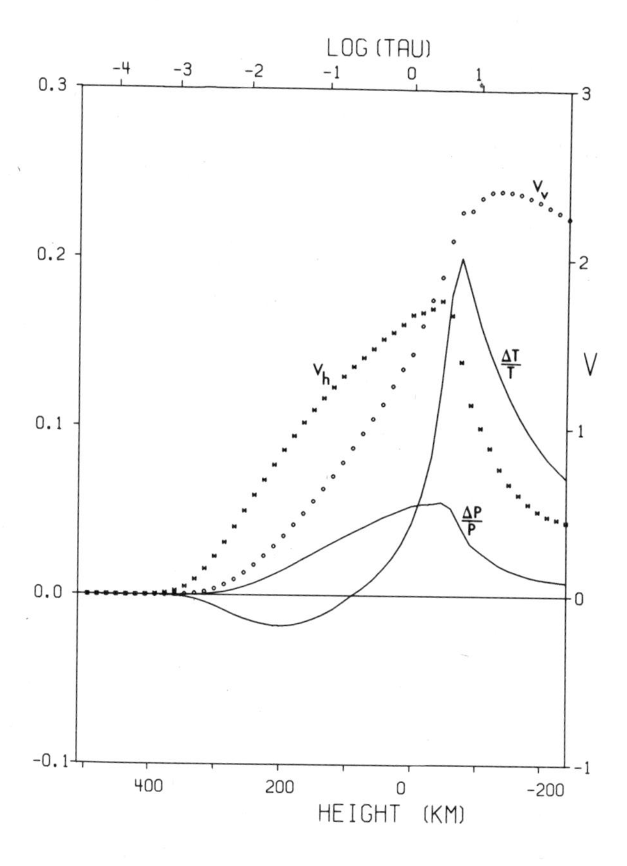

Figure 7. Model for solar granulation ΔT/T and ΔP/P and the horizontal and vertical velocities refer to amplitudes of sinusoidal variations (from Nelson, 1979).

substantial horizontal motions and a temperature inversion at τ_0 = 0.23 (as indeed has been observed in the wings of the Ca^+ H and K lines). This model has a cell size of 1060 km, somewhat below the observed value of 1500 km and it does not, of course, include the wide distribution of convective cell sizes actually observed on the sun.

Solar granulation, although mostly constant across the solar disk, does show some decrease in both contrast and size in active regions and perhaps at supergranule boundaries.

4. UNRESOLVED SOLAR MOTIONS

Depending on whether one accepts the results and models by Canfield (1976), Keil (1979) and Nelson (1979) or those by Durrant *et al*. (1979) and Nordlund (1976, 1977, 1978, 1979) there is or there is not a significant amount of unresolved solar motions (at scales $\leq$ 500 km). This is an unresolved question but I tend towards the Canfield and Keil results. That leaves most of the solar motion field at small scales. There are some clues as to the nature of this small scale velocity field:

4.1 SHORT PERIOD OSCILLATIONS

Deubner (1976a, b) suggests that the remaining line broadening and the micro-velocities are due to short period ($\leq$ 1 minute) propagating acoustic waves with spatial wavelengths along the line of sight comparable to or smaller than the width of the velocity weighting function. He supports the suggestion by observations of the temporal power spectrum of solar velocities which show small high frequency peaks (periods 30-100^s) which he contends are the result of the peculiarities of spatial filtering along the line of sight by the velocity weighting function. After a large correction for this filtering he derives $\sim$ 1 km/sec rms velocities at $\tau_0 \approx 0.1$ for these oscillations which comes close to explaining the entire line broadening. Cram *et al*. (1979), on the basis of a full dynamic model for a 30-second period oscillation, find that these rms velocities could be as small as 0.1-0.2 km/sec which is insufficient to explain the line broadening but sufficient for coronal heating.

4.2 UNRESOLVED CONVECTIVE MOTIONS FROM THE LIMB EFFECT

After correction for gravitational redshift solar lines show a blue shift with respect to their laboratory wavelength standard. Because of its peculiar center to limb variation this shift has been called the "limb effect." Of the three most likely causes for the limb effect, pressure shifts, wave shifts and convective shifts, only the latter remains as a possible one after Beckers and DeVegvar (1978) and Cram *et al*. (1979) showed that the former two are much too small and perhaps of the wrong sign. Convective wavelength shifts are caused by the correlation of intensity and velocities seen e.g. in the solar granulation. The different weighting of the out and in line-of-sight velocities by the intensities results in an apparent outward motion (blue shift) of solar lines even when there is no net mass flux. The origin of the limb effect as a convective line shift receives further support from the properties of the limb effect: (i) the decrease of the line shift with increasing line strength (e.g. Beckers and DeVegvar, 1978) as the consequence of the decrease of the convective overshoot with height, (ii) the absence of the line shift in sunspots (Beckers, 1977) as the consequence of the suppression of convection in sunspots, and (iii) the explanation of the center-to-limb variation as the consequence of the effect of both vertical and horizontal motions in convective elements (Beckers and Nelson, 1978).

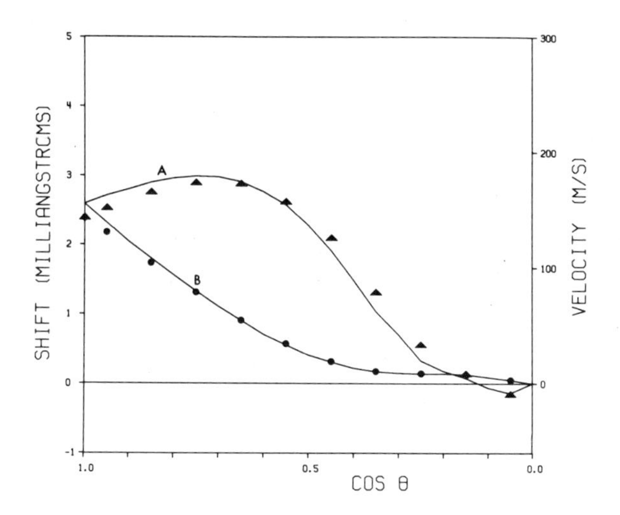

Figure 8. Calculation of the limb effect for a weak 2eV excitation potential line using the Nelson (1979, see also Figure 7) model of solar granulation. Curve B includes only vertical velocities in the granulation, curve A includes both vertical and horizontal velocities. Full lines are for line center Doppler shifts, dots and triangles are for line center of gravity displacements (from Beckers and Nelson, 1978).

Figure 8 shows the calculated center-to-limb variation for the Nelson (1979) model of solar granulation including both vertical and horizontal motions. The calculated wavelength variation is indeed very similar in shape to the observed limb shift but it is a factor of 2 to 3 smaller. If indeed the convective shift interpretation is correct, it implies that the Nelson model underestimates the velocities and/or temperature fluctuations by a factor of ~ 2. This may be the result of insufficient resolution of the spectra used to calibrate the model. The Nordlund model, on the other hand, would give results quite consistent with the observed limb effect. As the most likely state of affairs I will assume that the Nelson model holds for convective cell sizes between 500 and 2000 km and I will postulate at $\tau_0 \approx 0.1$ to 0.01 a smaller convective cell regime with sizes between 50 and 500 km and with velocities twice that of the granulation (see Table 1). With the addition of the granular convective shift these cells produce a total convective shift comparable to the observed shift.

TABLE 2

Contributions to the Total Non-Thermal Velocity Field

Structure	rms Velocity (m/s) τ_0 = 0.1 (h = 138 km)		rms Velocity (m/s) τ_0 = 0.01 (h = 238 km)	
	Vertical	Horizontal	Vertical	Horizontal
Supergranulation	30?	150	30?	150
5^m Oscillations	400	0	450	0
Resolved Granules	460	890	90	450
Unresolved Convection from limb effect	920	1780	180	900
Short Period Oscillations	250	200?	280	220?
Total	1130	2000	570	1080
Total Macro- and Micro- velocities from Line Width*	1460	2060	910	1550
Residual	920	500	710	1110
Total Microvelocities*	1160	1770	730	1100

*From Canfield and Beckers (1976).

Table 2 completes Table 1 with this assumed motion and compares the total motion thus derived with the total non-thermal motions and the microvelocities as summarized by Canfield and Beckers (1976). Figures 9 and 10 show the same data in the form of the power contained in the different size regimes. The residual power at horizontal scales below 50 km is shown, rather arbitrarily, as a Kolmogoroffian distribution. Also shown is the total velocity power above a given horizontal size as well as the power contained in microvelocities (μ). Table 2 and Figures 9-10, crude as they are, give a reasonable summary of our current understanding on the size distribution and the physical nature of the solar photospheric velocities.

5. CONCLUSION

Much, and probably most, of the solar non-thermal motions occur on scales smaller than the resolution limits of present observations. Although the sun teaches us something about the nature of velocity fields in stars of the solar type, it has

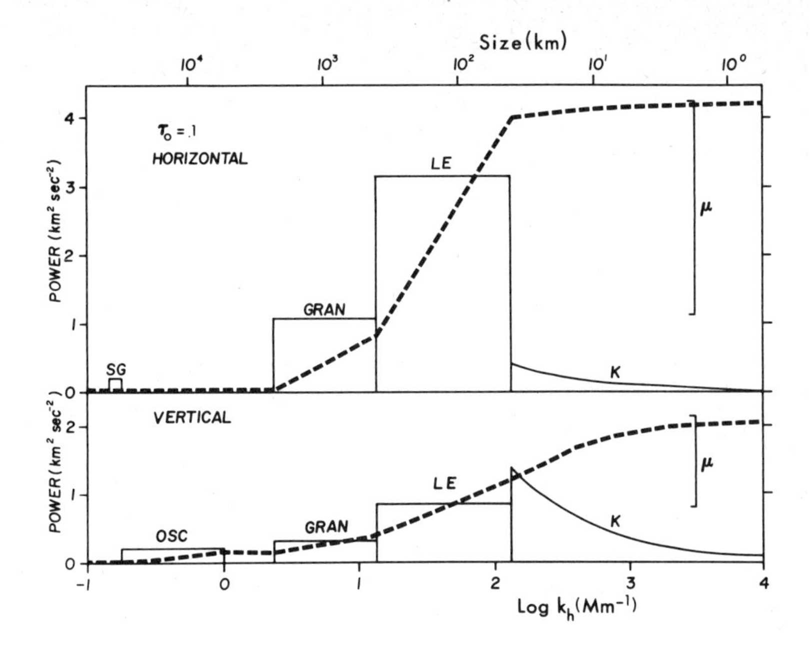

Figure 9. Power (rms velocities squared) associated with the different velocity regimes at $\tau_0 = 0.1$. SG = supergranulation; OSC = oscillations; GRAN = resolved granulation according to Nelson (1979); LE = unresolved convective motion derived from limb effect and K = assumed Kolmogoroffian velocity distribution for spatial scales less than 50 km with an amplitude to give the total non-thermal velocity as observed from line width. μ = microvelocity according to Canfield and Beckers (1976); the upper limit of the bar associated with μ is the total micro and macro velocity field. The dashed line represents the total velocity power at scales above the scale in the abcissa.

not told us all. In this paper I have not discussed any of the numerous interesting chromospheric and coronal motions which have been observed on the sun. Neither did I discuss motions in strong magnetic field regions like sunspots which are similar in total magnitude to those on the quiet sun but totally different in nature. The study of motions in the quiet photosphere by itself is however a most interesting topic because of the interesting astrophysical processes involved which include both small scale convection, large scale circulation, pressure waves, rotation, etc. The sun allows us to make more detailed measurements of these motions and to compare these with astrophysical predictions.

Comments by Drs. Cram, Keil and Zirker helped me in the preparation of this paper.

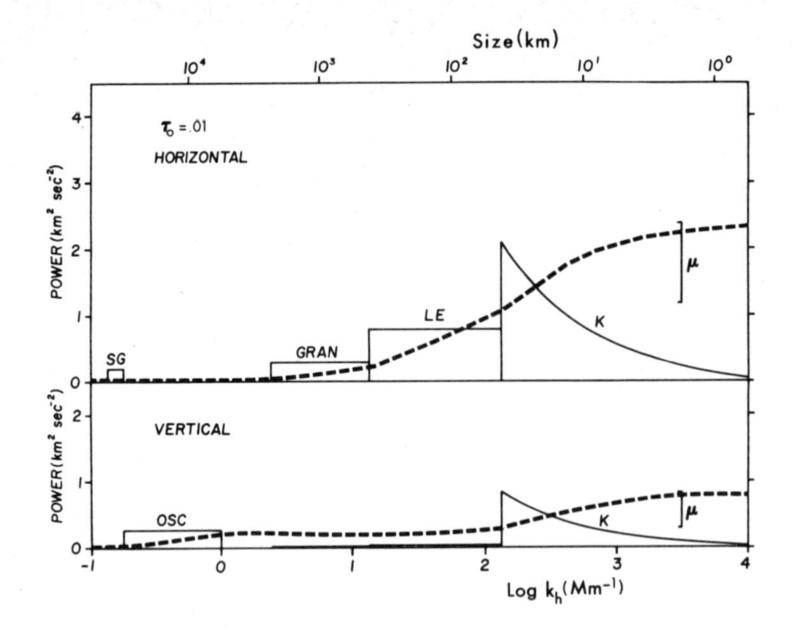

Figure 10. As Figure 9 but for τ_0 = 0.01.

REFERENCES

Ando, H. and Osaki, Y.,1975, P. A. S. Japan 27, 581.

Beckers, J. M., 1977, Astrophys. J. 213, 900.

Beckers, J. M., 1980, NASA/CNRS Series on "Non-Thermal Stellar Atmospheres," in preparation.

Beckers, J. M., Bridges, C. A. and Gilliam, L. B., 1976, AFGL Environmental Research Paper No. 565 = AFGL-TR-76-0126.

Beckers, J. M. and Canfield, R. C., 1976, CNRS Colloquium No. 250, p. 207 = AFCRL-TR-75-0592 part 1.

Beckers, J. M. and DeVegvar, P., 1978, Solar Phys. 58, 7.

Beckers, J. M. and Nelson, G. D., 1978, Solar Phys. 58, 245.

Canfield, R. C., 1976, Solar Phys. 50, 329.

Canfield, R. C. and Beckers, J. M., 1976, CNRS Colloquium No. 250 p. 291, = AFCRL-TR-75-0592 part 2.

Cram, L. E., Keil, S. L. and Ulmschneider, P., 1979, Astrophys. J. (in press).

Deubner, F.-L., 1975, Astron. and Astrophys. 44, 371.

Deubner, F.-L., 1976a, Astron. and Astrophys. 51, 89.

Deubner, F.-L., 1976b, IAU Colloquium No. 36, p. 45.

Deubner, F.-L., 1977, Mem. Soc. Astr. Italia 48, 499.

Deubner, F.-L., 1978, Sac Peak Proc. Symp. on Large Scale Motions on the Sun, (S. Musman, ed.), p. 77.

Deubner, F.-L., Ulrich, R. K. and Rhodes, E. J., 1979, Astron. and Astrophys. 72, 177.

Durrant, C. J., Mattig, W., Nesis, A., Reiss, G. and Schmidt, W., 1979, Solar Phys. 61, 251.

Goldberg, L., 1958, Astrophys. J. 127, 308.

Gray, D. F., 1977, Astrophys. J. 218, 530.

Howard, R., 1976, Astrophys. J. 210, L159.

Howard, R., 1978, Reviews of Geophys. and Space Phys. 16, 721.

Keil, S. L., 1979, preprint.

Keil, S. L. and Canfield, R. C., 1978, Astron. and Astrophys. 70, 169.

Livingston, W. C. and Duvall, T. L., 1979, Solar Phys. 61, 219.

Nelson, G. D., 1979, Solar Phys. 60, 5.

Nordlund, Å., 1976, Astron. and Astrophys. 50, 23.

Nordlund, Å., 1977, IAU Colloquium No. 38, Problems in Stellar Convection, p. 237.

Nordlund, Å., 1978, Astronomical Papers Dedicated to B. Strömgren, (A. Reiz, T. Anderson, eds. (in press).

Nordlund, Å., 1979, preprint.

Paternó, L., 1978, Proc. Catania Workshop on Solar Rotation, p. 11.

Rhodes, E. J., Ulrich, R. K. and Simon, G. W., 1977, Astrophys. J. 218, 901.

Schröter, E. H. and Wöhl, H., 1978, Proc. Catania Workshop on Solar Rotation, p. 35.

Smith, M. A., 1978, Astrophys. J. 224, 584.

Stenholm , L. G., 1977, Astron. and Astrophys. 61, 155.

Ulrich, R. K., 1970, Astrophys. J. 162, 993.

White, O. R. and Cha, M. Y., 1973, Solar Phys. 31, 23.

Wilcox, J. M. and Howard, R., 1970, Solar Phys. 13, 251.

Zirker, J. B., 1979, Proc. 17th Aerospace Sciences Mts., AIAA, preprint.

DIAGNOSTIC USE OF FE II H & K WING EMISSION LINES

Lawrence E. Cram and Robert J. Rutten
Sacramento Peak Observatory
Sunspot, NM 88349, U.S.A.
Bruce W. Lites
Laboratory for Atmospheric and Space Physics (University of Colorado)
Boulder, CO 80369, U.S.A.

The Fe II λ3969.4 line is one of the weak lines in the wings of Ca II H and K that appear in emission near the solar limb, and in the flux spectra of cool giants. In spatially resolved solar spectrograms the line shows very pronounced small-scale spatial intensity variation, which is strongly correlated to the line structure of the local H-wing background, and not at all to the chromospheric structure seen in the H & K cores. A 15-level atomic model computation for iron shows that this behaviour is due to pumping by photons in the wings of the strong Fe II resonance lines near 2600 Å, in the deep photosphere. The λ3969.4 line is therefore deeply controlled, with large sensitivity to photospheric inhomogeneities, while its background is formed much higher. This makes the line a useful diagnostic of stellar photospheric line structure, in contrast to the adjacent H core for which emission indicates chromospheric line structure.

TURBULENCE IN MAIN SEQUENCE STARS

Thomas Gehren
Max-Planck-Institut für Astronomie
Königstuhl, 6900 Heidelberg, Germany

I. Introduction

It has for long been realized that the interpretation of stellar spectral lines normally requires broadening velocities well in excess of purely thermal motions. The origin and structure of such velocity fields, which have usually been summarized as turbulence, still appear to be a subject of controversy.

In the Sun, according to Worrall and Wilson (1973), there is no evidence for small scale velocity fields with amplitudes large enough to account for the observed line broadening. Subsequently, Wilson and Guidry (1974) tried to explain the center-to-limb variation of the solar Na D lines by temperature fluctuations associated with a two-stream model. More recently, Nelson and Musman (1977) have constructed a model of the solar granulation which predicts granular motions with a scale height appreciably smaller than indicated by the unresolved small scale velocities in the solar photosphere. As Deubner (1976) has shown, further out in the solar chromosphere short period acoustic waves may account for the observed line broadening velocities.

The acoustic flux generated in the convection zone of main sequence stars has been estimated by Hearn (1974) to be insufficient to maintain microturbulence against viscous dissipation, if turbulence is due to progressive sound waves. His suggestion that the broadening of spectral lines results from convective overshoot into the photosphere has been investigated in detail by Nordlund (1978), who confirms that even a laminar velocity field describing the up- and downward motions of penetrating convective eddies can reproduce line strengths as well as their center-to-limb variation in the solar photosphere. The amplitude of such a velocity field is chosen to have a maximum on granular scales, and Nissen and Gustafsson (1978) suggest that parametrization of small scale velocity fields derived from analysis of horizontally homogeneous model atmospheres probably leads to an overestimate of the true microturbulence.

II. Micro- and Macroturbulence

The two-component model atmosphere approach has not yet been systematically applied to the interpretation of spectral lines in main sequence stars. Except for the Sun, our knowledge of stellar velocity fields is restricted by the assumption of plane-parallel horizontally homogeneous atmospheres in which the concept of random motions on two different scales (micro- and macroturbulence) is used.

Small scale motions are usually described by a gaussian velocity distribution, where the microturbulence parameter ξ represents the most probable amplitude. Some authors have tried to derive its dependence on continuum optical depth, $\xi(\tau)$, from the application of the Goldberg method. For Procyon (F5IV-V) Sikorski (1976) obtained a microturbulence parameter smoothly increasing from 2.5 km/s at $\tau = 0.3$ to 3.5 km/s at $\tau = 0.001$. A similar increase with height, derived for Arcturus, is in accordance with the work of Mäckle et al. (1975). Contradicting these results, Stenholm (1977) has determined $\xi(\tau)$ for the Sun and the subgiants η Cep (KOIV) and γ Cep (K1IV) using Goldberg's method. His solar microturbulence shows a distinct decrease from 1.5 km/s at $\tau = 0.3$ to 0.5 km/s at $\tau = 0.001$, and a similar behaviour of $\xi(\tau)$ is found for η and γ Cep.

It has been emphasized by Holweger et al. (1978) that the horizontally homogeneous approach requires the existence of an anisotropic small scale velocity field in the Sun, since the limb strengthening of photospheric lines cannot be explained entirely by a depth-dependent isotropic parameter. Moreover, a solar microturbulence, increasing with height in the fashion derived for Procyon, is incompatible with the observed width of faint lines originating in the upper photosphere at $\tau < 0.01$. These lines require very small velocity amplitudes, probably below 0.5 km/s (cf. Canfield and Beckers, 1976, for a review on unresolved solar motions). Thus, the most plausible depth variation seems to be represented by a smooth decrease of photospheric small scale velocities with height, followed by an increase in chromospheric layers.

Large scale motions (macroturbulence) in the atmosphere of Procyon have been estimated by Evans et al. (1975). From a line shape analysis they rule out microturbulence velocities larger than 1 to 2 km/s, the dominant line broadening process being large scale motions with most probable velocities of about 2 km/s. This agrees roughly with the ratio of micro/macro amplitudes in the Sun derived by Holweger et al. (1978) as well as with that in giants (Reimers, 1976).

III. Microturbulence and Basic Stellar Parameters

Further information on small scale velocity fields in stars along the main sequence is constrained by the assumption of an isotropic gaussian velocity distribution with a depth-independent microturbulence parameter ξ. The determination of the microturbulence parameter has been carried out according to one of the following methods:

(a) Narrow band photometry, calibrated with synthetic spectra from model atmospheres (Gustafsson et al., 1974, Nissen and Gustafsson, 1978)
(b) Curve-of-growth analyses of different degrees of sophistication as compiled by Glebocki (1973) and Morel et al. (1976)
(c) Fourier transform analysis of line profiles (Gray, 1973)

Since microturbulence values taken from the catalogues of Glebocki and Morel et al. are often contradictory, the principal sources of systematic errors must be elucidated.

Frequently the microturbulence is only a by-product of abundance analyses, and some authors do not seem to have recognized that their element abundances are severely degraded by an improper determination of the turbulence parameter. It should be emphasized that the one-layer curve-of-growth analysis is particularly unsuited to derive small scale velocities, because the choice of the "best fit" theoretical curve is often biased by the large scatter in weak line strengths. Curve-of-growth results may additionally be affected by unrecognized blends and uncertainties in the determination of the continuum level.

Besides these general statements some sources of systematic errors deserve a more detailed investigation (cf. Gray and Evans, 1973):

(1) Effective temperature and gravity of model atmospheres used to compute saturation parameters or synthetic spectra are usually determined from observed colours by calibration relations. This procedure often involves the Sun as a reference point, although the solar colours are a subject of controversy. Frequently the result may have been a significant underestimate of the temperatures of solar-type stars.

(2) The temperature distribution predicted by theoretical line-blanketed model atmospheres of solar-type stars displays strong backwarming effects which cannot be reconciled with observations of the solar continuum (Gehren, 1979, see also Fig. 2 of Gustafsson and Bell, 1979). The model temperature gradients in the region of line formation are too steep and, when applied to the interpretation of observed line strengths, may simulate a low microturbulence.

(3) Many abundance analyses have been based on absolute oscillator strengths. Because of systematic errors in these measurements prior to

1970, most of the related microturbulence data are unreliable (Reimers, 1976). A rediscussion by Andersen (1973) revealed a considerable change of the earlier microturbulence determinations.

(4) The problem of absolute f-values may be overcome by analyses using solar f-values. However, such a procedure requires the knowledge of solar equivalent widths as well as the solar microturbulence parameter (the average small scale velocity amplitude in the solar photosphere). Differential curves-of-growth for long were based on the Utrecht atlas which, as shown by analysis based on modern photoelectric measurements (Gehren, 1978), may lead to a larger scatter of line strengths around the curve-of-growth and to microturbulence values systematically low. Furthermore, broadening velocities obtained for stars depend critically on the solar value. Some stellar data (e.g. Gustafsson et al., 1974) refer to $\xi_{\odot} = 0.5$ km/s , as derived by Foy (1972) from a disc center solar curve-of-growth. Recent determinations group around 0.9 km/s for the disc center and 1.3 km/s for the integrated disc (Blackwell et al., 1976, Holweger et al., 1978).

(5) Magnetic intensification and hyperfine structure may play a role in special classes of stars or lines, respectively.

(6) Departures from LTE have been shown to be of minor importance as far as line strengths are considered (Holweger, 1973). However, they may affect Fourier transform analysis of line profiles. Smith and Gray (1976) suggest to avoid these difficulties by analyzing lines of different strength. This approach has been challenged by Durrant (1979) who finds that the microturbulence derived from the Fourier transform analysis strongly depends on line strength.

(7) A final source of uncertainty is due to the damping constants (Blackwell et al., 1976). Since the contributions of microturbulence and damping can be separated in a line strength vs. abundance diagram (cf. Garz et al., 1969, Fig. 1), however , correction factors for the simple hydrogenic treatment of van der Waals interaction may be found.

From this discussion it is evident that, in order to study microturbulence , we have to restrict ourselves to results presumably unaffected by systematic errors. The data compiled by Glebocki (1973) and Morel et al. (1976) have been disregarded on this account. The microturbulence values displayed in Figs. 1 and 2 are from the following sources: narrow band photometry of Nissen and Gustafsson (1978, o), curve-of-growth analyses of Chaffee (1970, corrected by Andersen, 1973,+) and Clegg (1977,×), model atmosphere analyses of Gehren (1977, Δ , 1978, 1979, ✱), and Fourier transform analysis (Gray, 1973,□). Small symbols denote stars for which contradictory results have been found

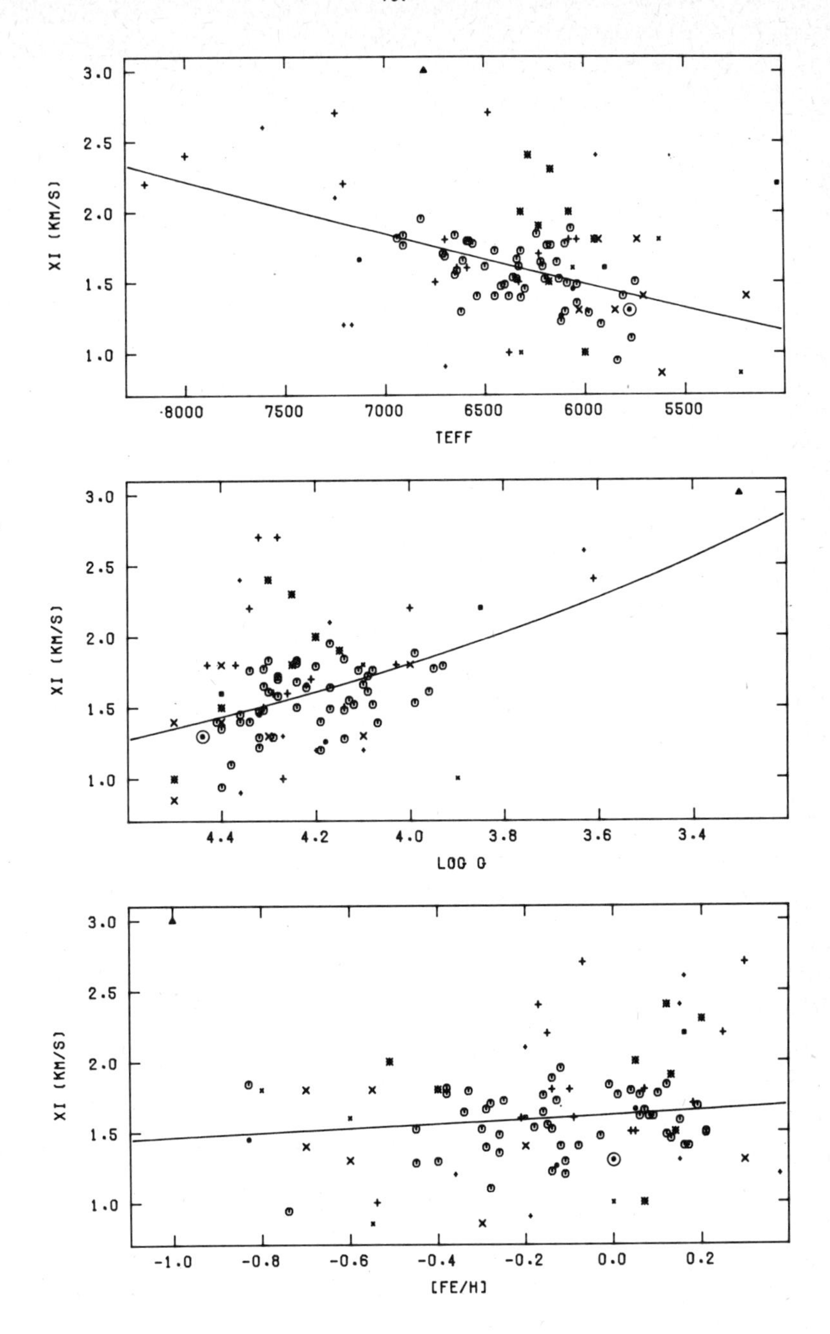

Figure 1: Microturbulence velocities as a function of the basic stellar parameters, temperature, gravity, and metal abundance. Solid curves represent least square power laws. See text for an explanation of the symbols

among the sources just mentioned and the compilations of Glebocki and Morel et al. The corresponding microturbulence values have been given lower weight.

In Fig. 1 is shown the degree of correlation between microturbulence and the three basic stellar parameters, T_{eff}, log g, and [Fe/H], where [Fe/H] is the logarithmic metal abundance relative to the Sun and (Fe/H) = $\exp_{10}$[Fe/H] . Since for some of the stars either gravity or metal abundance is not given by the authors, these data have been estimated from calibration relations of δc_1 and δm_1 (cf. Nissen and Gustafsson, 1978). Least square solutions assuming a power law dependence yield

$$\xi = 8.6\ 10^{-6}\ T_{eff}^{1.386} \pm 0.352 \text{ km/s},$$

$$\xi = 18.\ g^{-0.249} \pm 0.345 \text{ km/s, and}$$

$$\xi = 1.6\ (Fe/H)^{0.044} \pm 0.395 \text{ km/s},$$

while the multivariate solution

$$\xi = 2.5\ 10^{-3}\ T_{eff}^{0.945}\ g^{-0.175}\ (Fe/H)^{0.023} \pm 0.344 \text{ km/s}$$

is only a marginal improvement.

The considerable scatter displayed in Fig.1 is due to a few stars with comparatively high microturbulence , as β Vir, 9 Com, 37 UMa and μ Her A. Since for β Vir and 9 Com the error in the determination of ξ is probably within ± 0.3 km/s (Gehren, 1978, 1979), it is felt that there are significant departures from the mean relation that deserve special explanation. An independent determination with narrow band photometry would be also desirable in those cases. Excessively high velocities for A-type stars as noted by Baschek and Reimers (1969) and Smith (1971) are probably the result of using inaccurate absolute f-values (cf. Andersen, 1973). The same holds for the Hyades dwarfs of Chaffee et al. (1971). The Hyades stars VB 14 and VB 47 included here do not show a high microturbulence.

Whereas least square solutions confined to the more homogeneous data of Nissen and Gustafsson (1978) yield a considerable reduction of the scatter , the functional dependence found above is not changed. Thus the least square solutions are not excessively weighted by the few stars with high temperature or low gravity. From Fig. 1 it is seen that the microturbulence velocity is not an independent stellar parameter, in agreement with the results of Nissen and Gustafsson.

IV. Discussion

The overall variation of the microturbulence parameter with T_{eff} and log g is remarkably similar to the one found by Reimers (1973) for giants and supergiants covering a considerably larger interval in gravity. Moreover, it closely follows the temperature and gravity dependence of the Wilson-Bappu effect, which yields for the Ca II K emission widths $W_o \sim T_{eff}^{1.25}\ g^{-0.22}$. This seemingly uniform behaviour of stellar velocity fields, extending over a large part of the Hertzsprung-Russell diagram, led Böhm-Vitense (1975) and Reimers (1976) to suggest that microturbulence is correlated with the velocities in the upper layers of stellar convection zones. Even within the restricted range in temperature and gravity considered in Fig.1, a correlation between microturbulence and convective velocities does indeed exist. This is shown in Fig. 2 where we have drawn microturbulence velocities against maximum convective velocities computed by de Loore (1970), since photospheric velocities due to penetrative convection are not yet available for a grid of temperatures and gravities. Although the maximum convective velocities depend critically on the mixing-length parameter, their variation with T_{eff} and log g appears to be reasonably well defined. The linear least squares solution shown in Fig.2 is

$$\xi = 0.31\ v_{max} + 0.79 \pm 0.328 \text{ km/s} .$$

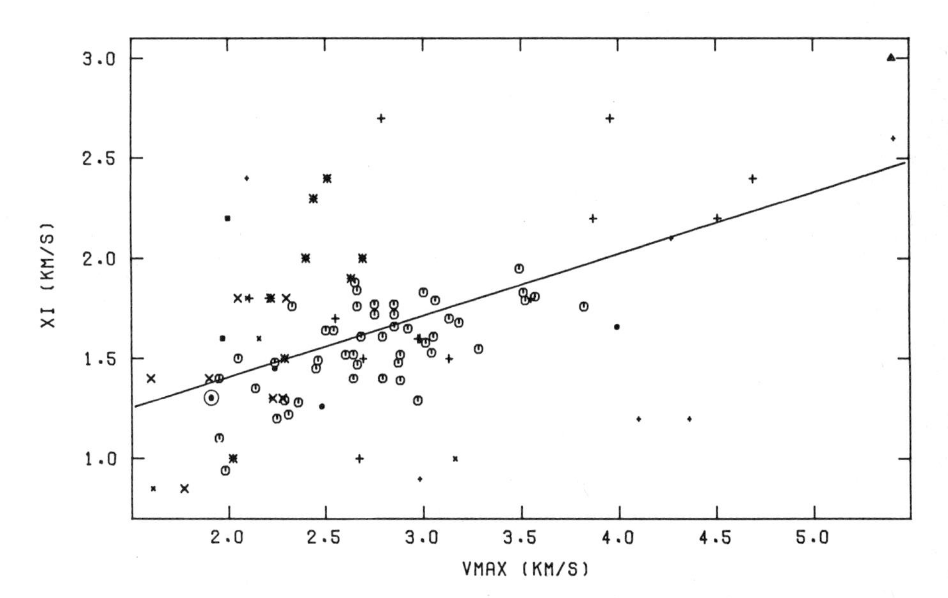

Figure 2: Microturbulence vs. maximum convective velocities, computed by de Loore (1970). The solid line denotes the linear least squares solution

Since photospheric velocity amplitudes due to convective overshoot may be expected to vary smoothly with maximum convective velocities, it is perhaps not surprising that this solution corresponds to

$$\xi \simeq v_{conv}(\tau = 1) ,$$

when $v_{conv}(\tau = 1)$ and v_{max} are taken from Nordlund's two-component models (Nordlund, 1976 , Nissen and Gustafsson, 1978). Photospheric velocities of this amplitude may be also derived from mixing-length theory if allowance is made for a non-local computation of velocities (Parsons, 1969, Nordlund, 1974).

Even though the convective origin of the observed broadening velocities seems to be sufficiently well established, the nature of the transport mechanism cannot yet be ascertained. Regular streaming patterns on granular scales may resemble microturbulence as shown by Nordlund (1978). His results are most convincing because they explain the total velocity amplitudes (micro- and macroturbulence) as well as the solar center-to-limb broadening of photospheric lines. There still remains the possibility that a fraction of the convective energy is redistributed to motions on smaller scales (< 100 km), which are unresolvable even in the Sun and would look more like the classical microturbulence. Progressive sound waves generated in the convection zone may be another transport mechanism (Edmunds, 1978). However, the corresponding wave velocity amplitudes appear to be too small in the photosphere of a main sequence star. Short period acoustic waves propagating on top of stellar atmospheres are more likely to account for chromospheric motions (Oster and Ulmschneider, 1973, Deubner, 1976).

The similar dependence upon temperature and gravity of both regular convective motions and progressive sound waves provides a simple explanation for the correlation between small scale velocities observed in stellar photospheres and chromospheric Ca II K emission widths.

V. Concluding Remarks

In order to discriminate between different line broadening velocity fields, future work will have to concentrate on the Sun, where the obtainable spatial and temporal resolution is orders of magnitude higher than for any other star.

A crucial test of the relation outlined here between convective and line broadening velocities may be supplied by a reexamination of the microturbulence in A-type stars. If convection in main sequence stars fades out beyond effective temperatures of 8500 K, the microturbulence velocities should also diminish.

The cool branch of the main sequence deserves further attention. No reliable microturbulence values are available for stars later than spectral type G5V.

The microturbulence parameter as a measure of random motions on small scales thus will probably remain one of the main goals of spectrum analysis for years to come.

References

Andersen, P.H. 1973, Publ.Astron.Soc.Pacific 85,666
Baschek, B., Reimers, D. 1969, Astron.Astrophys. 2,240
Blackwell, D.E., Ibbetson, P.A., Petford, A.D., Willis, R.B. 1976, Monthly Not.Roy.Astron.Soc. 177,227
Böhm-Vitense, E. 1975, in "Problems in Stellar Atmospheres and Envelopes", eds. B. Baschek, W.H. Kegel, G. Traving, Heidelberg, p. 21
Canfield, R.C., Beckers, J.M. 1976, in "Physique des Movements dans les Atmosphères Stellaires", eds. R. Cayrel, M. Steinberg, Paris, p. 291
Chaffee, F.H. 1970, Astron.Astrophys. 4,291
Chaffee, F.H., Carbon, D.F., Strom, S.E. 1971, Astrophys.J. 166,593
Clegg, R.E.S. 1977, Monthly Not.Roy.Astron.Soc. 181,1
Deubner, F.-L. 1976, Astron.Astrophys. 51,189
Durrant, C.J. 1979, Astron.Astrophys. 76,208
Edmunds, M.G. 1978, Astron.Astrophys. 64,103
Evans, J.C., Ramsey, L.W., Testerman, L. 1975, Astron.Astrophys. 42,237
Foy, R. 1972, Astron.Astrophys. 18,26
Garz, T., Holweger, H., Kock, M., Richter, J. 1969, Astron.Astrophys. 2,446
Gehren, T. 1977, Astron.Astrophys. 59,303
Gehren, T. 1978, Astron.Astrophys. 65,427
Gehren, T. 1979, Astron.Astrophys. 75,73
Glebocki, R. 1973, Acta Astronomica 23,135
Gray, D.F. 1973, Astrophys.J. 184,461
Gray, D.F., Evans, J.C. 1973, Journ.Roy.Astron.Soc.Canada 67,241
Gustafsson, B., Kjærgaard, P., Andersen, S. 1974, Astron.Astrophys. 34,99
Gustafsson, B., Bell, R.A. 1979, Astron.Astrophys. 74,313
Hearn, A.G. 1974, Astron.Astrophys. 31,415
Holweger, H. 1973, Solar Physics 30,35
Holweger, H., Gehlsen, M., Ruland, F. 1978, Astron.Astrophys. 70,537
de Loore, C. 1970, Astrophys.Space Science 6,60
Mäckle, R., Holweger, H., Griffin, R., Griffin, R. 1975, Astron.Astrophys. 38,239
Morel, M., Bentolila, C., Cayrel, G., Hauck, B. 1976, in "Abundance Effects in Classification", eds. B. Hauck, P.C. Keenan, Dordrecht, p. 223
Nelson, G.D., Musman, S. 1977, Astrophys.J. 214,912
Nissen, P.E., Gustafsson, B. 1978, in "Astronomical Papers Dedicated to Bengt Strömgren", eds. A. Reiz, T. Andersen, Copenhagen, p.43
Nordlund, Å. 1974, Astron.Astrophys. 32,407
Nordlund, Å. 1976, Astron.Astrophys. 50,23
Nordlund, Å. 1978, in "Astronomical Papers Dedicated to Bengt Strömgren", loc.cit. p. 95
Oster, L., Ulmschneider, P. 1973, Astron.Astrophys. 29,1
Parsons, S.B. 1969, Astrophys.J.Suppl. 18,127
Reimers, D. 1973, Astron.Astrophys. 24,79
Reimers, D. 1976, in "Physique des Movements dans les Atmosphères Stellaires", loc.cit. p. 421

Sikorski, J. 1976, Acta Astronomica 26,1
Smith, M.A. 1971, Astron.Astrophys. 11,325
Smith, M.A., Gray, D.F. 1976, Publ.Astron.Soc.Pacific 88,809
Stenholm, L.G. 1977, Astron.Astrophys. 61,155
Wilson, A.M., Guidry, F.J. 1974, Monthly Not.Roy.Astron.Soc. 166,219
Worrall, G., Wilson, A.M. 1973, in "Vistas in Astronomy", ed. A. Beer, Vol. 15, p. 39

OBSERVATIONAL ASPECTS OF MACROTURBULENCE IN EARLY TYPE STARS

Dennis Ebbets
McDonald Observatory
University of Texas at Austin

I. Introduction

The basic premise of this paper is that the atmospheres of supergiants of spectral type O, B, and A are not homogeneous static and stable. Observations show effects in the spectrum which seem to indicate velocity fields of many different scales, which probably are variable with depth and are certainly variable with time. Some effects can be understood in terms of global, steady, symmetric velocities, namely rotation and expansion. Those effects which still cannot be accounted for are generally attributed to turbulent velocities.

The concept of curve of growth microturbulence dates back to Struve and Elvey (1934), but will not be discussed at length here. A wide ranging review was presented by Wright (1955) and lengthy discussions of problems of interpretation are found in Thomas (1960), and Mihalas (1979). The concept of spectroscopic macroturbulence may also have originated with Struve (1952), and was developed and clarified by Huang and Struve (1953). Theoretical arguments, based on the outward force of radiation pressure, led Underhill (1949) to predict that supergiant atmospheres would be mechanically unstable. In fact, irregular, semi-periodic radial velocity variations were discovered in A type supergiants by Abt (1957). Indirect evidence for turbulent velocities was discussed by Slettebak (1956), and Abt (1958). Reviews of these arguments were presented by Underhill (1960) and Rosendhal (1970).

I will summarize the results of observational studies since then, which have been carried out with new and higher quality data. For the most part these programs incorporated more sophisticated statistical and analytical techniques, and often made reference to modern model atmosphere calculations. I will discuss primarily OBA supergiants, and say little about main sequence stars, Wolf-Rayet stars, or binary systems. Theoretical and computational techniques are discussed in the volume by Cayrel and Steinberg (1978), and in other papers in this colloquium.

II. Photometric Results

During the past decade photometric studies have provided a great deal of new information about the magnitudes, colors, and polarization properties of the continua of early type stars. Time resolved studies have detected and analyzed variability of several interesting and complicated types. The most extensive study of O type stars was that by Morrison (1975), in which u-v-b-y-Hβ photometry of over one hundred stars was analyzed and compared to the predictions of the static non-LTE model atmospheres of Mihalas (1972). Maeder and Rufener (1972) surveyed the photometric properties of 80 non-supergiant OB stars, and 34 supergiants of type B2-G8, and described the variability of their color indices. Sterken (1977), Burki et al. (1978), and Feitzinger (1978) reported the results of statistical studies of the variability of a smaller number of extremely luminous galactic and large Megallanic Cloud supergiants. Schild and Chaffee (1975) investigated the spectrophotometric properties of the Balmer discontinuity in O and B type supergiants, and Serkowski (1970) and Hayes (1975) have studied the intrinsic linear polarization of the continua of O type stars. The results of these and other similar reports can be summarized as follows:

The most luminous stars of all spectral types are variable in both light and color. A rough lower limit to the luminosity at which variability is detected is at an absolute visual magnitude of $M_V \sim -5$. Included therefore are most of the O stars earlier than about O6, all O supergiants and Of stars, and all Ia and Iab supergiants of spectral types B, A, and F. Amplitudes of variation range from the limit of certain detection ($\lesssim$ .1 mag.) in color indices (U-B or v-y). The amplitude correlates fairly well with spectral type and luminosity class. For a given luminosity, the hottest stars show the largest range of variation. Similarly, at a given spectral type, the more luminous stars vary with a larger amplitude.

The variations have been observed with a wide range of timescales. Long term trends are present in monthly and yearly mean magnitudes for all stars, but owing to a paucity of observations, few stars have been studied in detail. At the other extreme, only in the most luminous Ia supergiants, and those with the largest amplitudes, can statistically significant variations within a single night be confirmed. On the other hand, all of the supergiants vary significantly in several days. Unique periods do not exist, but characteristic timescales, sometimes called "semi-periods" can be found in a well defined manner. A harmonic analysis is performed in which a model light curve is represented by the lowest order terms of a Fourier series. A number of trial periods are tried, each time adjusting the amplitudes until the χ^2 deviation between the model and the observations is minimized. The correlation coefficient is computed for each model and that with the highest correlation is taken to be the best singly periodic representation of the observations. In general, these solutions are good, but not perfect for B8 Ia - A2 Ia stars, with correlation coefficients of about .95, and semiperiods of 20-30 days. Among the earlier types, B0 Ia - B2 Ia, a singly periodic solution is a much less satisfactory model, with correlations of .5 to .6. The semi-periods are shorter, with values of 5-10 days being typical. Among samples of stars with similar luminosity, say $-8.2 \leq M_V \leq -7.2$, the semi-period expressed as log P increases nearly linearly with spectral type, in the sense that the later type stars have longer periods. Similarly, at a given spectral type, the more luminous stars have longer periods - see Figure 1.

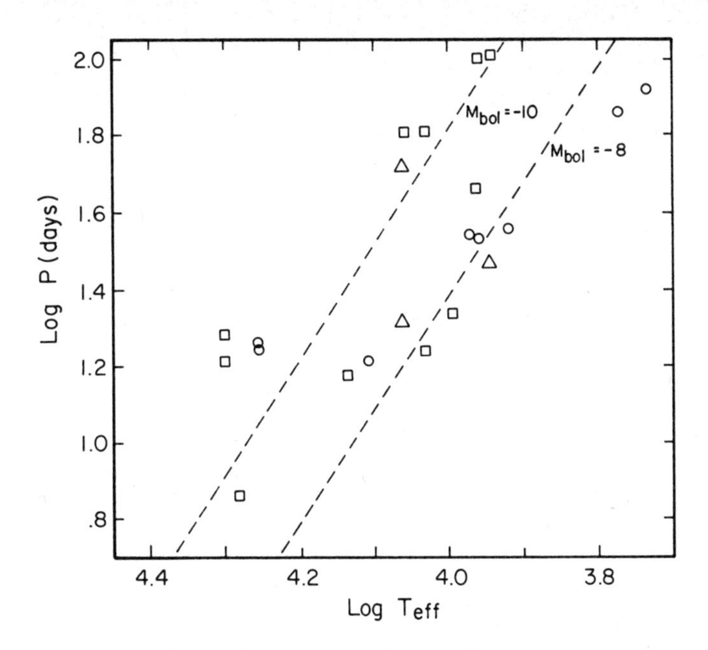

Fig. 1 - The semi-periods of photometric variation compiled from Maeder and Rufener (1972) Δ, Sterken (1977) □, and Burki et al. (1978) O, are plotted as a function of temperature. The temperature calibration of Barlow and Cohen (1977) was used. The dashed lines represent equation 4 with Q = .1 day.

It is thought that the observed variability is basically a pulsation phenomenon, possibly the simultaneous excitation of two or more non-radial modes. Despite the lack of strict periodicity the characteristic timescales and their correlation with temperature and luminosity are suggestive of pulsation. For gravitational oscillations, the period is related to the mean density as

$$P = Q(\rho/\rho_{\odot})^{-\frac{1}{2}} , \quad \text{Log } P = \text{Log } Q - .5 \text{ Log } (\frac{M}{M_{\odot}}) + 1.5 \text{ Log } (\frac{R}{R_{\odot}}) \tag{1}$$

Taking the radius from the definition of effective temperature

$$\frac{L}{L_{\odot}} = (\frac{R}{R_{\odot}})^2 (\frac{T}{T_{\odot}})^4 , \quad \text{Log } (\frac{L}{L_{\odot}}) = 2 \text{ Log } (\frac{R}{R_{\odot}}) + 4 \text{ Log } (\frac{T}{T_{\odot}}) \tag{2}$$

the mass from the supergiant mass-luminosity relation

$$\text{Log } (\frac{L}{L_{\odot}}) \simeq 2.5 \text{ Log } (\frac{M}{M_{\odot}}) + 1.38 \tag{3}$$

and the solar values $M_{\odot} = 4.77$ and $T_{\odot} = 5760$ K, gives

$$\text{Log } P = \text{Log } Q - .22 M_{bol} - 3 \text{ Log } T_{eff} + 12.61. \tag{4}$$

Even this simple expression demonstrates the observed correlations as can be seen in Figure 1. The slopes of the relationships are approximately correct when a vlue of Q ~ .1 day is adopted.

III. Radial Velocities

The measurement of precise absorption line positions, expressed as radial velocities, provides the most direct evidence that large scale velocity fields are present in the line forming regions of early type supergiants. The observations can be grouped into two basic categories; those that indicate a radial velocity gradient through the atmosphere (almost always an expansion velocity), and those that show velocity fluctuations.

The radial velocity study of O and Of stars by Conti, Leap, and Lorre (1977) showed that expansion velocities of 20-30 km sec^{-1} are present in the deepest visible layers of Of stars. Bohannon and Garmany (1978) found a strong gradient present in the Balmer line velocities of 25 O type stars. Ebbets (1979) measured strong gradients in O type supergiants, but found no such effect in main-sequence stars (Figure 2). The effect found in all of these studies was that the more opaque Hydrogen lines, Hα, Hβ, etc., have systematically more negative radial velocities than the higher Balmer lines H10-H15, or the helium, silicon, and CNO lines. Hβ is typically 20-30 km sec^{-1} more negative, and Hα may be shifted by from 50 to 200 km sec^{-1} in the same sense.

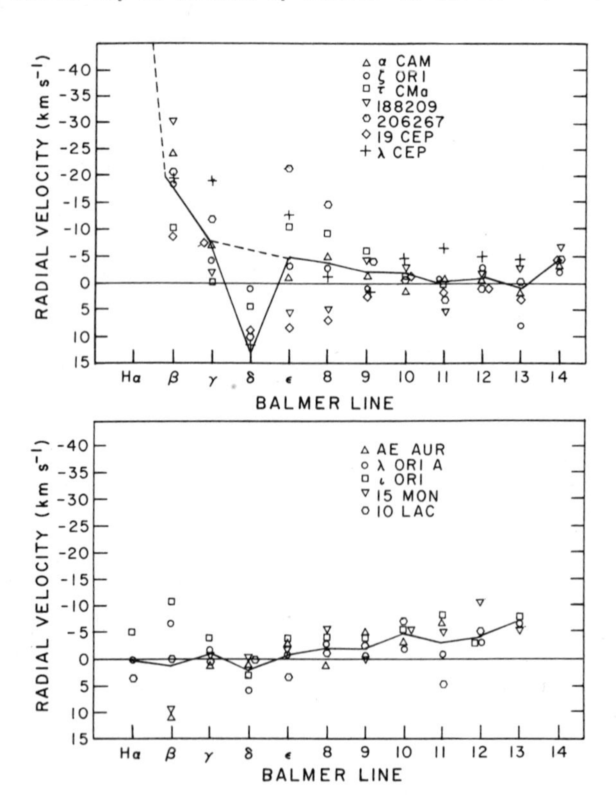

Fig. 2 - The Balmer lines in O supergiants show a systematic trend in velocity which indicates that the lines are formed in an expanding atmosphere which accelerates over the line forming region. The non-supergiants show no such effect. The zero point is the mean velocity of lines of C, N, O, and Si. These figures were adopted from Ebbets (1979), courtesy of Ap. J.

This phenomenon is not restricted to O stars but has been well studied in B types by Hutchings (1976) and in the A type supergiants by Aydin (1972), and Wolf et al. (1974). In these stars, the lines of many different ions can be measured, and a strong correlation between ionization potential and radial velocity is found. The sense of the correlation is almost always that the lower excitation lines have persistent and systematically more negative velocities.

The radial velocity variability of hot supergiants has been known for over twenty years, but has been the subject of few thorough studies. The range of variations is usually small, from the limit of confident detection (~ 1 km sec^{-1}) to about 40 km sec^{-1} in extreme cases. As was the case with photometric variations, the largest amplitudes tend to be found in the most luminous stars. The timescale of variations ranges from days to years. Despite careful statistical searches (Lucy 1975) significant changes within one night have not been found. Long term changes - timescales longer than a year - are similarly difficult to confirm. The best documented variations occur on timescales of 4 to 30 days. Often the changes are fairly regular, but again no stable, single period is ever present. No convincing correlation has yet been demonstrated between the characteristic timescale, and spectral type or luminosity class, since most of the studies to date have analyzed a limited range of spectral types.

The earliest spectral type star for which a comprehensive, modern radial velocity study exists is α Cam -O9.5 Ia. The absorption lines of this massive ($M_* \sim 30\ M_\odot$) and luminous ($M_{bol} \sim -9.75$) supergiant show both line to line, and time to time variations in radial velocity. This complication demands that a careful statistical procedure be used to separate these effects. Such an analysis was initiated by Bohannon and Garmany (1978) in which a two-way analysis of variance, accompanied by an F test, confirmed the highly significant nature of both effects. The gradient across the Balmer line is about 25 km sec^{-1} at Hβ, and itself varies significantly. Systematic variations of the entire spectrum with an amplitude of about 20 km sec^{-1} are also confirmed with a timescale of about thirty days. In a follow-up study with more and better data, Tryon and Garmany (1979) were not able to represent these motions with an orbital solution. The motions are definitely real, with a well established amplitude and timescale, but do not allow a simply periodic description.

An even more thorough analysis of the radial velocity variations of α Cyg - A2 Ia - by Lucy (1975) produced a remarkable insight into the nature of supergiant variability. Lucy was able to represent a long series of 447 radial velocity measurements (Paddock 1935) with a harmonic series containing sixteen terms. Sophisticated statistical tests demonstrated that the periods, amplitudes, and phases of all terms were significant, and stable for at least the six years during which the observations were made. The periods of the individual terms range from seven to one hundred days with no significant power present at shorter timescales. The individual amplitudes range from .4 to 1.0 km sec^{-1}, but add together in such a way as to produce a total range of variation of 10 km sec^{-1}. It is hypothesized that these terms represent many discrete modes of non-radial pulsation, and that other modes are present, but either undetected or unresolved.

IV. Absorption Line Profiles

The inference of turbulent velocities in early type supergiant atmospheres was first made many years ago on the basis of absorption line strengths and shapes. Equivalent widths of all lines in O star spectra were stronger than LTE model atmospheres predicted, and the profiles were much too broad to be attributed to the thermal and stark broadening. Microturbulent velocities in excess of 20 km sec^{-1} were required to account for the observed equivalent widths, and rotation and/or macroturbulence often in excess of 100 km sec^{-1} was implied by the line shapes. Considerable progress has been made in these areas in the past decade, with the availability of static non-LTE model atmospheres, and several new observational studies. For the most part, the predictions of line strengths by the new models are entirely successful at reproducing the observations (Auer and Mihalas 1972, Conti 1973). Most of the hydrogen, neutral helium, and ionized helium equivalent widths are consistent with zero microturbulence, however some important differences do remain. The neutral helium lines at λ5876 Å and λ6678 Å are still observed to be considerably stronger than the models allow. Conti (1974) and Rosendhal (1973b) suggest that some small classical microturbulence included in the calculations would ameliorate these discrepancies. Most of the lines of He II are adequately accounted for by the non-LTE models. A possible discrepancy was found by Snijders and Underhill (1975) for lines having a lower level of n = 4. In a detailed analysis of the He II spectrum of ζ Pup - O4 f - they found these lines to be systematically stronger than expected on the basis of a model atmosphere which describes the other lines well. They suggest that the n = 4 level of He II is over-populated by absorbing Hydrogen Lyman Alpha photons in the n = 2-4 transition. In a quiescent medium these transitions are not exactly coincident in wavelength (Auer and Mihalas 1972) but if a velocity field (the authors suggest macroturbulence) of about 50 km sec^{-1} were present, the broadened lines could overlap sufficiently to produce the necessary pumping. The only other lines whose characteristics are seriously in disagreement with the static, plane parallel models are Hα and He II λ4686 Å and He II λ10124 Å, whose formation almost surely occurs in the higher, extended and expanding regions of the atmosphere.

Two studies of the absorption line widths in O stars by Slettebak (1956) and Conti and Ebbets (1977), were intended to measure rotation velocities, but in addition outlined the statistical properties of what is called macroturbulence. Both studies started out with line profiles thought to be free of any macroscopic broadening, either the observed profiles from sharp lined stars, or the theoretical profiles from model atmospheres. These sharp lines were then numerically broadened, using the convolution approximation with a limb darkened rotation function, and the relationship between full width at half maximum vs. V sin i was calibrated. The line widths of about two hundred program stars were then expressed as rotation velocities. Finally, the distribution with spectral type and luminosity class was studied (Figure 3). The conclusions of both studies regarding line broadening were similar, and can be summarized as follows: The O stars show a wide range of apparent rotation velocities, from essentially zero to some in excess of 400 km sec^{-1}. Regardless of how the stars are classified, the mean and modal velocity is near 100 km sec^{-1}. Only the stars of type O9 V have any members which show lines sufficiently sharp to be consistent with zero macroscopic broadening. Stars hotter than type O9, and with luminosity brighter than class V have no members with unbroadened lines. The number of observations is sufficiently large that if the broadening were purely rotation, some sharp lined - i.e. pole on stars - would be expected. This residual broadening increases with increasing temperature and increasing luminosity. Among the earliest types which are well observed - say O6 V - the minimum broadening is equivalent to 50 km sec^{-1} of rotation. Among the Ia supergiants of spectral type O8 and O9, the non-rotational component of line broadening is about 30 km sec^{-1}. These velocities are referred to as macroturbulence. They describe line broadening which cannot be attributed solely to rotation, and refer only to groups of stars in a statistical sense.

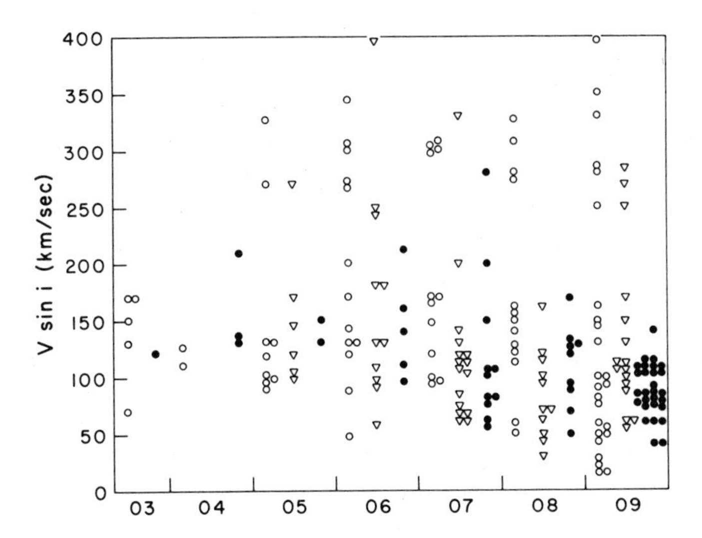

Fig. 3 - Apparent rotational velocities for O stars as catalogued by Conti and Ebbets (1977). Open circles represent main sequence stars, triangles are giants; filled circles indicate supergiants. This figure originally appeared in the Ap. J.

An attempt to separate the effects of rotation and macroturbulence in individual stars was made by this author (Ebbets 1979) using the Fourier Transform analysis described by Smith and Gray (1976). In the wavelength domain the effects of different velocity structures are subtle and difficult to separate, buth given sufficiently precise data, the differences are more pronounced in the Fourier transform domain (Figure 4). Adopting an isotropic Gaussian representation of the macroturbulence, the rotation function of Huang and Struve (1953) and the unbroadened profiles of Auer and Mihalas (1972), I derived rotation and macroturbulence parameters from the He I and He II lines of 16 O supergiants and non-supergiants. The rotation velocities were consistent with the previous values, and macroturbulence ranged from less than the detection threshold of 10 km sec^{-1} in O9 V stars to 30 km sec^{-1} in O9.5 Ia stars. In addition the macroturbulence parameter shows a strong correlation with the Balmer lines' velocity gradient.

Variations in the absorption line profiles are also detected in some supergiants. Perhaps the most dramatic example is that reported by Wolf et al. (1974) in the A2 Ia

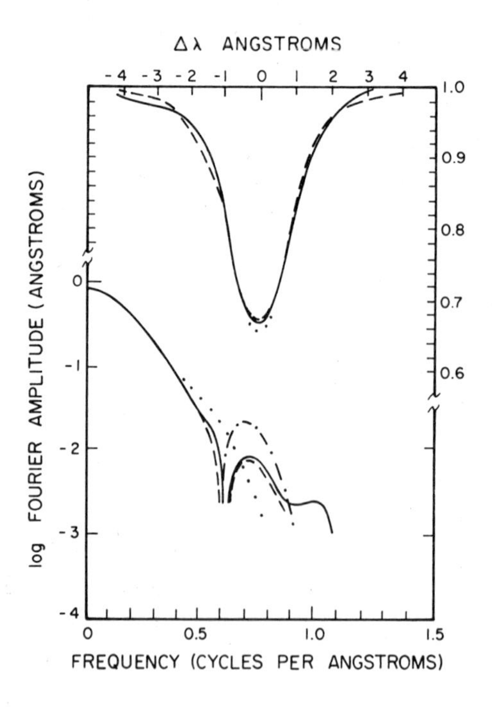

Fig. 4 - A line profile and Fourier transform analysis was performed to measure macroturbulence in O type stars. The line shown is He I λ4471 in 19 Cep. Models with various combinations of rotation and turbulence can reproduce the profile well, but only one set, V sin i = 75 km sec^{-1}, V_{turb} = 20 km sec^{-1}, matches the observation in both domains simultaneously. The dot-dashed transform is from a model with no turbulence but V sin i = 80 km sec^{-1}. The dotted profile and transform model zero rotation, but V_{turb} = 55 km sec^{-1}. From Ebbets (1979).

star HD 160529. In addition to the systematic and variable displacements discussed in the previous section, the metallic lines (Fe II, Si II, etc.) are variable in profile and equivalent width. In extreme instances, the lines actually split into two distinct components sometimes by as much as 40 km sec^{-1}. This splitting is irregular in time, the relative strengths of the components vary, and the effect does not appear to be due to a binary companion. The authors believe that very large sections of the atmosphere moving with quite different velocities are responsible for the phenomenon. If this interpretation is correct, it is a paradigm example of the concept of macroturbulent motions in a stellar atmosphere. Similar, but less dramatic variations are observed in other stars. Smith (private communication 1979) has observed rapid changes (often within one night) in the Si III lines of ρ Leo B1 Ib, and Rosendhal (1972) reports variation of the Si II lines in α Cyg A2 Ia on timescales of days. The interpretation of these observations is unclear at present, but must imply some kind of substantial change in the visible hemisphere on short timescales.

V. Emission Line Profiles

In addition to their absorption lines, the spectra of many O stars, and almost all O, B, and A supergiants, contain emission lines of hydrogen, neutral helium, and ionized helium. It is generally thought (Beals 1951 for example) that the emission lines arise in the atmospheric layers well above the photosphere, in the extended and expanding envelope of the star. The strength, profiles, and variability of emission lines therefore provide information about the state of motion of the gas in layers much higher in the atmosphere than the regions where absorption lines are formed. Not surprisingly, careful observations have revealed that the emission lines are variable in strength and profile on a variety of timescales. The best example is the Hα profile of the B3 Ia star, 55 Cygni, whose changes over about 25 years were sketched by Underhill (1966). At various times the profile was strong and symmetric in absorption, very weakly in absorption, or strongly in emission with a pronounced P-Cygni profile. Since the time coverage was very spotty, a characteristic timescale is difficult to assign to these changes. Rosendhal (1973a) studied the variability of Hα in B and A type supergiants, again showing large changes in spectra taken several years apart. Conti and Niemela (1976) observed a gradual weakening of Hα in ζ Pup over three years, and suggested a global decrease in the envelope density, and therefore the rate of mass loss.

On the more rapid timescales - several days - Conti and Frost (1974), Conti and Leep (1979), Hutchings and Sanyal (1976) documented variations in the Hα and He II λ4686 Å emission lines in the O6 ef star λ Cep. A possible periodicity (~ 3.3 days) consistent with the rotation of an inhomogeneous envelope was suggested. Similar nightly, and often hourly, changes are present in the Hα profile of α Cam, O9.5 Ia (Figure 5, from Ebbets 1980a). Again, the recurrence of similar profiles at multiples of the estimated

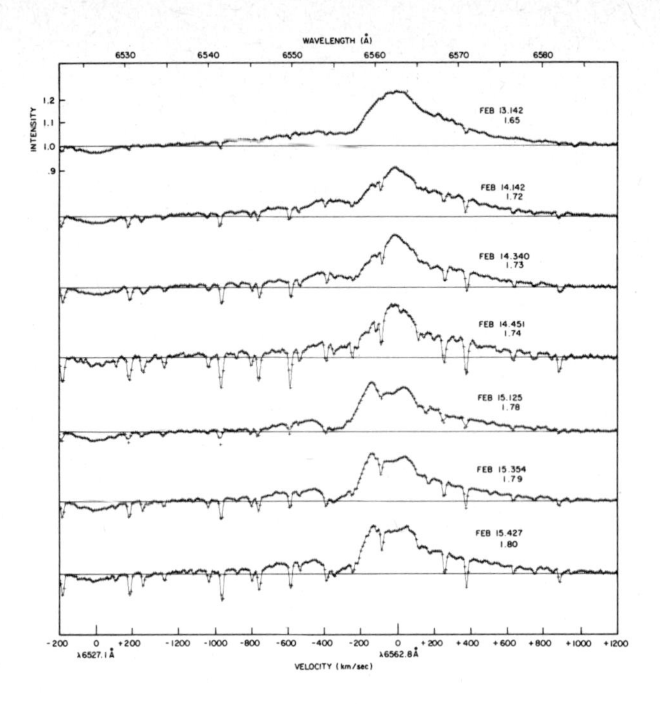

Fig. 5 - The Hα profile of the O9.5 Ia star α Cam shows a Beals Type III P Cygni profile. The width of the line, and the extensive, symmetric wings are due to emission and scattering in the lower regions of the stellar wind. The narrow features are atmospheric water vapor.

rotation period (15 days) is suggestive of a non-axisymmetric envelope. Occasionally a star which ordinarily does not show emission lines in its spectrum experiences a short episode of activity. Such an episode occurred in the O9 V star ζ Oph in 1973. Emission at Hα was first reported by Niemela and Mendez (1974), and its development and dispersal was followed by Barker and Brown (1974). This outburst was presumably related to the extremely rapid rotation ($V \sin i > 450$ km sec^{-1}) but is cited as an example of a transitory large scale velocity field.

A final relevant observation is the presence of extremely broad but faint wings flanking the Hα emission line in most O type supergiants. These features extend to at least ±1500 km sec^{-1} in α Cam and ζ Ori, and to over ±2000 km sec^{-1} in ζ Pup. Their presence was detected in high signal-to-noise coude reticon observations, and are probably due to a combination of emission in the rapidly accelerating lower envelope, plus the effect of non-coherent scattering by free electrons (Ebbets 1980b). Scattering by electrons in the expanding atmosphere may also produce some of the line broadening observed in photospheric absorption lines.

VI. Ultraviolet Observations

The most opaque transitions in early type stars are the ultraviolet resonance lines of the dominant ionization stages of carbon, nitrogen, and oxygen. Observations of these lines therefore probe the highest atmospheric regions accessible to the spectroscopist. By now the message of these lines is well known - the outer atmosphere of all O type stars, and most B and A supergiants are expanding at supersonic velocities (often in excess of 2500 km sec^{-1}) and returning mass to the interstellar medium at rates of several times 10^{-6} solar masses per year. (See Snow and Morton 1976, Conti 1978, Cassinelli 1979 for extensive reviews of stellar winds.) Detailed analyses of ultraviolet spectra, in which the resonance line profiles are modeled and interpreted, have investigated the temperature, density, and velocity distribution in the envelope. All studies to date (for example Lamers and Morton 1976, Olson 1978) have assumed spherically symmetric, laminar steady flow, and have been reasonably successful at reproducing the observed profiles. Turbulence was not explicitly included in these studies, and there are no glaring inconsistencies which cry out for its introduction. Two classes of models postulate a temperature in the wind higher than that which strict radiative equilibrium would allow, and may implicitly require the dissipation of some form of mechanical energy as a heating mechanism. The "warm wind" model and the coronal models are reviewed by Cassinelli, Castor, and Lamers (1978).

Since ultraviolet observations can only be made from spacecraft, time series observations have been difficult and rare. There is, however, a growing literature reporting variability in the ultraviolet spectrum. Snow (1977) reobserved fifteen O and B stars which had shown evidence of massive stellar winds in the first Copernicus survey. Most stars showed only marginal changes over timescales of two to four years. At least two stars, ζ Pup, O4 f, and δ Ori A, O9 I, seemed to have less material in the expanding envelope when reobserved, implying a decrease in the mass loss rate. The profiles in ζ Ori A O9 I suggest a slightly more dense stellar wind and thus an increased mass loss. The only observation searching for variations on timescales of days were reported by Conti and Leep (1979), who found little significant change in the spectrum of λ Cep. Very rapid changes in the fine structure of the ultraviolet absorption lines (timescale ~ hours) were reported by York <u>et al.</u> (1977). The nature and duration of the fluctuations were such that these authors suggested density perturbations - waves - propagating upward as a transient disturbance in the atmosphere. Prolonged ultraviolet observations of several hot supergiants with time resolutions of several hours to a day would clearly be an interesting and worthwhile project.

Summary and Conclusions

Observations of many different kinds demonstrate that the outer layers of early type supergiants are variable. Photometric and radial velocity studies indicate semi-regular fluctuations of low amplitude and timescales of days to weeks. The nature of these variations suggests that pulsation contributes at least somewhat to the phenomenon. Brightness and color fluctuations may reflect changes in the depth dependent velocity field. Variations in the electron pressure in the continuum forming layers may contribute to the erratic behavior of the hydrogen Balmer discontinuity. The combined effect of the velocities of many low amplitude modes of oscillation may help explain the width of absorption lines and their occasional asymmetry and splitting. Such turbulent velocities could also allow the n = 4 level of ionized helium to overlap with hydrogen Lyman alpha enough to slightly overpopulate that level in the deeper layers, accounting for the anamolous strength of the absorption lines of that series.

All of these effects occur in the deeper layers of the stellar atmosphere. What effect they have on the properties of the stellar wind is an unsettled question. Observations have not yet provided a strong case for short term variability in the wind. Do photospheric disturbances merely damp out in the lower layers, do they propagate outwards quiescently, or do they shock and act as a source of heat for a chromosphere corona structure? Purely radiative models (Lucy and Solomon 1970, Castor, Abbot, and Klein 1975) indicate that the mass loss can be driven without a mechanically heated coronal push. At the same time, models which do incorporate a thin corona are more successful at explaining the observed ionization balance, while not interfering with the basic radiatively driven wind (Cassinelli, Olson, and Stalio 1978). On the other hand, the imperfect flow model (Cannon and Thomas 1977) contends that the propagation and dissipation of m chanical energy is of fundamental importance in determining the properties of the expanding atmosphere.

Observations leave no doubt that variations are present in even the deepest visible layers. A good deal of progress has been made in characterizing the fluctuations, and identifying possible physical mechanisms. Further progress demands that we seek and clarify the connection between photospheric "turbulence" and the dynamics and energetics of the expanding supergiant envelope.

References

Abt, H.A. 1957, *Ap. J.*, **126**, 158.

________ 1958, *Ap. J.*, **127**, 658.

Auer, L.H., and Mihalas, D. 1972, *Ap. J. Suppl.*, **24**, 193.

Aydin, C. 1972, *Astron. & Astrophys.*, **19**, 369.

Barker, P., and Brown, T. 1974, *Ap. J.*, **192**, 111.

Barlow, M., and Cohen, M. 1977, *Ap. J.*, **213**, 737.

Beals, C.S. 1951, *Pub. Dom. Ap. Obs.*, **9**, 1.

Bohannon, B., and Garmany, K.D. 1978, *Ap. J.*, **223**, 908.

Burki, G., Maeder, A., and Rufener, F. 1978, *Astron. & Astrophys.*, **65**, 363.

Cannon, C.J., and Thomas, R.N. 1977, *Ap. J.*, **211**, 910.

Cassinelli, J.P. 1979, Ann. Rev. Astron. Astrophys., 17:
Cassinelli, J., Olson, G., and Stalio, R. 1978, Ap. J., 220, 573.
Cassinelli, J.P., Castor, J.I., and Lamers, H.J.G.L.M. 1978, Pub. A.S.P., 90.
Castor, J.I., Abbott, D.C., and Klein, R.I. 1975, Ap. J., 195, 157.
Cayrel, R., and Steinberg, M. 1978, ed. Physique Des Mouvements Dans Les Atmospheres Stellaires, Centre National de la Recherche Scientifique, Paris.
Conti, P.S. 1973, Ap. J., 179, 161.
__________ 1974, Ap. J., 187, 539.
__________ 1978, Ann. Rev. Astron. Astrophys., 16: 371-392.
Conti, P.S., and Frost, S.A. 1974, Ap. J., 190, L137.
Conti, P.S., and Niemela, V.S. 1976, Ap. J., 209, L37.
Conti, P.S., and Ebbets, D. 1977, Ap. J., 213, 438.
Conti, P.S., and Leep, E.M. 1979, preprint.
Conti, P.S., Leep, E.M., and Lorre, J.J. 1977, Ap. J., 214, 759.
Ebbets, D.C. 1979, Ap. J., 227, 510.
__________ 1980a, Ap. J., in press.
__________ 1980b, Ap. J., in press.
Feitzinger, J.V. 1978, Astron. & Astrophys., 64, 243.
Hayes, D.P. 1975, Ap. J., 197, L55.
Huang, S.S., and Struve, O. 1953, Ap. J., 118, 463.
Hutchings, J.B. 1976, Ap. J., 203, 438.
Hutchings, J.B., and Sanyal, A. 1976, Pub. A.S.P., 88, 279.
Lamers, H.J., and Morton, D.C. 1976, Ap. J. Suppl., 32, 715.
Lucy, L.B. 1975, Ap. J., 206, 499.
Lucy, L.B., and Solomon, P.M. 1970, Ap. J., 159, 879.
Maeder, A., and Rufener, F. 1972, Astron. & Astrophys., 20, 437.
Mihalas, D. 1972, Ap. J., 176, 139.
__________ 1979, preprint, Curves of Growth and Line Profiles in Expanding and Rotating Atmospheres, submitted to Ap. J.
Morrison, N.D. 1975, Ap. J., 200, 113.
Niemela, V., and Mendez, 1974, Ap. J., 187, L23.
Olson, G.L. 1978, Ap. J., 226, 124.
Paddock, G.F. 1935, Lick Obs. Bull., 17, 99.
Rosendhal, J.D. 1970, in Stellar Rotation, ed. A. Slettebak, Reidel, Dordrecht, 122.
__________ 1972, Ap. J., 178, 707.
__________ 1973a, Ap. J., 182, 523.
__________ 1973b, Ap. J., 183, L39.
Schild, R.E., and Chaffee, F.H. 1975, Ap. J., 196, 503.
Serkowski, K. 1970, Ap. J., 160, 1083.
Slettebak, A. 1956, Ap. J., 124, 173.
Snijders, M.A.J., and Underhill, A.B. 1975, Ap. J., 200, 634.
Snow, T.P. 1977, Ap. J., 217, 760.
Snow, T.P., and Morton, D.C. 1976, Ap. J. Suppl. Series, 32, 429.
Smith, M.A., and Gray, D.F. 1976, Pub. A.S.P., 88, 809.
Sterken, C. 1977, Astron. & Astrophys., 57, 361.
Struve, O. 1952, Pub. A.S.P., 64, 118.
Struve, O., and Elvey, C. 1934, Ap. J., 79, 409.
Thomas, R.N. 1960, Aerodynamic Phenomena in Stellar Atmospheres, Nicola Zanichelli, Bologna.
Tryon, P.V., and Garmany, C.D. 1979, preprint
Underhill, A.B. 1949, M.N.R.A.S., 109, 562.
__________ 1960, in Aerodynamic Phenomena in Stellar Atmospheres, ed. R. N. Thomas, Bologna.
__________ 1966, The Early Type Stars, New York: Gordon and Breach.
Wolf, B., Campusano, L., and Sterken, C. 1974, Astron. & Astrophys., 36, 87.
Wright, K.O. 1955, Transactions of the I.A.U. Vol. IX, p. 739.
York, D.G., Vidal-Madjar, A., Laurent, C., and Bonnet, R. 1977, Ap. J., 213, L61.

PHOTOSPHERIC MACROTURBULENCE IN LATE-TYPE STARS

Myron A. Smith
Department of Astronomy
University of Texas at Austin

I. Why Study Macroturbulence?

It is intimidating to attempt a review of the subject of late-type stars macroturbulence which follows so closely on the heels of David Gray's (1978) fine review in Solar Physics. Therefore this paper will avoid many emphases of his review while filling in some areas where some progress has emerged in the last 1-1/2 years. My slant will come mainly from the standpoint of line broadening analysis.

Each of us has his own reasons for being interested in macroturbulence of late-type stars. A short list of motivations might look like the following:
1) The relation of spectroscopic "macroturbulence" to the general atmospheric turbulence spectrum. Even in the Sun we do not know yet how the resolved velocity fields add up to "spectroscopic macroturbulence". The relationship of this macro-field is still not well defined in terms of microturbulence, and it is far from settled as to how important and unique a mesoturbulent description is. Finally, the cause of macroturbulence, whether convection, granules, nonradial pulsation, or even rotation, is still unspecified and it may be different in different regions of the H-R Diagram.
2) The relevance of macroturbulence to energy dissipation in the chromosphere. Here, at least, it appears that some tentative answers are beginning to emerge (§ III C.).
3) The relationship to chromospheric parameters in stars. These promise to provide us with kinematical models for the chromosphere-corona-solar wind complex both in stars and in the Sun. Adopting the solar-physics-of-stars theme, consider that stars of varying observable characteristics (T_{eff}, log g, composition, rotation, age, are the usual quintet) will help us to see the dependence of upper atmospheric phenomena on fundamental attributes of a star. Even though well observed, the Sun alone cannot provide this dependence. The Wilson-Bappu effect is an excellent historical example of what could have been a relationship between a global parameter and a turbulence, though now it appears that velocity fields are not the culprit after all (Ayres 1979).
4) The relationship of 5-minute-type oscillations to macroturbulence. Analysis of this oscillatory pattern, a consequence of nonradial p-modes (Deubner 1975, Rhodes et al. 1977), has already facilitated probes of interior properties such as the convection zone depth and differential rotation rate. The possibility exists that time-resolved analyses will provide similar information for other late-type stars.

II. Toward the Detection and Modeling of Radial-Tangential Macroturbulence.

In most stars the measurement of macroturbulence is hampered by the presence of a substantial rotational broadening. What makes at all possible the dissection of a profile's broadening into rotational and macroturbulent velocity fields is a combination of high-resolution, high S/N data on one hand and the sharply differing models for the two velocity distributions on the other. Whereas the Unsold "rotation function" is U-shaped, radial-tangential macroturbulence (the appropriateness of which is discussed below) produces profiles with extended wings and a deep, pointed core. Predictably, the Fourier transforms of these two functions are different and these are depicted in Figure 1. The rotation transform can be expressed approximately as a first order Bessel function. This function contains a series of regularly spaced zeroes and sidelobes. As one proceeds to stars of slower rotational velocities, the zeroes shift to higher Fourier frequencies until even the first zero recedes to unobservable frequencies and, ultimately, one is left only with a filtering due to the main lobe of the rotational transform. This lobe is rather square-shaped and for low rotational velocities, the interference from rotation quickly becomes negligible. It is for this reason that most of our information on macroturbulence comes from old, late-type stars. Let us now place these statements on a more quantitative basis: Lines of intermediate strength (~ 100 mÅ) generally offer recovery to the highest Fourier frequency at a fixed S/N (Smith and Gray 1976, Gray 1978). Such a line will have its first natural zero at about 0.14 s km^{-1}. As a practical matter the presence of a macroturbulent broadening agent can be best detected in stars having $V_R \sin i \lesssim 17$ km s^{-1} (Smith 1975, Kurucz et al. 1977). For broadening in these stars, e.g. those occurring in the middle of the H-R Diagram, almost nothing can be said concerning the turbulence distribution. When one passes to $V_R \sin i = 10$ km s^{-1} (see Figure 2), the rotational zero occurs near 0.06 s km^{-1}. One can then perceive a slightly less bowed Fourier main-lobe, which corresponds to extended wings of the profile in

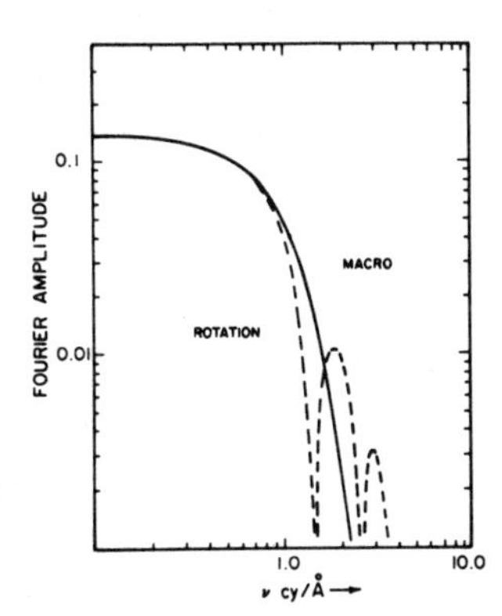

Fig. 1 - Fourier transforms of a model profile at λ6000 broadened by rotation ($V = V_R \sin i = 22.5$ km s^{-1}) and radial tangential macroturbulence (M = 15 km s^{-1}).

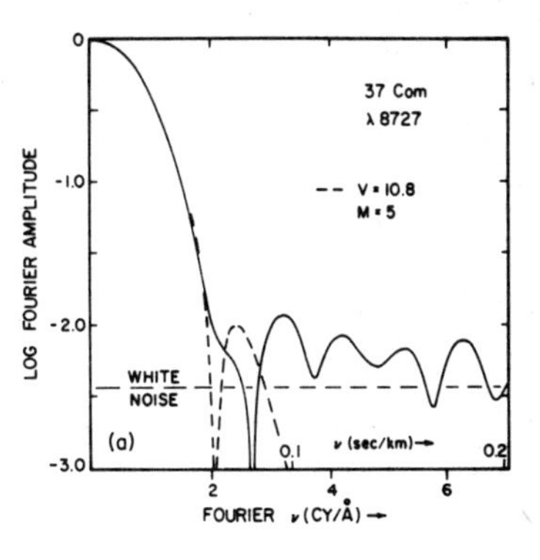

Fig. 2 - Transform of line modeled with rotation and macroturbulence. Rotation dominates, but one can see that R-T is a better turbulence model than is a gaussian.

the wavelength domain. At this point, corresponding to V_R sin $i/M_{RT} \approx 2$ in supergiants (Figure 2) and about 3 in giants, one is on the threshold of distinguishing radial-tangential macroturbulence from much different distributions like the gaussian. For V_R sin $i = 3$, i.e. at a velocity ratio near unity, the rotational effects become negligible and a host of other radiative transfer and instrumental parameters become more important (see Smith 1980 for a ranking of sources of error). Included in this velocity range are the ultra-slowly rotating dwarfs like the Sun and the red giants; most of our turbulence information is derived from these sources.

Older studies of macroturbulence used isotropic or single-stream gaussian representations for the lack of anything better. More recently, observers have been forced to a radial-tangential or exponential macroturbulence for certain sharp-lined early-type stars (Smith and Karp 1978), the Sun (Rutten et al. 1974, Gurtevenko et al. 1976, Smith et al. 1976, Gray 1977), solar-type dwarfs (Smith 1976, Smith 1978), and luminous K stars (Lambert and Tomkin 1974, Gray 1975, Luck 1977, Gray and Martin 1979 ("GM79"), Smith and Dominy 1979 ("SD79")). Gray (1975, 1976) first suggested the two-stream, radial-tangential model with internal gaussian dispersion (see Table 1 for definition) because of the similarity of this distribution to Benard cells envisioned in solar granules. Now if these eddies actually followed a circular and not a square pattern, the resulting macroturbulence would be described by an isotropic gaussian distribution. This contrast demonstrates the importance of geometry in the modeling of these motions. Both Gray and Smith have used equal radial/tangential stream areas and velocity dispersions in their macroturbulence modeling. Beckers and Morrison's (1970) results on solar intragranular flow patterns suggest that these adopted equalities appear to be well within a factor of two of reality, but their results also make it clear that granular flows do not turn square corners! An alternate way of representing the observed distribution is with a depth-dependent isotropic gaussian distribution of velocities (Smith et al. 1976). This model is not too physically unreasonable because of the strong depth dependences shown by granular and 5-minute oscillation patterns. In his solar flux-profile study Gray (1977) indeed found a 0.6 km s^{-1} increase in macroturbulence or weak lines formed in the lower photosphere. In sum, both the Benard cell and depth-dependent assumptions seem to have some basis in fact and together contribute toward the radial-tangential model. This conceptual agreement with solar observations also implies that "macroturbulence" as the spectroscopist observes it may well be due to the granulation pattern, perhaps with some residual help from 5-minute-type oscillations.

This reviewer would be remiss in not alluding to other independent techniques of measuring large-scale velocities. One consists of measuring radial velocity shifts (see W. Buscombe's paper) for high and low excitation lines preferentially formed in updraft/downdraft regions. Using this technique Dravins (1974) finds the same result that emerges from line broadening studies: the Sun and α Boo have similar macroturbulent velocities. Line shift studies ought to be encouraged and have a continuing place in photographic work. Finally, Traub et al. (1978) have broken ground in searching for five-minute-type oscillations in spectra of other late-type stars. This work can be most easily extended by time-resolved observations of red giants. These stars are bright and the oscillations are expected to have large amplitudes and periods. Cram and Smith have in progress a pilot investigation of α Boo. The strategy here will be to use a conventional coude system with a Reticon detector and to minimize ultra-small instrumental wavelength shifts by referencing stellar lines with nearby terrestrial features. It is hoped that future reviews on this subject will have less emphasis on line broadening and more on resolved oscillatory patterns.

III. Recent Results.

A. Macroturbulence and the Turbulence Spectrum.

Mesoturbulence - Despite the work done on broadening by finite-sized eddies, there still seems to be disagreement on whether model profiles computed with a correlation-length formulation fit center-to-limb observations, (cf. Auvergne et al. 1973, Frisch 1975 vs. Canfield and Beckers 1976), or the detailed shape of a given profile (Smith and Frisch 1976 vs. Gray 1977), better than a micro-/macro-model does. Most of us observers hope that a micro-/macro-description is an adequate representation of stellar profiles but a decisive answer is not yet forthcoming. Consider that although the solar intensity profile analysis of Smith and Frisch, and the flux analysis of Gray 1977, were each done "correctly", the two studies led to conflicting results. The latter could find no evidence for narrow sidelobes indicative of the dominance of mesoturbulence. Perhaps the fault lies in the assumptions going into the analysis of one or both studies (e.g. inaccurate eddy size distribution, or use of the convolution approximation for rotation and flux profiles) or in systematic errors in the atlases; the matter is not resolved. It may be, as implied from other solar observations (Beckers and Parnell 1969), that mesoturbulence is physically important in a stellar atmosphere but that operationally one can ignore it in a stellar profile, as Gray suggests. Future studies would do well to look for abnormally narrow or broad sidelobes in transforms of intermediate-strong symmetrical lines (~ 200 mÅ). However this approach requires a Fourier noise amplitude of -3 or less in the log. Therefore, only high S/N, completely unblended profiles can be used for studying mesoturbulence.

Microturbulence - The relationship between macro- and micro-turbulence is equally unclear. In his review (Gray 1978) shows an impressive-looking ξ_t - M_{RT} correlation. Contrariwise, our work has not turned up yet any such correlation. For example, the microturbulence values determined by Smith (1978) and Smith and Dominy (1979) increase from 0.5 km s^{-1} to 2.0 km s^{-1} going from dwarfs to giants, but there is no such increase in the macroturbulence values. Perhaps a correlation in broadening does exist, particularly if supergiants are included, but there are reasons to question a turbulence interpretation for any such relation. The quoted values of microturbulence in supergiants (which may include non-LTE effects and/or velocity gradients), on which the Gray correlation largely rests, cannot yet be considered well-documented turbulences. In some cases these values are actually supersonic. In short the turbulence measured in dwarfs and supergiants may arise from totally unrelated phenomena. Thus we seem to be in the same state here as a decade ago. A study of particular lines including effects of non-LTE and asymmetry in the analysis needs to be done to test the turbulence interpretation.

In one important respect progress has been made: the microturbulence values from both profile and curve of growth (e.g. Lambert and Ries 1977) analyses of both dwarfs and giants are finally in good agreement.

B. Macroturbulences of Main Sequence and Luminous Late-Type Stars.

General Survey -- Table 1 is a compilation of recent rotational and macroturbulence velocities obtained by photoelectric means for G type dwarfs and luminous K stars. Note in the table the definition of the M_{RT} velocity-scale used herein; we have scaled Gray's values downwards by $\sqrt{2}$ to put all numbers on a common system.

Where comparison is possible between observers, the agreement in this table is excellent. For the Sun the comparison agrees to within 0.1 km s^{-1}. Note that these turbulence values, even when corrected for velocity scale differences, are perhaps a factor of two larger than the sum of all the resolved solar velocities (e.g. Edmonds 1967). A second comparison can be made from the mean macroturbulence for a group of four giants studied by GM79 and SD79 studies. The mean value in both studies is 3.0 km s^{-1}. Finally, SD79 report a macroturbulence difference of only 0.2 km s^{-1} if the same line is observed and analyzed independently by two investigators (Gray and Smith). This suggests that most errors in the red giant analyses derive from different lines being used in the analyses; that is, errors involving treatment of line blending and radiative transfer are most severe. As main sequence stars tend to have fainter apparent magnitudes, the errors in their analyses are dominated by spectrophotometric errors (Smith 1978).

Main Sequence Stars -- A pair of mid-F and K stars in Table 1 show smaller turbulence values than the G stars do. However, using a scaled solar $T(\tau)$-relation, Smith (1978) found no evidence for a change in the mean macroturbulence values from G0 to K0. It is certainly premature to make a statement on the dependence of turbulence with T_{eff}. Even when more F and K stars are observed, it will be imperative to find a trustworthy model $T(\tau)$-relation in order to carry out an analysis before any results can be related to G stars.

TABLE 1

Recent Fourier Determined $V_R \sin i$ and M_{RT}* Values for Late-Type Stars

Dwarfs --

Star	Sp. Type	$V_R \sin i$	M_{RT}	Ref.
Sun	G2-4 V(?)	---	3.1	S76
		2.2 ± 0.7	3.2 ± 0.2	S78
		1.9	3.1	GM78
Procyon	F5 IV	3.5 ± 0.5	1.2 ± 0.3*	WJ78
59 Vir	F8 V	5.5 ± 0.4	3.8 ± 0.4	S78
β Vir	F8 V	4.1 ± 1.0	4.2 ± 0.5	S78
π^1 UMa	G0 V	10.8 ± 0.4	4.8 ± 0.4	S78
HR 3625	G0 V	6.0 ± 0.2	3.8 ± 0.4	S78
β Com	G0 V	5.8 ± 0.2	3.5 ± 0.3	S78
η Cas A	G0 V	4.3 ± 0.7	2.8 ± 0.7	S78
47 UMa	G0 V	3.3 ± 0.7	3.4 ± 0.5	S78
β CVn	G0 V	4.7 ± 0.7	3.1 ± 0.4	S78
λ Ser	F0 V	5.4 ± 0.2	1.8 ± 0.2	S78
51 Peg	G5 IV	5.0 ± 0.7	2.4 ± 0.5	S78
τ Cet	G8 V	2.2 ± 1.1	2.6 ± 0.5	S78
Gmb 1830	%8 VI	1.0 ± 1.0	1.8 ± 0.5	S78
70 Oph A	K0 V	3.3	4.1	S78
δ Eri	K0 IV	2.2 ± 0.9	2.9 ± 0.2	SD79
ν^2 CMa	K1 IV	1.5 ± 0.7	3.0 ± 0.1	SD79
γ Cep	K1 IV	2.4 ± 0.4	2.8 ± 0.6	SD79
61 Cyg A	K5 V	2.0	1.8	VF79

*All numbers have been normalized to the following definition of M_{RT}:

$$f^*(\Delta\lambda) = \frac{1}{2\sqrt{2\pi}\, M_{RT} \sin\theta} \exp(-(\Delta\lambda/\sqrt{2}\, M_{RT} \sin\theta)^2) + \frac{1}{2\sqrt{2\pi}\, M_{RT} \cos\theta} \exp(-(\Delta\lambda)/\sqrt{2}\, M_{RT} \sin\theta)^2)$$

(cf. Gray 1976, eqn. 18-12). My M_{RT} scale is smaller by $\sqrt{2}$ than the "ζ_{RT}" values quoted by Gray; it is 1.5 times larger than those of Wynn-Jones et al.

Giants --

Star	Sp. Type	$V_R \sin i$	M_{RT}	Ref.
γ Leo B	G7 III	1.0 ± 1.3	2.8 ± 0.7	SD79
β Her	G8 III	3.5	4.2	G79
η Dra	G8 III	0.0	4.2	G79
ε Vir	G9 III	2.7 ± 1.1	2.9 ± 0.8	SD79
β Gem	K0 III	0.8 ± 1.3	3.3 ± 1.0	SD79
		2.2	2.8	GM79
α UMa	K0 III	3.2	2.8*	GM79
ε Cyg	K0 III	2.1	2.9*	GM79
β Cet	K1 III	3.3 ± 0.8	4.2 ± 0.2	SD79
β Oph	K2 III	---	4.2*	Gray 1975
α Ari	K2 III	2.9 ± 0.8	2.7 ± 0.8	SD79
		0.0	2.8	
β Oph	K2 III (SMR)	3.0	2.7*	GM79
μ Leo	K2 III (SMR)	2.4	2.8*	GM79
α Boo	K2 III	2.7 ± 0.5	3.2 ± 0.5	SD79
	2.8	3.3*		GM79
α Ser	K2 III (SMR)	2.0 ± 0.3	2.7 ± 0.3	SD79
		2.3	2.8	GM79
α Tau	K5 III	2.7 ± 0.2	3.3 ± 0.5	SD79
γ Dra	K5 III	3.5 ± 0.7	2.7 ± 0.5	SD79
Supergiants --				
37 Com	G9 II-III	10.4 ± 0.3	5.8 ± 0.5	SD79
56 Peg	K0 Ib	3:	5.0 ± 0.3	SD79
α Cas	K0 II_III	3:	6.0 ± 1.2	SD79
ε Peg	K1 Ib	≥ 3	7.4 ± 1.0	SD79
ν^1 And	K3 II	---	4.1*	Gray 1975
ι Pup	K3 Ib	3:	5.3 ± 0.7	SD79
γ Aql	K3 II	---	5.3 ± 0.4*	Gray 1975
ξ Cyg	K5 Ib	3:	6.1 ± 1.3	SD79
ζ Aur	K5 II	3:	4.9 ± 0.6	SD79
HR 8726	K5 Ib	3:	6.9 ± 1.4	SD79

In the same study (Smith 1978), we reported an anticorrelation between macroturbulence and age for G dwarfs. Although this relation is significant only at the 90% level, this relation is still more significant for this sample of stars than the (venerable) correlation between rotation and age. It is likely that age uncertainties have introduced scatter into the turbulence-age relation. A physical explanation for the age correlation effect has not yet been advanced.

Smith and Dominy (1979) found that the mean macroturbulence velocity of early K stars remains constant for five magnitudes along the red giant branch. At $M_{Bol} \simeq -2$, corresponding to the appearance of class II and Ib stars, M_{RT} jumps suddenly from 3 to 6 km s^{-1}. For still brighter supergiants the macroturbulence values appear to increase with luminosity (Luck 1977, Imhoff 1977). This sudden jump is not easily understood, but it is helpful at least to know that stars having the larger macroturbulence values also have larger masses than the fainter K giants. In any case, this step-relation between macroturbulence and luminosity violates the expectation that the Wilson-Bappu effect is caused by a chromospheric velocity field. These results and the boundary temperature results of Desikachary and Gray (1978) support the radiative damping model advanced by Ayres (1978).

Of particular significance is the Kl III star β Cet. This star shows a macroturbulence of about 5σ larger than do other K giants of its luminosity. Anomalous broadening for this star has also been noticed by O'Brien (priv. comm.) in the chromospheric λ10830 line (seen in absorption). Moreover, an analysis of I.U.E. spectra of several K giants by Linsky and Haisch (1979) singles out this star as having an anomalous coronal temperature distribution. While it is unclear yet what makes this star so unique, one has to notice that this correlation of peculiar phenomena argues for photospheric and chromospheric velocity fields being correlated. Another argument for correlated velocities, though less convincing, concerns the well-known relation between photospheric and Hα line widths (Bonsack and Culver 1966, Imhoff 1977).

C. The Chromospheric Connection.

Returning to point #2 in § I, one asks: do macroturbulent motions dissipate their energy in the chromosphere? If so, one expects a correlation between chromospheric heating and photospheric macroturbulence. (This assumes the velocities in the chromosphere and photosphere _are_ related, a proposition for which, as just shown, there is some evidence.)

There are several ways to test this idea. Desikachary and Gray (1978) have found that Ca II Kl-minima indicate a higher temperature minimum for normal K-type stars than for super-metal rich ("SMR") stars. However, both stellar groups have higher temperature-minima than radiative equilibrium model atmospheres indicate. Therefore one expects dissipation, e.g. by waves, to be more noticeable in the normal stars than in the SMR stars. Finally, one also expects these waves to be observable as (or to be indirectly related to) macroturbulence. Following this reasoning, Gray and Martin (1979) searched for macroturbulence differences between members of these two stellar groups, but found none.

Another test of this concept can be constructed by plotting the macroturbulence values of a large number of luminous K stars against spectral type. Blanco _et al._ (1976) find that chromospheric K2 emission fluxes peak at spectral type Kl. Therefore, in Figure 3 I have plotted the M_{RT} values of the SD79 data against the spectral type difference, (Sp. Type of Star - Kl). For reference, the _relative_ K2 emission flux observed by Blanco _et al._ is included in the diagram. If one omits β Cet from consideration, one sees that neither the turbulence distributions of the giants nor of the supergiants follow the K2-emission relation. Both of these tests imply a lack of observable dissipation of macroturbulent motions in the lower chromosphere.

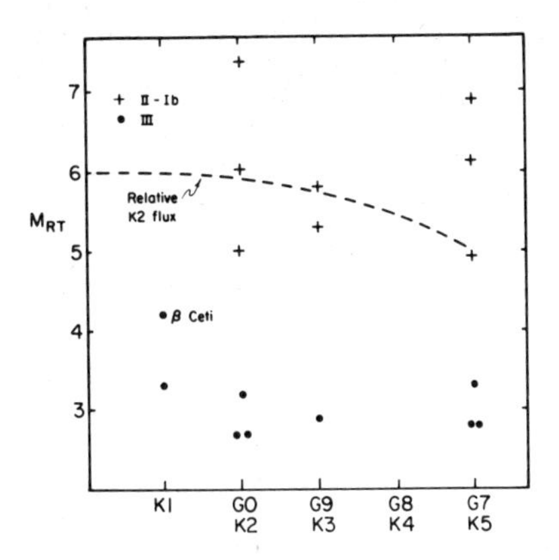

Fig. 3 - The macroturbulence vs. spectral type relation for early K III-IV stars. The K2 emission flux determined by Blanco et al. is shown for comparison.

While neither of the above tests may be sensitive enough, there are additional arguments that M_{RT} and chromospheric emission are related. For example, the K2-emission flux increases by more than a factor of ten between dwarfs and giants (Blanco et al. 1976), but there is no corresponding increase in macroturbulence among them, possibly to within 10%. In the same vein, macroturbulence does not increase from G0 V to K0 V types, whereas the K2-emission flux does. Finally, the macroturbulence value in the Sun does not argue for a detectable turbulence dissipation. One comes to this conclusion because the Sun has been recently reported to be an anomalously slowly rotating and chromospherically quiet star for its age (Smith 1980; Blanco et al. 1974). Despite this reported abnormality, the Sun manages to have a normal macroturbulent value for its age group (Smith 1978). What appears to be a normal photospheric macroturbulence is associated with a quiet chromosphere in the same star; therefore, the turbulence value must not be related to chromospheric activity. In sum, while it cannot yet be stated conclusively that macroturbulent velocities do not correlate with chromospheric heating, "the mood of the jury is perhaps becoming evident". These conclusions support theoretical arguments (e.g. Ulschneider 1974) that the heating of the lower chromosphere is due to dissipation of short-period (~ 30-sec) wave energy. Such waves have too small a wavelength to manifest themselves as macroturbulence. However, the possibility still exists that such waves may be identified in future analyses of strong lines as mesoturbulence, or through wavelength-shifted cores (Gray, priv. comm.).

IV. Prognosis.

Why study macroturbulence in late-type stars in the next few years? Here is one list of problems awaiting our attention:

1) A solar weak-line/strong-line analysis across the limb. Such an analysis has not yet been carried out using all the velocity signatures available in the Fourier domain.

It begs attention if the relationship between mesoturbulence and large/small scale-length turbulences is ever to be addressed properly. Such a study may also explain why intensity and flux studies can arrive at different answers.

2) Macroturbulence studies in new regions of the H-R Diagram. Values of M_{RT} are needed for KM stars because of their tendency toward complete convective equilibrium in the envelope. The question of M_{RT} increases in supergiants must be investigated too, though only after the non-LTE excitation effects are described for each line and "moving atmosphere" aspects are evaluated.

3) Macroturbulence studies in young clusters. Young stars provide the final, perhaps decisive, test in the search for a correlation between M_{RT}, age, and chromospheric dissipation. One search would be provided by finding a sharp-lined (near pole-on) young star in a cluster. Another would be to dissect velocities in dMe stars. So far one attempt on BY Dra, has led to negative results; the rotational velocity is too large (Vogt and Fekel 1979).

4) Macroturbulence and chromospheric variability. Two chromospheric diagnostics, K2 emission (Wilson 1978) and He I λ10830 (O'Brien and Lambert 1979) are particularly informative chromospheric signatures. Especially with the latter study nearing completion, it should be possible to search for abnormalities in the chromospheres of certain stars (e.g. β Cet) that also betray themselves in the photospheres.

5) The search for 5-minute-type oscillations. If successful, a search for rapid, periodic radial velocity excursions (either of the total line or the line core) promises to lead to a breakthrough in probing the convective envelopes of red giants.

References

Auvergen, M., Frisch, H., Frisch, U., Froeschle, C., and Pouquet, A. 1973, Astr. Ap., 29, 93.
Ayres, T.R. 1979, Ap. J., 228, 509.
Beckers, J.M., and Morrison, R.A. 1970, Solar Phys., 14, 280.
Blanco, C., Catalano, S., Marilli, E., and Rodono, M. 1974, Astr. Ap., 33, 257.
__________ 1976, Astr. Ap., 36, 297.
Bonsack, W.K., and Culver, R.B. 1966, Ap. J., 145, 767.
Canfield, R.C., and Beckers, J.M. 1976, Colloq. Intern. C.N.R.S. No. 250, ed. R. Cayrel and M. Steinberg, Paris, p. 291.
Desikichary, K., and Gray, D.F. 1978, Ap. J., 226, 907.
Deubner, F.L. 1975, Astr. Ap., 44, 371.
Dravins, D. 1974, Astr. Ap., 36, 143.
Edmonds, F.N., Jr. 1967, Contr. Bilderberg Conf. on the Photosphere, Apr. 12-16, unpub.
Frisch, H. 1975, Astr. Ap., 40, 267.
Gray, D.F. 1975, Ap. J., 202, 148.
__________ 1976, The Observation and Analysis of Stellar Photospheres, New York: John Wiley and Sons.
__________ 1977, Ap. J., 218, 530.
__________ 1978, Solar Phys., 59, 193.
__________ 1978, 4th Trieste Conf. on High Res. Spectrometry, (ed. M. Hack), p. 268.
__________ 1979, priv. comm. ("G79").
Gray, D.F., and Martin, B.E. 1979, Ap. J., 231, 139 ("GM79").
Gray, D.F., Smith, M.A., Wynn-Jones, I., and Griffin, R. 1979, Pub. A.S.P., in press.
Gurtovenko, E.A., de Jager, C., Lindenbergh, A., and Rutten, R.J. 1976, Colloq. Intern. C.N.R.S. No. 250, (ed. R. Cayrel and M. Steinberg), Paris, p. 331.
Kurucz, R.L., Traub, W.A., Carleton, N.P., and Lester, J.B. 1977, Ap. J., 217, 771.

Lambert, D.L., and Ries, L.M. 1977, Ap. J., 217, 508.
Lambert, D.L., and Tomkin, J. 1974, Ap. J., 194, L89.
Linsky, J.L., and Haisch, B.M. 1979, Ap. J., 229, L27.
Luck, R.E. 1977, Ap. J., 218, 752.
O'Brien, G., and Lambert, D.L. 1979, Ap. J., 229, L33.
Rhodes, E.J., Ulrich, R.K., and Simon, G.W. 1977, Ap. J., 218, 901.
Rutten, R.J., Hoyng, P., and de Jager, C. 1974, Solar Phys., 38, 321.
Smith, M.A. 1975, Ap. J., 203, 603.
___________ 1978, Ap. J., 224, 584 ("S78").
___________ 1980, Pub. A.S.P., submitted; I.A.U. Joint Meeting on Stellar Rotation, Montreal, August, 1979.
Smith, M.A., and Dominy, J.F. 1979, Ap. J., 231, 477 ("SD79).
Smith, M.A., and Frisch, H. 1976, Solar Phys., 47, 461.
Smith, M.A., Testerman, L., and Evans, J.C. 1976, Ap. J., 207, 308, ("S76").
Traub, W.A., Mariska, J.T., and Carleton, N.P. 1978, Ap. J., 223, 583.
Ulmschneider, P. 1974, Solar Phys., 39, 327.
Vogt, S.S., and Fekel, F. 1979, Ap. J., in press ("VF79").
Wilson, O.C. 1978, Ap. J., 226, 379.
Wynn-Jones, I., Ring, J., and Wayte, R.C. 1978, 4th Trieste Conf. on High Res. Spectrometry, (ed. M. Hack), p. 512 ("WJ78").

DEPTH-DEPENDENCE OF TURBULENCE IN STELLAR ATMOSPHERES

Robert E. Stencel
Laboratory for Astronomy & Solar Physics
NASA - Goddard Space Flight Center

Greenbelt, Maryland 20771/USA

I am very pleased to have this opportunity to present some recent observational work bearing on the depth dependence of motions in stellar atmospheres. I thank Dr. Gray for this invitation and also wish to acknowledge the patronage of the NAS-NRC.

Photospheres

The bulk of the work published on the depth dependence of turbulence, v(h) in stellar photospheres has tended to concern microturbulence alone. The solar case is perhaps the most well studied as both horizontal and vertical components of both micro- and macro-turbulence can be examined with depth. The recent summaries by Beckers (1975) and Canfield (1975) disclose a nearly constant 1 - 2 km/s turbulence through the solar photosphere, possibly rising slightly into the upper chromosphere. Other late type stars can be compared against this best case.

Arcturus is probably the next best studied case. Figure 1 collects several of the efforts to determine depth dependence of turbulence in its photosphere. A distinction among the types of motions measured should be noted: some techniques deliver microturbulence, while others deliver a total non-thermal velocity (including micro, meso and macro-velocities). Classical curve of growth studies, such as that by Griffin and Griffin (1967), yield a mean microturbulence, a depth-averaged value probably applicable to around log optical depth (5000A) of -1, as deduced from moderate strength metal lines. The Goldberg-Unno (1958, 1959), which uses pairs of lines in multiplets, has been applied by Sikorski (1976) and Stenholm (1977) to deduce the variation in v(h). Curiously, despite the similarity of spectroscopic materials and method, these solutions diverge somewhat. This may suggest a limitation of the Goldberg-Unno method in non-solar cases. A thoughrough discussion of the accuracy

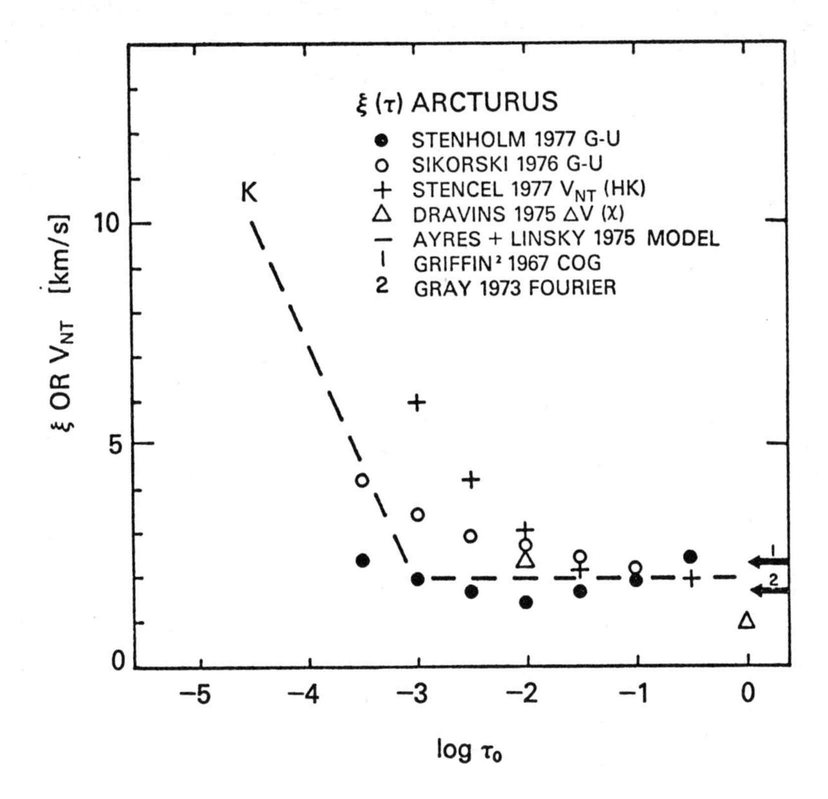

Figure 1. Turbulence variations in the photosphere of Arcturus.

to be expected with this method has been given by Teplitskaja and Efendieva (1975), and the conclude that uncertainties of 10 to 40 percent are likely. Thus the differences may be largely due to the inherent accuracy of the method. Dravins (1974) found, by an entirely different technique, that differential Doppler shifts of 2 km/s were typical of the Arcturan photosphere. If convective cells dominate, higher excitation lines will be formed in the hotter, rising (blueshifted) convective regions, while lower excitation lines will be formed in cooler, sinking (redshifted) elements. Thus, systematic wavelength shifts, similar to those known for the Sun (Glebocki and Stawikowski, 1971) should be observable. Dravins finds evidence for this, which reveals a macro-motion. Another technique, developed for the solar case by Canfield (1971) and applied to Arcturus in my own work (Stencel, 1977) uses the widths of weak emission lines occuring in the wings of the Ca II H & K lines. Their formation against the H & K line wings enables a probe of the total non-thermal line broadening to be evaluated with depth in the stellar photosphere. The result is again an indication for increasing motions with height through the photosphere. As new equipment for high resolution and improved signal-to-noise spectroscopy becomes available, application of Gray's (1976) Fourier techniques for deconvolution of line broadening mechanisms will help to reduce the discepancies. Gray's analysis of photospheric lines in the Arcturus spectrum leads to a mean microturbulence in agreement with curve of growth results, plus a small contribution due to stellar rotation. Finally, progress has been made in modeling of the upper photosphere of cool giant stars via spectral synthesis of the H & K lines. The statistical equilibrium methods of Ayres and Linsky (1975) seem to require a microturbulence which rises from 2 km/s in the photosphere, to 10 km/s in the upper chromosphere, which is again in the sense of the previous results.

In summary, to the extent that Arcturus can be considered typical of cool giants and comparable to the Sun, it seems that: 1. both the Sun and Arcturus show evidence for an increase in the total micro- plus macro-velocities with height, starting at mid-photospheric levels, and, 2. no single technique appears to possess an inherent lack of uncertainty, be it due to limitations of the data or the physical approximations and simplifications. Thus, only in the agreement of independent determinations of v(h) can we find hope that we have an idea of the atmospheric state.

Chromospheres

Of the few diagnostics of stellar chromospheric velocity fields available in the visible region, the Ca II K line may be the most valuable. In G, K and M stars, the central emission is formed over heights covering the low to mid-chromosphere. This maps the temperature rise until collisional coupling of the source function is diminished by the low density in the upper chromosphere. Historically, the influence of velocity fields on the emission width to absolute magnitude correlation (Wilson-Bappu effect) has been the subject of a healthy debate. However, recent compelling theoretical arguments by Ayres (1979) strongly suggest that Wilson-Bappu operates more as a barometer (pressure sensitive) than a tachometer (velocity sensitive).

Although the effects of velocity fields may not appear in the emission core width of the K line directly, they might be strongly visible in the gross asymmetry of the doubly reversed emission profile, characteristic of the line in cool giants and supergiants (cf, Wilson and Bappu, 1957; Linsky, et al., 1979). As first shown by Hummer and Rybicki (1967) and subsequently verified in more general cases, differential atmospheric motions can shift the self-absorbed core (K3) and change the ratio of the peak fluxes of K2V to K2R ("V/R"). An outflow which increases with height would blueshift K3 relative to K2 and diminish the K2V peak, resulting in V/R less than one. This situation is as observed among the late K giants, where blueshifted K4 (circumstellar) concurs with this interpretation. The opposite, as seen in the solar flux spectrum, can be intrepreted as an inflow, probably due to cooler downward motions as part of convective circulation. We must note that some have questioned whether a velocity field uniquely determines a particular asymmetry.

The extensive observational efforts of O.C. Wilson have provided us with an extensive sample of V/R measures (Wilson, 1976). Previous work lead me to examine this data as a function of effective temperature among the cool giant stars. The result (Stencel, 1978) was the quantization of the asymmetries in the HRD: the G and early K giants exhibit the solar asymmetry, while late K and M giants show the outflow pattern. The Kl to K3 giants comprise the transition objects, as is shown in Figure 2. Note the uncertainty of 0.5 magnitude in M_V and 0.06 magnitude in the V - R color, which could distort the sharpness of the transition.

Is this same asymmetry transition seen in the analogous Mg II 2800A resonance doublet which also shows strong chromospheric emission cores? Due to its larger oscillator strength and abundance, the Mg II emission cores are formed about three scale heights above the Ca II H & K lines, into the upper chromosphere. Any differences in v(h) might be expected to show up as a displacement in the asymmetry transition line. Dermott Mullan and I have started to investigate this question with a survey of cool giants with the IUE, and some preliminary results are seen in Figure 3. While the Ca II survey contained over 500 stars, our present survey has but 50. Again, considering the smaller sample and the unfortunate choice of symbols selected by the draftsman,

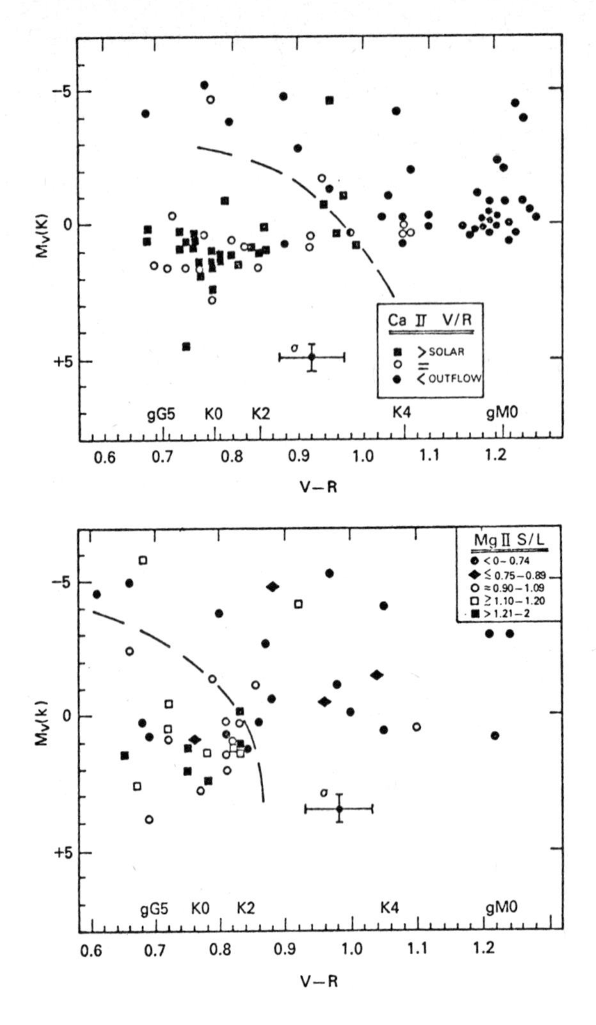

Figures 2 & 3. Ca II and Mg II asymmetry transitions. See text for details.

if you examine the overall trend, you find the Mg II asymmetry changes sense, much as does the Ca II. Our semi-objective judgement of the transition line is indicated and it is clearly shifted to smaller V-R color compared with the Ca II line. The addition of more data will help clarify the transition locus. It is plausible too that the upper chromosphere may be more 'volatile' than the Ca II line forming layers, so that the transition locus may be broadened by the number of stars undergoing asymmetry fluctuations due to upper atmospheric inhomogeneities. Overall, the G giants show the solar circulation pattern while the late K and M giants clearly exhibit outflow.
I also suspect some additional variations among the early G subgiants which should be

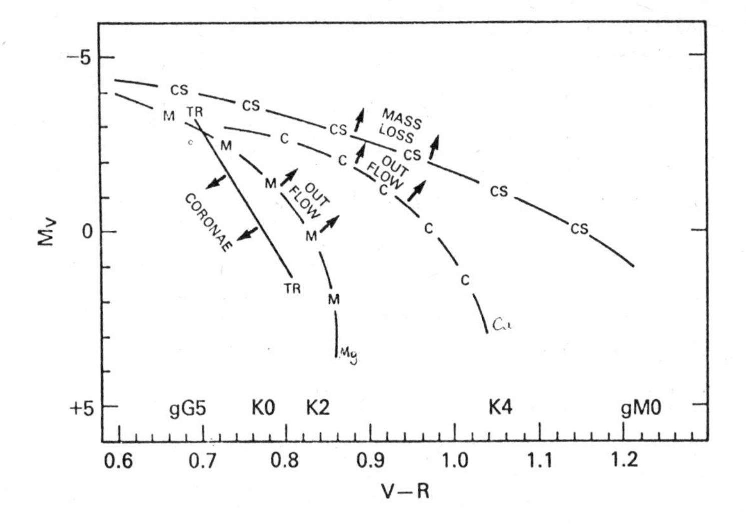

Figure 4. Collected transition loci in the HRD. See text.

explored. This Mg II data has been prepared for the Astrophysical Journal (Stencel and Mullan, 1979). In addition, we find very satisfactory agreement between the Wilson-Bappu M_V(Ca II K) and the Weiler and Oegerle (1979) M_V(Mg II k), except for a few peculiar objects like 56 Peg (cf, Basri, et al., 1980). In summary, the Mg II data points to changes once again in the hydrodynamic structure of the upper atmosphere. This fact should be considered in the current debate over the upper chromospheric pressure and velocity structure of cool evolved stellar atmospheres.

Coronae and Circumstellar (CS) Envelopes

The Ca II and Mg II asymmetry observations can be combined with additional studies to reveal an interesting description of the upper atmospheric structure of cool evolved stars. Reimers (1977) has delineated the occurance in the HRD of stars exhibiting either permanent or transient CS absorption (K4). This locus parallels the Ca II asymmetry transition (Figure 4). In addition, UV spectra of the presence or absence of high excitation lines (10^5K) in cool giants obtained by Linsky and Haisch (1979) lead them to postulate the existence of a thermal transition among the early K giants, such that warmer stars possess hot upper atmospheres (coronae, by implication), while slightly cooler stars do not have such coronae. This transition needs more observations in both UV and X-ray regions to be verified, but if it is correct, placing the proposed transition locus in the HRD leads to an interesting speculation: as a star evolves through the K giants en route to its helium flash, some hydrodynamic mechanism first dissipates the corona and uses the available energy to drive a stellar wind, as is seen in solar coronal holes. Initially the wind influences the upper chromosphere where Mg II emission is formed, and later penetrates to the Ca II line forming region and affects that emission asymmetry. Ultimately, as a consequence of the wind containing enhanced densities, CS features begin to become visible. The timescale necessary for crossing these transitions may position some stars "on the brink" such that patches of corona and of coronal holes alternately dominate the outer atmosphere. This is seen in flux as variable Ca II and Mg II emission core asymmetry changes, and collectively as a broadening of the transition loci. Variable far UV and X-ray flux might be expected. High resolution profiles of the far UV high excitation lines for a sample of stars would help reveal upper atmospheric motions as well, but these will be difficult to obtain even with the IUE.

Mullan's (1978) expanation for this involves the changing height at which motions become supersonic. This supersonic transition locus (STL) is assumed to occur in the coronae of G stars, and to impinge on a discontinuity between hot thin coronae and cooler denser chromospheres among the early K giants. This disrupts the corona and increases the mass loss rate. The Mg II asymmetry is first affected, then the Ca II asymmetry as the STL drops deeper into the atmosphere. While this theory is appealing, it does rely on a strict solar analogy, whereas a steadilly increasing mass loss rate _might_ produce comparable observed effects without requiring a chromosphere-corona discontinuity (Dupree, private communication). Resolution of this will require some sophisticated transfer calculations, but regardless of which theory one favors, the observations reveal a depth dependent effect of velocity fields among a wide range of cool evolved stars. A crucial test of the STL would be made if a non-transient case of Mg II level inflow with Ca II level outflow could be found in a flux observation.

Theoretical work on waves and inhomogeneities is crucial in this context. Recent work by MacGregor and Hartmann (1979) and by Haisch, et al. (1979) and Basri (1979) which consider mechanical and radiative fluxes, are excellent beginnings. This work combined with observations of the depth dependence of turbulence and winds will enable us to come to grips with details of atmospheric structure. At this point, observational advances in the visual wavelengths will come with application of the newest detectors to very high dispersion spectroscopy. Further improvement of statistics and resolution in the UV and X-ray regions will help. Finally, solar studies where the inter-relation of the velocity fields and the line formation can be examined in detail, are vital.

REFERENCES

Ayres, T. 1979 Astrophys. J. 228, 509.
Ayres, T. and Linsky, J. 1975 Astrophys. J. 200, 660.
Basri, G. 1979 Dissertation, Univ. Colorado.
Basri, G., Linsky, J. and Stencel, R. 1979 -- in preparation.
Beckers, J. 1975 "Physics of Motions in Stellar Atmospheres", Colloq. Int. du CNRS, No. 250, ed. R. Cayrel and M. Steinberg.
Canfield, R. 1971 Solar Phys. 20, 275.
Canfield, R. 1975 "Physics of Motions in Stellar Atmospheres", Colloq. Int. du CNRS, No. 250, ed. R. Cayrel and M. Steinberg.
Dravins, D. 1974 Astron. & Astrophys. 36, 143.
Glebocki, S. and Stawikowski, A. 1971 Acta Astron. 21, 185.
Goldberg, L. 1958 Astrophys. J. 127, 308.
Gray, D. 1976 "Observation and Analysis of Stellar Photospheres" (N.Y., Wiley).
Griffin, R. and Griffin, R. 1967 M.N.R.A.S. 137, 253.
Haisch, B., Linsky, L. and Basri, G. 1979 Astrophys. J., submitted.
Hummer, D. and Rybicki, E. 1968 Astrophys. J. 153, L107.
Linsky, J. and Haisch, B. 1979 Astrophys. J. 229, L27.
Linsky, J., Worden, S., McClintock, W. and Robertson, R. 1979 Astrophys, J. Suppl., -- in press.
MacGregor, K. and Hartmann, L. 1979 Bulletin A.A.S. 11, 448 (abstract).
Mullan, J. 1978 Astrophys. J. 226, 151.
Reimers, D. 1977 Astron. & Astrophys. 57, 395.
Sikorski, J. 1976 Acta Astron. 26, 1.
Stencel, R. 1977 Astrophys. J. 215, 176.
Stencel, R. 1978 Astrophys. J. 223, L37.
Stencel, R. and Mullan, D. 1979 Astrophys. J., submitted.
Stenholm, L. 1977 Astron. & Astrophys. 54, 127.
Teplitskaja, R. and Efendieva, S. 1975 Solar Phys. 43, 293.
Unno, W. 1959 Astrophys. J. 129, 375.
Weiler, E. and Oegerle, W. 1979 Astrophys. J. Suppl. 39, 537.
Wilson, O. 1976 Astrophys. J. 205, 823.
Wilson, O. and Bappu, W. 1957 Astrophys. J. 125, 661.

TURBULENCE IN THE ATMOSPHERES OF ECLIPSING BINARY STARS

K.O. Wright
Dominion Astrophysical Observatory
Herzberg Institute of Astrophysics

Abstract: A review of the orbits and dimensions of the ζ Aurigae systems is given, based on photometric and spectrographic observations. The Ca II K-line has been studied intensively to determine the extent and uneven structure of the chromospheres of these stars; the multiple structure of this line observed at several eclipses confirms the presence of large-scale clouds in the atmospheres. Measurements of line profiles and equivalent widths show that macroturbulent velocities up to 10 km/sec in the upper chromosphere, and up to 20 km/sec. in the lower chromosphere are present. Microturbulent velocities in the lower chromosphere are about 10 km/sec. Recent ultraviolet observations indicate that the B star in the 32 Cygni system may be within the outer chromosphere of the giant component and its radiation may affect the chromospheric structure more than had previously been suggested.

In the study of stellar atmospheres the composition and structure at increasing heights above the photosphere are required. The sun is the only star whose chromosphere can be observed directly. The other principal source for observing such structure is in the ζ Aurigae stars where, during ingress and egress at eclipse, the relatively small, hot, secondary component acts as a probe when its radiation passes through the atmosphere of the much-larger late-type primary star. Much information can be obtained from the light curves of these stars concerning the types of the components and their relative dimensions, while irregularities in the curves during ingress and egress phases may indicate disturbances in their atmospheres. However, it is important that both photometric and high-resolution spectrographic observations be combined in order to obtain the maximum knowledge about each system and the individual component stars. For many years I.A.U. Commission 42 on Close Binary Stars has stressed the need for such observations and a number of cooperative programs have been sucessfully carried out.

The spectrographic data can be used in a number of ways. The orbital elements of the system can be determined from the radial velocities when one or more cycles are covered. Although the lines in the secondary component are usually difficult to measure, radial velocities can be obtained and, knowing that the inclination must be near 90°, the masses can be determined. In some cases there is a circumstellar cloud surrounding the system and its structure can be inferred from changes in the spectra over the cycle. Close to eclipse major changes in the low-excitation lines of both neutral and ionized atoms can be seen. Detailed analysis of the intensities of these lines give numerical values for the temperature, pressure and composition of the outer atmospheres of the giant component of the system at various heights above the photosphere. The intensities and profiles of the lines in these spectra give indications of the turbulence and other mass motions in the atmosphere that we have come to discuss at this colloquium. It has usually been assumed that the atmospheres of these giant stars are similar to single stars since the components are not close, relative to most binary systems, as the periods are hundreds of days rather than a few days or less. However, there have been suggestions that the hot component does affect the structure of the giant companion, which is the principal object of many of our investigations.

Quite an extensive literature on the ζ Aurigae stars has been built up over the past forty years. We shall discuss only the principal systems, ζ Aurigae, 31 and 32 Cygni and VV Cephei since they are the brightest examples and therefore have been studied most intensively and with the highest spectrographic dispersion. The standard work on the detailed analysis of these systems was written by Wilson (1960). Wright (1970, 1973) up-dated the survey of the orbits and dimensions of the ζ Aurigae stars and reviewed the changes of the Ca II K-line that had been observed at several eclipses. Cowley (1969) surveyed the literature on the VV Cephei stars and noted that several of these stars, especially AZ Cassiopeiae, WY Geminorum and Boss 1985, may be eclipsing systems that would warrant more systematic study. Recently Sahade and Wood (1978) examined the close binary systems and included a section on atmospheric eclipses. Observational programs have been continued in recent years, especially by Hack and Faraggiana, by Kitamura and his associates and by Wright. New studies of these systems should become available in the relatively near future.

Since this colloquium has been organized to discuss turbulence in stellar atmospheres, this paper will be devoted to brief discussions of current data available for the four principal systems: their orbits and dimensions which are required to give the scale for determining the atmospheric structure; the observations of the Ca II K-line that can be observed higher in the atmosphere than most other lines; and the curve-of-growth analyses that determine the microturbulence as well as other atmospheric parameters.

TABLE 1. ORBITS AND DIMENSIONS OF THE ζ AURIGAE STARS

	ζ Aurigae		32 Cygni		31 Cygni		VV Cephei	
	Primary	Secondary	Primary	Secondary	Primary	Secondary	Primary	Secondary
Period P days	972.160		1147.8		3784.3		7430.5	
Eccentricity e	0.41		0.30		0.22		0.35	
Longitude of Periastron ω^{o}	336.0		218.2		201.1		59.2	
Periastron Passage, T JD 2,400,000+	34585.74		33141.80		37169.73		38461.0	
Mid-eclipse JD 2,400,000+	27692.82		40110.		37685.65		35931.	
Systemic Velocity V_o km/sec.	+12.9		-5.7		-7.7	-12.3	-20.2	-18.5
Semi-amplitude K km/sec.	24.6	31.4	17.0	34.	14.0	20.8	19.4	19.1
Semi-major axis a sin i ($x10^8$)km.	3.00	3.83	2.56	5.12	7.11	10.6	18.6	18.3
Inclination,°	90		82		88		77	
Spectrum	K4 Ib	B6 V	K5 Ib	B6 V	K4 Ib	B4 V	M2ep Ia	08
Luminosity, Mv	-3.5	-1.3	-3.3	0.1	-4.6	-2.0	-7.5	-3
Temperature, T_{eff}°	3550	15,400	3315	15,400	3550	17,600	3000	25,000
Diameter Θ	200	5	250	2.0	210	5.5	1600	13
Mass, M_{Θ} sin i	7.54	5.91	9.15	4.85	9.16	6.19	18.0	18.2

ORBITS AND DIMENSIONS

The orbits and dimensions of these eclipsing binary systems are not as well determined as might be desired, partly because the periods are long and few cycles have been covered, the light curves show some changes from eclipse to eclipse, and the secondary spectra are difficult to measure because the multi-lined K - or M - type spectrum dominates most of the wavelength range that is covered; data derived from light curves and spectrographic observations do not always agree. Nevertheless the results are sufficiently accurate to give an adequate scale of the atmosphere. The best available data are listed in Table 1. No more precise orbital data for ζ Aurigae, 31 and 32 Cygni have been published since Wright's (1970) review; those for VV Cephei are from Wright (1977) and are based on measurements of the Hα line for the O-type star and lines in the same region of the spectrum for the M-type star. Some revisions for the luminosities, masses and diameters have been considered: for ζ Aurigae, Kiyokawa and Kitamura (1973) revised the light-curve data for M_V, T_{eff} and the diameter of the K-type star; for 32 Cygni the Saijo and Saito (1977) analysis of the light curve indicated that the spectrum of the B-type star is probably B8 rather than B4 with corresponding revisions for the temperature and the diameters; they also suggested small changes in the masses. Koch et al. (1970) rank the photometric data for these systems in the category of "lowest reliability", chiefly because the light curves show some variability, and the eclipses are atmospheric and therefore the depth of eclipse varies with wavelength.

It is worth mentioning that, for ζ Aurigae, Kiyokawa (1967) discussed the light curves for five eclipses from 1935 to 1963-64 and found a small increase in the length of totality with time, which suggested a gradual expansion of the atmosphere of the primary star, but Kitamura (1974) found that narrow-band observations of the 1963-64 eclipse indicated a shorter time for total eclipse, which leaves the question of changes in the size of the K-type star still open.

It has been known for many years that the eclipse for 32 Cygni is grazing; the observations were discussed by Wright (1970). Wellmann (1957) decided that totality lasted for 9 days in 1949, and 12 days in 1952, while Scholz (1965) concluded that the eclipse was total for 6 days in 1962, and Herczeg and Schmidt (1963) considered that totality lasted between 1 and 3 days for the same eclipse. However, Saito and Sato (1972) analyzed the numerous observations made by the Japanese observers for the 1968 and 1971 eclipses and concluded that these eclipses were grazing and almost total, but showed no flat-bottomed minimum in the light curve that would indicate totality. Thus it appears probable that, for 32 Cygni, changes in the extent of the outer atmosphere do occur. It has been known, of course, that the opacity varies with wavelength and that the eclipses are much deeper in the ultraviolet region of the spectrum than in the red. Thus in the far ultraviolet region, the eclipses of 1968 and 1971 may have been effectively total. The Victoria spectrographic observations of Wright and Hesse (1969), based chiefly on the appearance of the K-line indicated that totality lasted 12 days in 1965 and 8 days in 1962.

RADIAL VELOCITY OBSERVATIONS OF THE CA II K-LINE

The K-line in the spectra of the ζ Aurigae stars has been studied intensively and many measures of radial velocity and intensity have been published. For ζ Aurigae the velocities of all lines, including the K-line, are more positive relative to the orbital velocity during the ingress phase and more negative at egress. However, the displacements, though in the same general direction as those that might be expected for a rotating star, are more random in character and do not fit the pattern predicted even for a slow rotation. Wilson (1960) concluded that the rotation effect must be small, that there is a chromospheric equatorial current and that the chromosphere contains sizeable concentrations of matter moving with different velocities. Saito (1970) studied the problem for ζ Aurigae and concluded that the hydrogen gas facing the B-type star should be ionized by the ultraviolet radiation of the hot star and an ionization front is formed as a boundary between the H I and H II regions. For ζ Aurigae Saito concluded that the velocity field of the K-type chromosphere consists of a non-steady field due to gas clouds moving with different velocities overlapping the main velocity field of gas expanding from the K star at an ejection rate of 1×10^{-8} $M_{\odot}$/year and a rotation rate of 5 km/sec. At some distance away from the K-type star in the direction of the B star velocities at egress are produced by the H I gas accelerated by Lyman quanta from the B star. Moderately good agreement between Saito's theoretical curve and the chromospheric observed radial velocities suggest that this proposed interpretation may have some validity. Velocities of chromospheric lines of other atoms, including Fe I and Ca II follow a pattern somewhat similar to the hydrogen lines.

On the other hand, observations of the Ca II K-line at the 1951 and 1961 eclipses of 31 Cygni (McKeller et al. 1959, Wright and Odgers, 1962) show that the radial velocities follow the orbital curve fairly closely with numerous negative displacements up to 20 km/sec during the ingress phases. In 1951 the K-line was clearly resolved into two components during the egress phases, with the mean velocity displaced above the orbital velocity curve; no observations could be made during the 1962 egress. The multiple structure of the K-line has been a feature of the chromospheric phases. This structure has been attributed to a small number of "clouds" possibly detached from the star in some cases where the displaced velocities are greater than the velocity of escape; the satellite line is usually sharper than the main component and can sometimes be observed for several days with nearly the same velocity. The velocities of the principal line during both ingress and egress seem to suggest a strong circulation in the K-star chromosphere moving in the opposite sense to the orbital motion and also in the opposite sense to that in ζ Aurigae.

For 32 Cygni the most extensive series of measurements are those of Wright and Hesse (1969) for the 1965 eclipse, and of Bisiacchi et al. (1974) for the 1971 eclipse. The trends of the radial velocities of the K-line are similar at each eclipse: There is considerable structure in the lines within about 40 days of mid-eclipse and the components can frequently be separated. The general trend is towards velocities slightly greater than the orbital velocity before mid-eclipse and slightly less after mid-eclipse. Bisiacchi et al. suggest that there may be some kind of cyclic oscillation in the atmosphere of the K-type star indicated by the measurements of the Fe I lines.

INTENSITY OBSERVATIONS OF THE CA II K-LINE

Wright's (1970) review of the intensity data for the chromospheric K-line observed in the ζ Aurigae stars still covers most of the data available. The K-line intensities can be measured two stellar diameters or more away from the photosphere (second contact) and the complex structure can often be seen at any phase as the radiation from the B-type star passes through the outer atmosphere of the K-type star.

Kawabata and Saito (1975) have published their data for the 1971-72 eclipse of ζ Aurigae. They obtained several plates on each of a number of nights during ingress and egress and found that the several satellite lines could nearly always be seen on each spectrogram obtained on a given night - thus confirming the reality of the components. The chromosphere of ζ Aurigae has now been observed for forty years. The K-line intensities may vary by as much as a factor of two from eclipse to eclipse; if they are weak during ingress they are also weak during egress. Thus the plots of the intensities at ingress and egress are quite similar, but are not necessarily mirror images. The intensities of the K-line at the eclipses of 1934, 1938 and 1971 are classified as weak. Those observed at Cambridge in 1937 are the strongest. The 1937 observations may be unusual, perhaps due to some kind of eruption at the time of the observations. The data for the 1947, 1950 and 1955-56 eclipses are listed as moderate. Kawabata and Saito suggest that the chromospheric activity of ζ Aurigae increased from 1939 until 1950-56, when it reached a maximum, and then decreased until 1971.

The K-line intensities in the spectrum of 31 Cygni for the eclipses of 1951 and 1961 have been studied by McKellar et al. (1959) and by Wright and Odgers (1962). The 1951 observations showed the component structure clearly, especially during the egress phases. The multiple structure was confirmed at the 1961 eclipse when numerous plates were obtained during ingress; a strong, negatively-displaced component was observed 5 stellar radii from the edge of the K-type star. The principal line increased in intensity from narrow to broad (1 Å) about one radius from the limb and then its width decreased for several days before it broadened again and showed the characteristic "square" appearance of a strong chromospheric line; the damping wings then began to appear as the probe passed through the innermost portions of the chromosphere. The decrease in breadth of the line so close to the limb seems to indicate that the density of the chromosphere is not uniform and, since the observations do not repeat at each eclipse, non-stable.

The chromospheric intensities for the 32 Cygni system for eclipses since 1949 have been discussed by Wright and Hesse (1969) and by Bisiacchi et al. (1975). The trends from eclipse to eclipse seem to be random. The intensities were least at the 1962 eclipse and greatest in 1965. At both the 1965 and 1971 eclipses the component structure is more evident during ingress than at egress, but the widths in 1965 were nearly fifty per cent greater than in 1971. However, these variations from one eclipse to another are not surprising in these cloudy, unstable atmospheres.

As the Ca II K-line shows the greatest changes during the eclipse phases, since it is the strongest line in the spectrum, its profile has been used to determine macroturbulent velocities. Perhaps the most careful determination of the profile has been made by Kitamura (1967) for ζ Aurigae. He made corrections for the effect of the B-type star on the observed continuum and for the K-line in the K-type spectrum. At a height of 16×10^6 km above the limb, Kitamura found a best fit between calculated and observed profile for a turbulent velocity of 15 - 19 km/sec. and a total number of $1 - 4 \times 10^{16}$ Ca II atoms in the line of sight. The fit with the theoretical curve is good but not excellent, partly because the observed profile has a small still-stand near half-intensity; the observed intensity change is more gradual than the theoretical profile at both zero intensity and at the continuum.

For 31 Cygni, McKellar et al. (1959) studied Victoria spectra obtained at the 1951 eclipse in some detail. They compared observed and computed profiles for both the principal lines and the satellite lines observed during egress. For the narrow lines far out in the chromosphere they found reasonable agreement with the observations by assuming a turbulent velocity of 10 km/sec. and 10^{12} atoms per unit cross section in the line of sight. For the broad lines in the inner chromosphere the turbulent velocity was 20 km/sec. and 10^{18} atoms. Underhill (1954) measured the profiles of a number of Fe I lines near K-line on the 1951 Victoria chromospheric spectra and concluded that the turbulent velocity for these lines was similar to that of the K-line for the inner chromosphere. The intensities of the Fe I lines decrease much more rapidly than those of the K-line and no estimates could be made for the outer chromosphere.

CHROMOSPHERIC INTENSITIES OF METALLIC LINES

Spectrograms obtained during ingress and egress for the ζ Aurigae giant eclipsing systems contain a wealth of information that can be used to determine the parameters of the chromosphere. No detailed analyses have yet been published using model-atmosphere calculations chiefly because the outer atmosphere of these stars seems to be unstable. Since condensations (prominences, clouds, etc.) are observed for the ionized calcium atoms, similar features may affect the intensities of other atoms. Since the radiation of the B-type secondary spectrum is effectively the continuum of the chromospheric lines, though some chromospheric continuum may also be present and the presence of the K-type spectrum must also be taken into account, curve-of-growth methods using a Schuster-Schwarzschild model with exponential absorption (van der Held, 1931, Unsöld, 1955) have been adopted in most of the published investigations. The standard work on the analysis of chromospheric line intensities is still that of Wilson (1960) who discussed the data available up to 1959. Wright's (1959) analysis of the 1951 eclipse data for 31 Cygni was published just after Wilson's survey and agrees with Wilson's principal conclusions although slightly different methods of reduction were used.

In order to determine the chromospheric intensities, plates obtained well outside of eclipse showing the normal composite spectrum and also plates taken during totality are required. The effect of the K-type spectrum must be removed and the chromospheric equivalent widths relative to the B-type probe are determined as discussed by Wilson (1960) and by Wright (1959).

In the curve-of-growth analysis of the data, the ordinate of the plot is log (W/λ) for the observed quantities and log $(W.c/\lambda\ .2v)$ for the theoretical data. Thus, when the plots of observed and theoretical curves of growth are superposed, the difference in the ordinates is a measure of the turbulence in the chromosphere. For the abscissa the observational data are usually laboratory f-values when they are available; theoretical intensities calculated from the sum rules can be used if necessary. In the theoretical plot the abscissa, log X_0 contains terms involving the abundance of the atom being studied, the lower excitation potential, the excitation temperature and the turbulent velocity. Plots are made for all lines of a given atom, such as Fe I, having approximately the same lower excitation potential. Each plot is superposed on the theoretical set of curves of growth. The mean displacement in the ordinates of these plots is a measure of the turbulent velocity. The displacement in the abscissa is a function of the excitation potential from which the excitation temperature can be derived.

For many years the best laboratory f-values for Fe I were considered to be the measures of King and King (1938). However, doubts have been cast on the temperature the Kings used for their electric furnace and it is now believed that the values can be improved. In order to check some of the data for the chromospheres of the eclipsing stars, the Fe I f-values used by Foy (1972), May et al. (1974) and by Bridges and Kornolith (1974) were plotted against Wright's (1959) data for 31 Cygni. The plots for each excitation potential were combined and mean observational curves of growth for each spectrum were fitted to the theoretical curves. While the results were not exactly the same as those originally published and the new f-values did not give an exact 1 : 1 correspondence with King's data, the differences were relatively small over the wavelength range covered (3700 - 4500 Å) and the conclusions concerning turbulence and temperature based on the King f-values do not seem to require major revision.

Published turbulent velocities agree that the chromospheric lines observed in the atmospheres of the ζ Aurigae stars show considerable turbulence. For ζ Aurigae, Wilson's (1960) summary notes that the microturbulence is between 5 and 15 km/sec., though some of the earlier, lower-dispersion data give values up to 20 km/sec. There is some evidence that the turbulence and also the excitation temperature increases with height in the atmosphere, and also that the value for ionized lines may be a few km/sec. greater than for neutral atoms. Effectively the same conclusions were reached by Wright (1959) for 31 Cygni.

A preliminary study of the Victoria high-dispersion spectrograms of 31 Cygni obtained at the 1971 eclipse has been begun (Morbey et al. 1975). A computer program was prepared to derive the spectrum of the B-type star by subtracting the spectrum of the K-type star, obtained during totality, from the composite spectrum observed far from eclipse. This program proved to be quite satisfactory and several helium lines could be observed on computer-constructed tracings in the spectrum of the B-type star that had not been detected previously. Some work has been done on the spectra obtained during the chromospheric phases of this eclipse, but the intensities have not been studied yet.

THE STRUCTURE OF THE CHROMOSPHERE

Several theories concerning the structure of the chromosphere of the ζ Aurigae stars have been suggested but the final solution probably has not yet been found. The Mount Wilson observers considered that the radiative flux from the B star is likely to be the chief source of the chromospheric excitation but found that, with a smooth density distribution, the B star should ionize the metals much more than has been observed; they concluded that the chromosphere must contain condensations sufficiently dense to prevent excessive ionization by the B-star radiation. Later Magnan (1965) discussed deviations from local thermodynamic equilibrium for ionized calcium with a three-level model atom and a continuum; he considered that the observations for 31 Cygni could be reconciled, taking electron collisions into account, with a homogeneous atmosphere where the average number of electrons is 10^{10} and the effective temperature is 10,000° K. As noted above, Saito discussed the effects of hydrogen Lyman quanta from the B star on ionization and shock-wave fronts in the K-type chromosphere and concluded that the hydrogen velocities observed near eclipse could be explained by such fronts combined with a slow rotation of the K-type star. Groth (1957) earlier had suggested that the source of super excitation required to produce the excitation and ionization in the chromosphere lies in the ultraviolet radiation from the transition zone between the chromosphere and the corona where the turbulent energy is dissipated. During the past few years Bernat, Boesgaard, Linsky, Reimers and others have been studying the chromospheres and circumstellar matter surrounding late-type stars and their model calculations may well be adaptable to conditions in the atmospheres of the giant components of the ζ Aurigae stars.

Very recently Stencel et al. (1979) were able to obtain observations of the 32 Cygni system with the IUE satellite in the region 1150 - 3200 Å at phase 0.2, a few months after eclipse. They observed numerous ionized lines of silicon, iron, magnesium, etc. with P Cygni type profiles (emission peaks at the redward edge of the absorption lines). The profiles of the strong lines showed absorption components of - 200 km/sec. and even up to - 400 km/sec., similar to, but with higher velocities than those previously observed only for the Ca II line near eclipse. They suggest that the Fe II emission lines would probably dominate the 2330 - 2630 Å range during the chromospheric phases and this emission would explain the excess radiation observed at that time by Doherty et al. (1974). The observed width of the Mg II lines and the strength of the Fe II emission lines combined with the lack of high-excitation lines of C IV and He II lead the authors to suggest that the underlying chromosphere of 32 Cygni may be qualitatively similar to the outer atmospheres of late-type supergiants like α Orionis. They conclude, from the observed strong P Cygni-type profiles, that the B star lies within the upper chromosphere of the K-type supergiant, and that the ionizing radiation from the B star penetrates into the supergiant atmosphere and sets up moving shock fronts within the chromosphere.

We have mentioned very little about VV Cephei in this paper although it clearly shows chromospheric lines at the time of eclipse. Although Goedicke (1939) and Peery (1966) made some analyses of the atmosphere, their data were not sufficient to give better than approximate results. Peery did estimate a turbulent velocity of 48 km/sec from his curve-of-growth data, but the value has not been verified.

At Victoria, spectra of VV Cephei have been obtained over a full cycle. The analysis of the Hα profiles and a determination of the orbit has been completed by Wright (1977). Most of the plates have been measured for radial velocity to determine which lines arise in the two stars and which are produced in the circumstellar envelope, but it has not yet been possible to analyze the intensities. Prior to the 1957 eclipse McKellar et al. (1957) studied the changes in the Ti II lines at 3759 and 3761 Å. These lines showed at least two components at -38 and +16 km/sec. with a suggestion of emission between them, especially close to totality. A preliminary study of these lines from 1975-78 show that these lines are present at the next eclipse - but they also appear with somewhat different velocities and intensities throughout the cycle. Thus it would appear that they are at least partially circumstellar in origin. The Victoria observations of the chromospheric phases of the 1976-77 eclipse were obtained in the region 3400 - 4250 Å in order to minimize the effect of the M-type star. Many of the plates could be stronger but they should be useful for the analysis of some phases of the chromosphere.

Much more information remains to be gained from observations of the ζ Aurigae stars. With modern techniques it is possible to obtain profiles of individual lines, such as the K line, much more accurately and probably in less time than with photographic plates, but when large regions of the spectrum are to be examined at a given phase, as is desirable for these stars, the photographic method would still seem to be most useful.

In this review we have examined the pertinent data obtained from plates of the ζ Aurigae stars observed near the time of their eclipses. Radial-velocity measurements show that there is probably some kind of chromospheric equatorial current that is superposed on the slow rotation of the star. The ionized calcium lines show that there must be large clouds or prominences, sometimes moving faster than the escape velocity, that can be observed for several days. Satellite data may well show that the early-type star that we have become accustomed to call a "probe" may be close enough to seriously affect conditions in the late-type supergiant star whose atmospheric structure we have hoped to determine - hence this structure may not be that of a normal single star. Modern theories of radiative transfer and line formation may give new clues concerning this structure based on the observational data obtained for these stars. It has been suggested that the microturbulent velocities assuming curves of growth based on pure absorption are merely the result of inadequate theory. I cannot answer this question, but it does seem that, for giant stars such as these, velocities up to 10 or more km/sec. would seem to be reasonable for the small-scale motions in the atmosphere, and the macroturbulent velocities for large-scale motions derived from the profiles of up to twice that value also seem to be plausible.

REFERENCES

Bisiacchi, G., Flora, U., and Hack, M. 1974, Astron. Astrophys. Suppl., **13**, 109.

Bridges, J.M., and Kornblith, R.L. 1974, Astrophys. J., **192**, 794.

Cowley, A.P. 1969, Publ. Astron. Soc. Pacific, **81**, 297.

Doherty, L., McNall, J., and Holm, A. 1974, Astrophys. J., **187**, 521.

Foy, R. 1972, Astron. Astrophys., **18**, 26.

Goedicke, V. 1939, Publ. Univ. Michigan Obs., **8**, 1.

Groth, H.R. 1957, Zeits. Astrophys., **43**, 185.

Kawabata, S., and Saito, M. 1975, Astrophys. Space Sci., **36**, 273.

King, R.B., and King, A.S. 1938, Astrophys. J., **87**, 24.

Kitamura, M. 1967, Publ. Ast. Soc. Japan, **19**, 194.

Kiyokawa, M., and Kitamura, M. 1973, Ann. Tokyo Astron. Obs. 2nd Ser., **13**, 243.

Koch, R.H., Plavec, M., and Wood, F.B. 1970, Publ. Univ. Pennsylvania Astron. Ser., **11**, 1.

Magnan, C. 1965, Ann. Astrophys., **28**, 512.

May, M., Richter, J., and Wichelmann, J. 1974, Astron. Astrophys. Suppl., **18**, 405.

McKellar, A., Aller, L.H., Odgers, G.J., and Richardson, E.H. 1959, Publ. Dom. Astrophys. Obs., **11**, 35.

McKellar, A., Wright, K.O., and Francis, J.D. 1957, Publ. Astron. Soc. Pacific, **69**, 442.

Morbey, C.L., Wright, K.O., and Carlberg, R.G. 1975, J. Roy. Astron. Soc. Canada **69**, 40.

Peery, B.F. 1966, Astrophys. J., **144**, 672.

Sahade, J., and Wood, F.B. 1978, Interacting Binary Stars, (New York: Pergamon), p. 120.

Saijo, K., and Saito, M. 1977, Publ. Astron. Soc. Japan, **29**, 739.

Saito, M. 1965, Publ. Astron. Soc. Japan, **17**, 107.

Saito, M. 1970, Publ. Astron. Soc. Japan, **22**, 455.

Saito, M. 1973, Astrophys. Space Sci., **22**, 133.

Saito, M., and Sato, H. 1975 Publ. Astron. Soc. Japan, **24**, 503.

Stencel, R.E., Kondo, Y., Bernat, A.P., and McCluskey, G.E. 1979, Astrophys. J., (in press).

Underhill, A.B. 1954, Mon. Not. Roy. Astron. Soc., **114**, 558.

Unsöld, A. 1955, Physik der Sternatmosphären, 2nd Edition, Springer pp. 288, 430.

Van der Held, E.F.M. 1931, Zeits. Astrophys., **70**, 508.

Wilson, O.C., and Abt, H.A. 1954, Astrophys. J. Suppl., **1**, 1.

Wilson, O.C. 1960, Stellar Atmospheres, ed., J.L. Greenstein, (Chicago: Univ. Press) p. 436.

Wright, K.O. 1970, Vistas in Astronomy, ed., A. Beer, (Oxford: Pergamon Press), p. 147.

Wright, K.O. 1973, Extended Atmospheres and Circumstellar Matter in Spectroscopic Binary Systems, ed., A.H. Batten, (Dordrecht: Reidel), p. 117.

Wright, K.O. 1977, J. Roy. Astron. Soc. Canada, **71**, 152.

Wright, K.O., and Hesse, K.H. 1969, Publ. Dom. Astrophys. Obs., **13**, 301.

Wright, K.O., and Odgers, G.J. 1962, J. Roy. Astron. Soc. Canada, **56**, 149.

DIFFERENTIAL LINE-SHIFTS

William Buscombe
Northwestern University
Evanston, IL 60201, U.S.A.

In the beginning of this review, I paraphrase three statements from Paul W. Merrill (1960). (1) In the spectra of long-period variables, absorption lines of neutral metals show a shortward displacement increasing algebraically with excitation potential, probably due to atoms moving outward in the star's upper atmosphere with an expansion velocity of the order of 20 km s^{-1}. (2) About 6 weeks after maximum light emission components appear within the broad absorption at H and K of Ca II, but displaced about -100 km s^{-1}. A similar, but less extreme, behaviour is known for classical cepheids. (3) In quasi-constant red giants, the resonance lines of Ca II, Ca I, and SR II show circumstellar components displaced shortward relative to the normal stellar photospheric lines by a velocity which is correlated with the spectral type, typically -8 km s^{-1} for M6 II to -25 km s^{-1} for class M0 III.

From the very few specific citations in the literature, Table 1 lists examples of much larger shifts detected on high-dispersion spectrograms of extremely luminous stars. The stars named are somewhat fainter than any yet studied in detail in the far ultraviolet. Morton (1976) quotes a velocity of -2260 km s^{-1} for O VI λ1031 in the upper chromosphere of ζ Pup.

It is noticeable that, in addition to the progression by excitation potential, already discussed by Merrill and the cited authors of the recent papers on supergiants, there appears a progression with photospheric temperature among stars of the highest bolometric luminosities. Imhoff (1976) demonstrated among G, K, and M stars respectively, such a progression with luminosity class exists. For the profiles of Hα with "reverse P Cygni" emission, the trend to larger velocity displacements from the chromospheres of more luminous M stars is seen in Figure 1. Some of the displacements change with time; Smolinski et al. (1979) have noted an outburst from recent spectra of HD 217476 in which ions with lower excitation potential have velocities similar to Hα . *

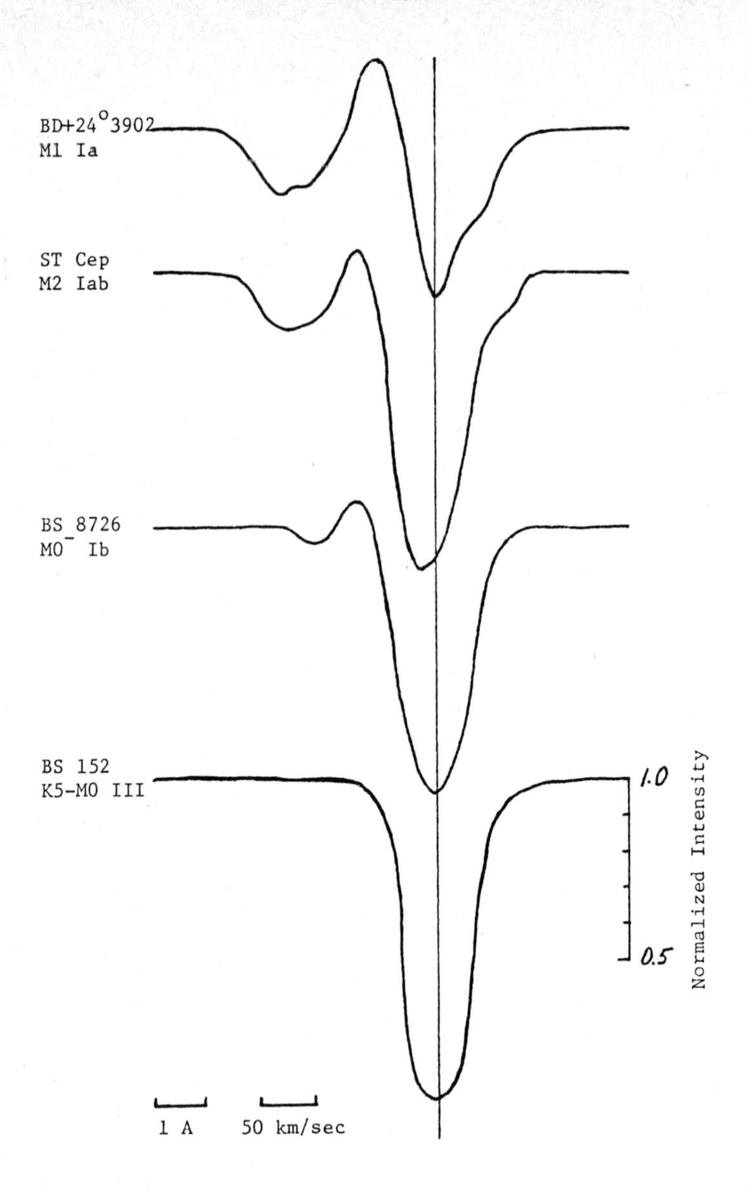

Fig. 1 Hα line profiles of early M stars of increasing luminosity. Uncorrected for instrumental profile. Vertical line marks rest wavelength of Hα . Wavelength increases to right.

(From Imhoff 1976)

Table 1

HD Name			MK Type	ΔV	Ion	Fig.	Reference
206936	=	μ Cep	M2 Ia	-52	Hα	1	Imhoff 1976
212466	=	RW Cep	K0 Ia-0	-71	Hα		"
217476	=	V 509 Cas	G5 Ia-0	-81	Hα		"
163506	=	V 441 Her	F2 Ia	-150:	Hα		Sargent 1969
223385	=	6 Cas	A3 Ia-0	-110	Hα		Aydin 1972
152236	=	ζ^1 Sco	B1 Ia-0	-210	SiIV	2	Hutchings 1968
2905	=	κ Cas	B0.7 Ia	-325	Hα		Rosendhal 1973
148937			O6 If	-340	Hα	3	Conti 1977

For supergiants of classes A and B, a progression with excitation potential has been discussed by several authors. Figure 2 summarizes the measures by Hutchings (1968) on the brightest member of the cluster NGC 6231. Conti (1977) has presented similar evidence for the Of star HD 148937 (see Figure 3).

It seems reasonable to suggest that the outward acceleration of matter escaping from stellar chromospheres is related to the age and possibly the mass of the star concerned. Each of the stars named in Table 1 has a bolometric luminosity of about two million suns.

I am indebted to R.E. Taam for the suggestion that this type of chromospheric activity may even be detectable as X-ray emission from very luminous stars.

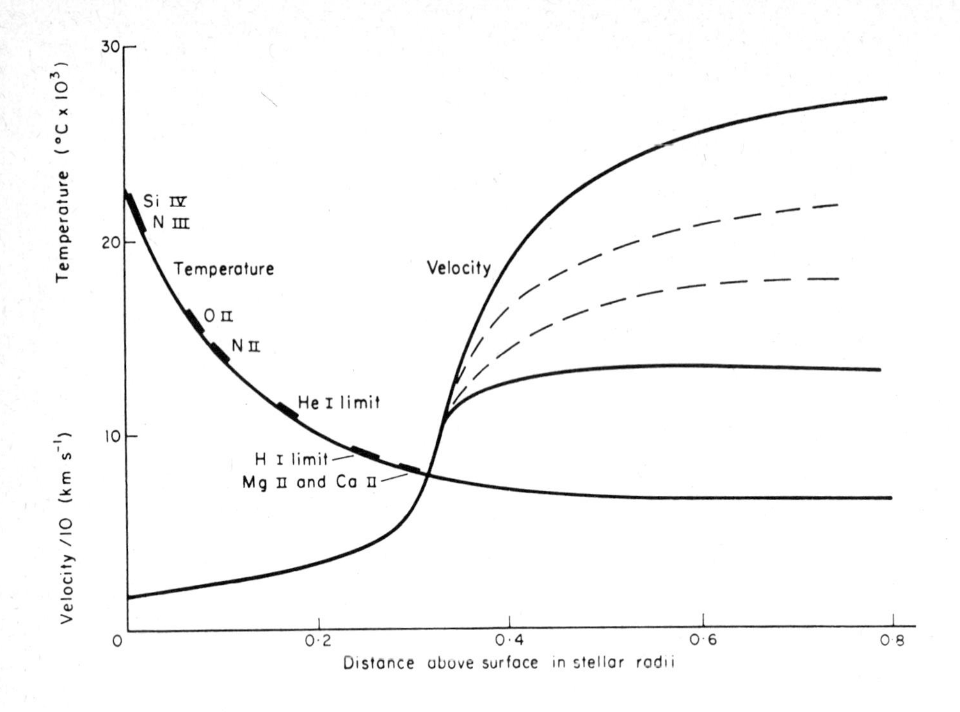

Fig. 2 Progression of velocity displacements for absorption lines in the spectrum of HD 152236 (ζ^1 Sco).

(From Hutchings 1968.)

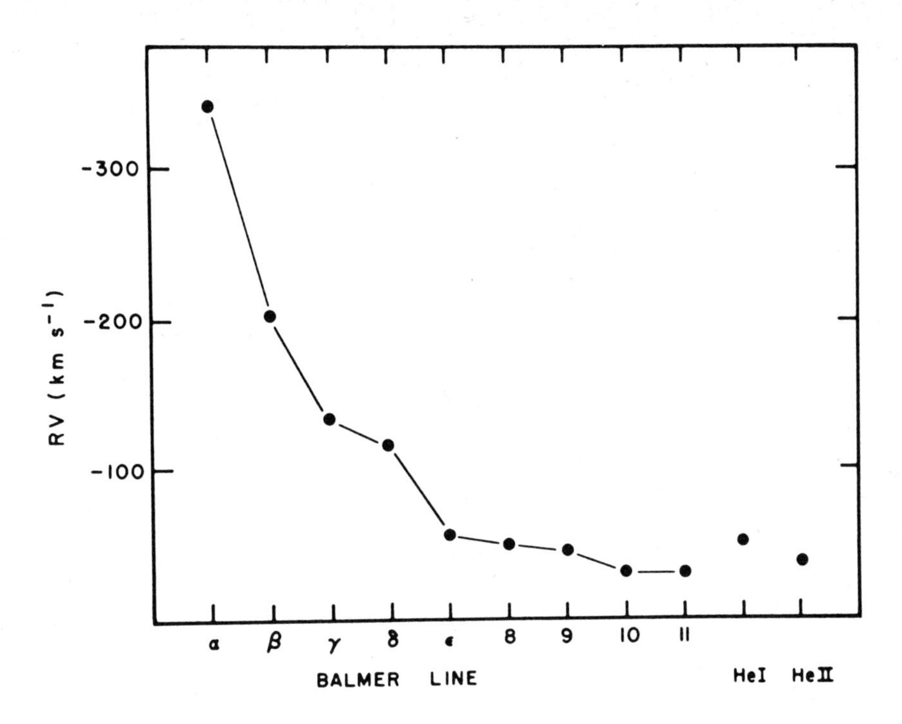

Fig. 3 Velocity progression of the Balmer, He I and He II absorption lines in the spectra of HD 148937.

(From Conti 1977.)

REFERENCES

Aydin, C. 1972, Astron. Astrophys. 19, 369.

Conti, P. 1977, Astrophy. J. 215, 561.

Hutchings, J.B. 1968, Mon. Not. R. Astron. Soc. 141, 219.

Imhoff, C.L. 1976, thesis, Ohio State University.

Merrill, P.W. 1960, in Stellar Atmospheres (ed. J.L. Greenstein), U. of Chicago Press, p. 516.

Morton, D.C. 1976, Astrophys. J. 203, 386.

Rosendhal, J.D. 1973, Astrophys. J. 186, 909.

Sargent, W.L.W. and P.S. Osmer 1969, in Mass Loss from Stars (ed. M. Hack), Dordrecht: Reidel, p. 57.

Smolinski, J.; J.L. Climenhaga, H. Funakawa, and J.M. Fletcher, 1979 Internat. Astronom. Un. Circular no. 3382.

TIME DEPENDENCE OF BALMER PROGRESSION IN THE SPECTRUM OF HD 92207

H.G. Groth
Inst. für Astronomie und Astrophysik, Universität München
München 80, West Germany

Abstract

The radial velocities measured in the spectrum of HD 92207 (AO Ia^+) show a Balmer progression, which varies considerably within one week and reverses within one month (or less). These observations give strong evidence, that at least in some parts of the atmosphere the velocity field reverses with time. (To be published in Astronomy and Astrophysics).

MICROTURBULENCE : AGE DEPENDENCES

R. Foy
Observatoire de Paris
92190 Meudon/France

The study of microturbulence in stars for itself, and not as a by-product of abundance determinations, is relatively recent. The first comprehensive study on this topic was carried out in 1966, by Bonsak and Culver. They found general trends of the behaviour of microturbulence ξ in the HR diagram. A decisive advance occured when Garz and Kock (1969) revised the scale of oscillator strengths of iron and, as a consequence, the value of the microturbulent velocity in the solar photosphere. The use of microturbulent velocities determined before 1969 should be avoided.

Before describing what we know about the behaviour of microturbulence with stellar age, I shall make some statements, and describe how the microturbulent velocity is found, from observations, to depend on the atmospheric parameters.

First, we shall be concerned here with stars in the spectral type range A5 - K5. Second, I shall distinguish the changes in microturbulence as a function of mass at fixed evolutionary stage, namely the main sequence, from the changes on microturbulence during the stellar evolution at fixed mass. Lastly, I shall consider the so-called microturbulent velocity as a physical atmospheric parameter, and not as an artificial parameter.

I determined microturbulent velocities for a sample of nearly 120 stars, from the vertical shift of the neutral iron curve of growth; I used our equivalent widths in Meudon, or published data; thus we considered large sample of stars measured in a homogeneous way. Equivalent widths were rescaled, using stars studied in common in at least two data sets.

Abscissae of the curves of growth were computed with model atmospheres, rescaled either from Gustafsson et al (1975) for giants or from Peytremann (1974) for dwarfs. Oscillator strengths were taken from my compilation (1972) or determined in the same scale from the weak line part of my solar curve of growth by Cayrel et al (1977).

DWARFS

In the case of dwarfs, the splitting of the damping part of the curve of growth is already present on the plateau of the curve of growth. The resultant dispersion of the flat part of the curve of growth has been interpreted in terms of random errors in the measurements either of the line equivalent widths, or from the abscissae of the curves of growth : this also led to systematic enhancements of the microturbulent velocity. This effect is dependent on the stellar gravity, as shown by Cayrel et al (1977), so that microturbulence measurements in dwarfs are often biased with respect to those for giants.

So the determination of ξ in dwarfs appears to be difficult for four reasons. First, the plateau is not at all horizontal; second, a slight error in the damping constant affects the level of the plateau; it means that in differential analysis, a difference in gas pressure does the same; third, the low excitation lines, which are less affected by damping, are not numerous; finally fourth, the microturbulent velocity is small as compared to the thermal velocity. Consequently noise is a severe problem.

I reanalysed published equivalent width data of dwarfs, observed with a reciprocal dispersion ranging around 3 Å/mm. The splitting in the damping part of the curve of growth is clearly visible, but this is not the case on the flat part, because of experimental errors. Therefore only an upper limit to the microturbulent velocity can be proposed.

The determination of the microturbulent velocity in solar type dwarfs does not appear very reliable to me. Note that this conclusion applies to the currently available spectroscopic material, but not to the very high resolution data obtained with modern techniques.

Consequently, the wellknown increase of microturbulence with increasing temperature along the main sequence should be reconsidered. Andersen (1973) found an increased microturbulence near spectral type A5, which is of particular interest. If it is real, it could be interpreted in terms of ages. The microturbulent velocity would be decreasing with increasing age, since A type stars are much younger on the average than G type stars. But I suspect that this trend could be interpreted in another way. Indeed a lot of Am stars which are in fact subgiants or giants are included in ξ versus T_{eff} relations, so that these relations could provide a valuable check of the change in microturbulence with stellar evolution.

GIANTS

Such a change is now well established for yellow stars : the microturbulent velocity in G and K giants is significantly larger than in dwarfs, provided the Sun is not an exception.Figure 1 shows the distribution of

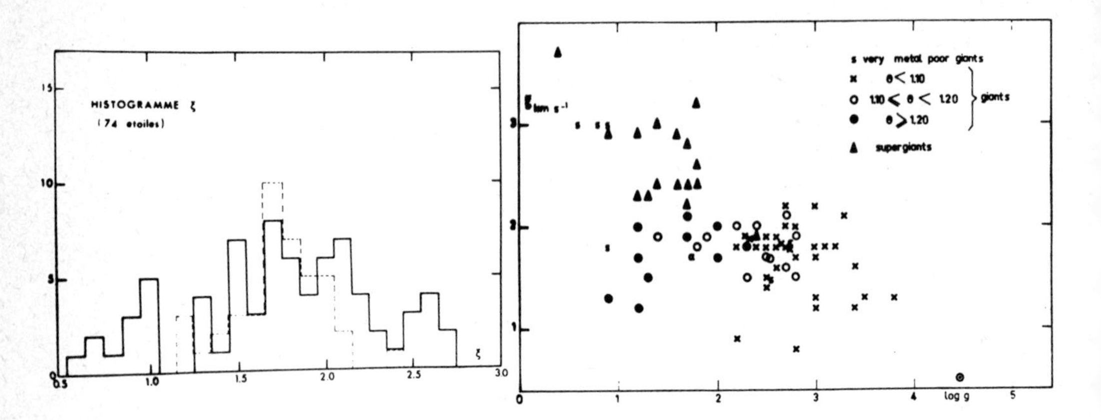

Figure 1. Distribution of microturbulence among G - K giants.

Figure 2. Microturbulent velocity versus surface gravity for giants.

the microturbulent velocity among giants : it is systematically larger than the solar value.The dotted line shows the distribution of microturbulence among giants obtained by Gustafsson et al (1974),from a sample having a narrower gravity range,using narrow band photometry.

The microturbulence appears to be strongly correlated with the surface gravity (Figure 2).However I do not venture to propose an analytical expression for this dependence.It tends to disappear when we remove supergiants and the Sun,i.e. when we reduce the gravity range of the sample.Inversely,it is better defined when removing only stars cooler than $\theta_{eff} = 5040/T_{eff} = 1.20$.The dependence of the microturbulence on the effective temperature is also better defined when removing the coolest stars (Figure 3).Could this be due to model effects? Or to a more complicated relationship?

Figure 4 shows that this is likely the case.Here we plot the gravity versus the effective temperature,with three microturbulent velocity ranges.Evolutionary tracks for 2 and 2.5 solar masses are shown for the asymptotic branch phase.From this diagram,the proportion of low values

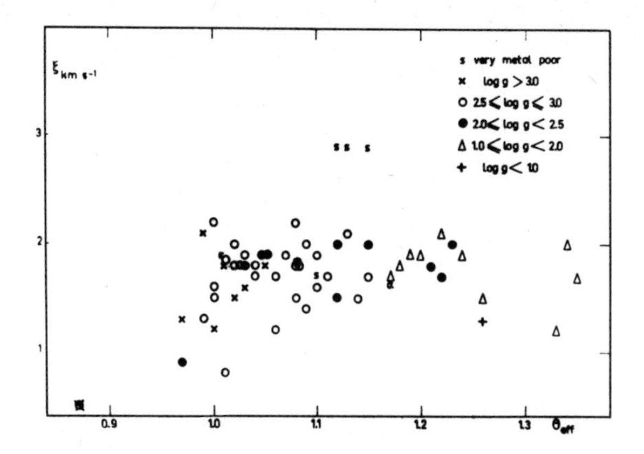

Figure 3. Microturbulent velocity versus effective temperature for giants.

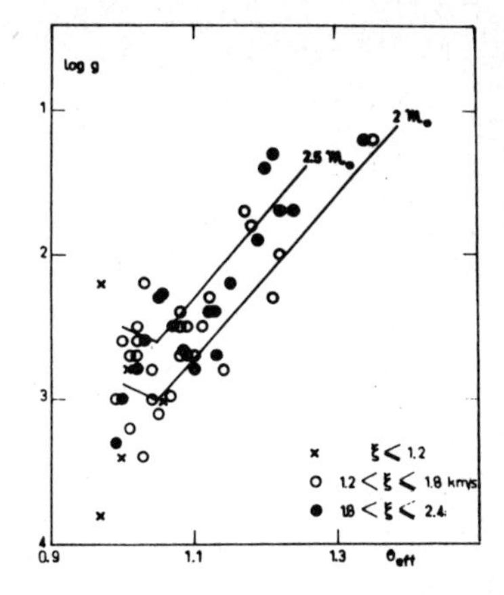

Figure 4. Behaviour of microturbulent velocity as compared to evolutionary tracks in the log g versus θ_{eff} plane.

of ξ relative to high values is smaller near the red giant tip than near the base of the asymptotic branch.Note that here we are concerned with a relatively small gravity range,but with a quite large effective temperature range.In this case no relation is clear on figures 2 and 3.

The same increase in ξ with stellar evolution occurs on the HR diagram (Figure 5).The less evolved giants have,on the average,lower microturbulent velocities than giants ascending the first asymptotic branch.There is a relatively clear cut division just at the beginning of the asymptotic branch,corresponding to the abrupt decrease in the atmospheric opacities and the rapid inward growth of convection.

It is also possible that microturbulence decreases after stars leave the red giant tip when they burn Helium in their core;during the Helium burning core,the efficiency of convection is rapidly decreasing (Iben,1967).I found (Foy,1978) indications that field giants with low microturbulent velocities would be evolving in the Helium burning core phase.An exemple is HD 71369,a G5 giant member of the Hyades group.Eggen (1972) found that it is one magnitude brighter than other giants having the same value of the R-I colour index;he suggested that its evolutionary state is different from that of the other giants in the Hyades group : it could evolve on a loop following the Helium core ignition.I found its microturbulent velocity to be as low as 0.8 km/s.

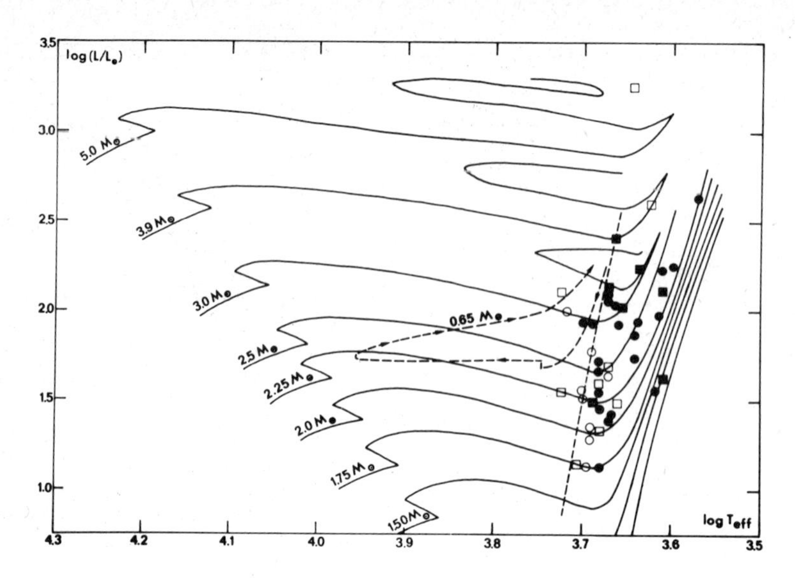

Figure 5. Behaviour of microturbulent velocity as compared to evolutionary tracks in the HR diagram.

To derive firm conclusions concerning this last relationship would require a comparison between the microturbulence in horizontal branch and asymptotic branch giants in a cluster;indeed the location on the horizontal branch is less questionable in the case of a cluster than for field stars.

Another program would be to use very high resolution techniques for a sample of bright field dwarfs,selected in two ways : firstly along the zero age main sequence,and secondly,perpendicularly to it.This program should provide information about the dependence of microturbulence on age,independently of the gravity and the temperature.

REFERENCES

Andersen,P.H.:1973,Publ.Astron.Soc.Pacific 85,666
Bonsack,W.K.,Culver,R.B.:1966,Astrophys.J. 145,767
Cayrel,G.,Cayrel,R.,Foy,R.:1977,Astron.Astrophys. 54,797
Eggen,O.J.:1972,Publ.Astron.Soc.Pacific 84,406
Foy,R.:1972,Astron.Astrophys. 18,26
Foy,R.:1978,Astron.Astrophys. 67,311
Garz,T.,Kock,M.:1969,Astron.Astrophys. 2,274
Gustafsson,B.,Bell,R.A.,Erikson,K.,Nordlund,A.:1975,Astron.Astrophys. 42,407

Gustafsson,B.,Kjaergaard,P.,Andersen,S.:1974,Astron.Astrophys. 34,99
Iben,I.Jr.:1967,Ann.Rev.Astron.Astrophys. 5,571
Peytremann,E.:1974,Astron.Astrophys. 33,203

HIGH LUMINOSITY F-K STARS MOTIONS AND Hα EMISSIONS

J. Smolinski, J.L. Climenhaga, and B.L. Harris
Department of Physics, University of Victoria
Victoria, B.C., Canada

Abstract

Changes and differences in radial velocities between neutral and ionized metals have been found for three F5-type supergiants: HD 231195, HD 10494, and HD 17971. Fifteen high dispersion coudé spectrograms (6 Å/mm) were used and 33 to 165 lines were measured on each. Semi-regular time variations up to about 8 km s^{-1} in radial velocity have been found. In addition, Hα line profiles for 8 high luminosity F-K stars have been analyzed. All of the stars show Hα emissions, variable in time, which is probably a common phenomenon in very luminous stars. Metallic emission lines with low excitation potentials, in particular the Ca I 6572.8 and the Fe I 6574.2 lines, are present in 5 of these stars.

TURBULENCE IN THE ATMOSPHERE OF B-TYPE STARS

Keiichi Kodaira
Department of Astronomy, University of Tokyo
Bunkyo-ku, Tokyo, Japan

Abstract

The stationary turbulent surface layer, whose depth is of the order of the pressure scale height in the subphotospheric layer, was investigated for B-type stars, using the momentum and the continuity equations with the inertia term neglected but the turbulence-viscosity term included. The mean velocity field is dominated by the horizontal component of the meridional circulation, driven by the pressure-density unbalance in the radiative envelope of the rotating star, and the differential rotation induced by the Coriolis force.

The model calculation for a B3IV-V star with the equatorial rotational velocity of 200 km/s led to the conclusions that the velocity field due to the differential rotation is of the order of 0.1-1 km/s, the velocity field of the meridional circulation itself is negligible, the velocity field of the three-dimensional shear turbulence is of the order of 0.1 km/s, and its scale is comparable to or less than the pressure scale height, the velocity field of the horizontal two-dimensional turbulence is of the order of 1-10 km/s, and its maximum scale ranges from a few times the pressure scale height to one-fiftieth of the stellar radius. If the index of the energy density law is close to 3.5, the turbulent surface layer may be dominated by the large scale (two-dimensional) barotropic eddies whose energy is fed by the small scale (three-dimensional) baroclinic turbulence in a way similar to the planetary atmospheres. In this case we may expect the tangential macroturbulence of the order of 1-10 km/s, though the interaction between this and the small-scale three-dimensional turbulence still remains to be investigated.

Mesoturbulence[+)]

G. Traving
Institut für Theoretische Astrophysik
der Universität Heidelberg
Heidelberg, FRG

[+)] In favour of a concise discussion of the basic concept of mesoturbulence we have refrained from presenting a comprehensive review of the work done in this field. Completeness, however, was aimed at in the references at the end of the contribution by E. Sedlmayr.

Abstract

The influence of a stochastic velocity field with a finite scale length l on the transfer of line radiation is described by means of a generalization of the transfer equation. Micro- and macroturbulence are contained in this mesoturbulence approach as limiting cases $l \to 0$ and $l \to \infty$ respectively.

Introductory Remarks

In hydrodynamics the term "turbulence" describes velocity fields which are dominated by inertial forces and of which only the statistical properties are controlled by initial- and boundary conditions. The meaning of the same word as used by spectroscopists is quite different. It encompasses any flow of unresolved pattern which - in addition to thermal motion - contributes to the Dopplerbroadening of spectral lines. So this term describes a situation which is characterized by a lack of information concerning the underlying velocity field. It is for this reason that the spectroscopist takes recourse to a description in statistical terms.

First of all, the basic information is contained in the mean square velocity $\langle v^2 \rangle = \sigma^2$, where v is the velocity component parallel to the ray. The second important parameter, the scale length l, is more difficult to determine.

Struve and Elvey (1934) discussed the limiting case $\kappa_{line} \cdot l \ll 1$ - with $\kappa_{line} [cm^{-1}]$ being the line absorption coefficient - where the hydrodynamic flow simply acts as an additional thermal broadening of the atomic absorption profile. In this case the saturation in the line decreases with the consequence that the curve of growth (say of stellar absorption lines) changes due to increasing equivalent widths. This is the microturbulence limit. Macroturbulence, on the other hand, is defined by $\kappa_{line} \cdot l \gg 1$. Then there is no velocity gradient along the ray and the radiative transfer is not affected. The profile in the radiation leaving the source has to be convoluted with the velocity distribution, a procedure which does not alter the equivalent widths.

The need for an approach based on the assumption of a finite l, which bridges the gap between these two limits, becomes obvious if one realizes that:

a) There is no doubt that "microturbulence" is a well established phenomenon in stellar spectroscopy. So l cannot have been large compared to κ_{line}^{-1}.

b) One has to exclude very small values of l since they would imply strong velocity gradients and hence excessive dissipation of kinetic energy.

Indeed, we note that under very general conditions the energy dissipation in a turbulent hydrodynamic flow is

$$\frac{dE}{dt} = 15\,\nu\,\langle v^2\rangle\,\lambda^{-2}\ [\mathrm{erg\ g^{-1}\ sec^{-1}}] \qquad (1)$$

if ν is the kinematic viscosity, which is roughly given by

$$\nu = v_{thermal} \cdot l_{free\ path} \qquad (2)$$

and λ the microscale of the flow defined by

$$\langle v^2\rangle/\lambda^2 = \langle(\frac{dv}{ds})^2\rangle. \qquad (3)$$

If the flow is stationary $\frac{dE}{dt}$ must be equal to $\langle v^2\rangle/2$ divided by a time which is characteristic for the renewal of the hydrodynamic energy. Let this time be the ratio of the equivalent height H of the atmosphere to the velocity of the flow; an assumption which seems to be reasonable either in case of buoyancy forces driving the turbulence or of convective transport of kinetic energy. If all velocities are of the same order of magnitude one finds that the smallest possible scale for hydrodynamic fields (in stellar atmospheres) is of the order of

$$\lambda = (30\,H\,l_{free\ path})^{1/2}\ . \qquad (4)$$

Inserting data for the solar photosphere yields $\lambda \sim 10^4..10^5$ [cm] which is of the same order of magnitude as the mean free path of a photon in an absorption line of medium strength.

If the effects of microturbulence are caused by sound waves of sawtooth form, Hearn (1974) has shown that the observed microturbulence velocities require a flux of mechanical energy which is about 100times the acoustic energy generated by the convection zones.

Hence, one cannot escape the conclusion that the naive interpretation of line broadening by small scale hydrodynamic flow as microturbulence has to be abandoned since it interfers with basic laws of hydrodynamics.

The Microturbulence Criterion Revisited

In the following we consider in more detail the condition for the validity of the microturbulence approach. Let

$$\frac{dI}{ds} + \kappa(v)\ I = 0 \tag{5}$$

be the transfer equation for the monochromatic intensity I in case of pure absorption. The inclusion of a non-zero source function S would make the formulae more involved without altering our conclusions. Note that I(s) depends on all values of v(s') for s'<s so that I(s) is not a function but a functional of v.

The microturbulent solution I_{mic} satisfies the equation

$$\frac{dI_{mic}}{ds} + \langle\kappa\rangle\ I_{mic} = 0 \tag{6}$$

with

$$\langle\kappa\rangle = \int_{-\infty}^{+\infty} \kappa(v)\ \mathbb{P}_1(v)\,dv \tag{7}$$

where $\mathbb{P}_1(v)$ is the one-point distribution function for the velocities along the ray. Defining u(s) by

$$I(s) = u(s)\cdot I_{mic}(s) \tag{8}$$

we obtain for u(s) the differential equation

$$\frac{du}{ds} + \Delta\kappa(v)\cdot u = 0 \tag{9}$$

with

$$\Delta\kappa(v) = \kappa(v) - \langle\kappa\rangle\ . \tag{10}$$

Solving eq. (9) with the initial value u(s=0)=1 by means of Picard's iteration one obtains

$$u(s)=1-\int_0^s ds_1\Delta\kappa_1+\int_0^s ds_1\int_0^{s_1} ds_2\Delta\kappa_1\Delta\kappa_2-\int_0^s ds_1\int_0^{s_1} ds_2\int_0^{s_2} ds_3\Delta\kappa_1\Delta\kappa_2\Delta\kappa_3\ldots \tag{11}$$

where

$$\Delta\kappa_i = \Delta\kappa(s_i , v(s_i)) \ . \tag{12}$$

Recalling that by means of the definition eq. (10) we have

$$\langle\Delta\kappa\rangle = 0 \tag{13}$$

we obtain for the lowest order deviation from unity of the expectation value of u(s)

$$\langle u(s)\rangle - 1 \approx \int_0^s ds_1 \int_0^{s_1} ds_2 \ \langle\Delta\kappa_1 \Delta\kappa_2\rangle \ . \tag{14}$$

Thus the lowest order deviation from the microturbulent case is determined by the two-point correlation of $\Delta\kappa$ which is calculated by means of the two-point velocity distribution $\mathbb{P}_2(v_1,s_1,v_2,s_2)$ according to

$$\langle\Delta\kappa_1 \Delta\kappa_2\rangle = \int_{v_1} dv_1 \int_{v_2} dv_2 \ \Delta\kappa_1 \Delta\kappa_2 \ \mathbb{P}_2(v_1,s_1,v_2,s_2) . \tag{15}$$

If $s_1 - s_2 > l$ the two point velocity distribution factorizes into two one-point distributions so that due to eq. (13) the contribution to the integral is zero. Taking this into account the rhs. of eq. (14) can be estimated as

$$\langle u(s)\rangle - 1 \approx \frac{l \cdot s}{2} \left(\langle\kappa^2\rangle - \langle\kappa\rangle^2\right) \ . \tag{16}$$

So we obtain the following condition for the validity of the microturbulence approach

$$\frac{\tau_l \ \tau_s}{2} \left(\frac{\langle\kappa^2\rangle}{\langle\kappa\rangle^2} - 1\right) \ll 1 \tag{17}$$

where the optical depths $\tau_l = l \cdot \langle\kappa\rangle$ and $\tau_s = s\langle\kappa\rangle$ have been introduced. Since in all cases of interest τ_s will be of the order one, we see that microturbulence is a good approach if $\tau_l \ll 1$ (the usual assumption) <u>and/or if $\langle\kappa^2\rangle/\langle\kappa\rangle^2 - 1$ approaches zero</u>. This is a condition which limits the amplitude of the velocity distribution. It should be mentioned that $\langle\kappa^2\rangle/\langle\kappa\rangle^2 - 1$ is zero at the line center for small $\langle v^2\rangle$ if the distribution functions are symmetric.

Approach to Mesoturbulence

The foregoing discusssion provides a first step towards a more general description of the radiative transfer which incorporates the to parameters $\langle v^2\rangle$ and l from the very beginning. Indeed, eq. (11) can be considered as the formal solution of such a transfer problem. One clearly sees that all higher order correlations of the velocity field enter.

An approach by means of a perturbation expansion - which bears at least some relation to the above formalism - has been formulated by Rybicki (1975). He developed all relevant quantities in orders of the perturbation by the velocity field.

$$\begin{aligned} \kappa &= \kappa^{(0)} + \kappa^{(1)} + \kappa^{(2)} + \dots \\ I &= I^{(0)} + I^{(1)} + I^{(2)} + \dots \\ S &= S^{(0)} + S^{(1)} + S^{(2)} + \dots \quad . \end{aligned} \tag{18}$$

S is the monochromatic source function which in case of scattering or non-LTE may depend on the velocity. Inserting these expansions into the transfer equation and equating terms of different order individually one obtains the perturbation expansion of the transfer equation. It turns out that the formal solution of the order (n-1) can be inserted as inhomogeneous term into the rhs. of the equation of order n. So there exists no closure problem since there is no dependence of the low order equations on the higher order solutions. It is by means of these rhs. terms that the correlations of the velocity field enter.

Apparently no attempts have been made to work out in more detail this formalism, into which correlations of all orders enter, but which - for practical reasons - is restricted to weak turbulence.

Obviously the problem is to determine the order of correlations which have to be taken into account. Apart from all theoretical considerations the answer to this question depends on the amount of information concerning the pattern of the velocity field which can be derived from the observed profiles. Apparently at present we can hardly expect to determine by such an analysis more than the influence of the lowest order correlations on the radiative transfer. Thus the assumption that all higher order correlations factorize into two-point correlations seems to be adequate.Approaches of this type which lead to simple formulae and which do not impose any constraints to $\langle v^2 \rangle$ have been followed independently by Auvergne et al. (1973) and by Gail et al. (1974).

The essentials common to the work of both groups can be presented in a very simple way: Let us assume LTE and a source function S which does not depend on the velocities. Then the transfer equation can be written as

$$dI = -\kappa(v)(I-S)ds \tag{19}$$

If the velocities v(s') for s'<1 are deterministic also I(s) is deterministic, but in case of random velocities I(s) will be random. Consider all I(s) which are compatible with the constraint that at s the velocity is in the interval v...v+dv, the probability of which is $\mathbb{P}_1(v,s)dv$. Let $q(v,s)$ be the mean value of the intensities subjected to this constraint. We now define

$$Q(v,s) = \mathbb{P}_1(v,s) \cdot q(v,s) \tag{20}$$

and obtain for the expectation value of the intensity

$$\langle I(s) \rangle = \int_v Q(v,s)dv \,. \tag{21}$$

We want to emphasize that in contrast to $I(s)$, which is a functional of v, the quantities $q(v,s)$ and $Q(v,s)$ are functions of v.

Aiming now at the transfer equations for $q(v,s)$ or $Q(v,s)$ we first restrict ourselves to the case of macroturbulence and assume constant v for all s. Then $q(v,s)$ and $I(s)$ have to comply with the same transfer equation since for v = const. both quantities are identical. Hence

$$dq(v,s) = -\kappa(v)(q(v,s)-S)ds \tag{22}$$

and by multiplication with $\mathbb{P}_1(v)$ - which we assume to be independent of s -

$$dQ(v,s) = -\kappa(v)(Q(v,s) - \mathbb{P}_1(v)S)ds \tag{23}$$

We now relax the macroturbulence condition (v=const.) by assuming a random field, the structure of which is dominated by two-point correlations or by the corresponding two-point distribution functions $\mathbb{P}_2(v_1,s_1,v_2,s_2)$, respectively. These can be written as the product of a one-point distribution function times a transition probability

$$\mathbb{P}_2(v_1,s_1,v_2,s_2) = \mathbb{P}_1(v_1,s_1) \cdot W(v_2|v_1,s_1,s_2-s_1) \tag{24}$$

where $W(v_2|v_1,s_1,s_2-s_1)dv_2$ is the probability of finding at s_2 the velocity v_2 in the interval dv_2 provided that at s_1 the velocity is v_1.

Any change of the velocities with the step ds will clearly affect the transfer equation (23). There will be an additional sink term for $Q(v,s)$ which is $Q(v,s)$ times the probability that with the step ds the velocity changes to any other velocity v'. Also an additional source term occurs, given by $Q(v",s)$ times the probability of a transition from any v" to v. It is by means of these transition probabilities that the scale length l is introduced.

Transitions of v which lead to

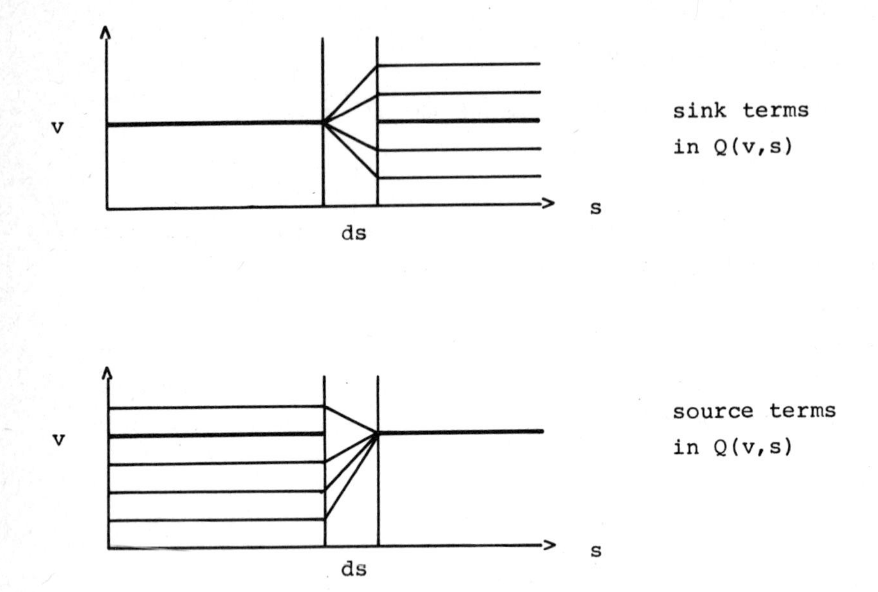

So, with W independent of s, eq. (23) turns into

$$dQ(v,s) = -\kappa(v)(Q(V,s) - \mathbb{P}_1(v)\cdot S)ds - \int_{-\infty}^{+\infty} Q(v,s)W(v'|v,ds)dv' + \int_{-\infty}^{+\infty} Q(v'',s)W(v|v'',ds)dv'' \quad (25)$$

This is the central equation of the two-point correlation approach. It has some resemblance with the transfer equation in case of scattering of line radiation. Indeed, if one relates the transition probability W with the redistribution function in case of scattering one can interpret eq. (25) as describing the transfer of monochromatic line radiation subjected to a scattering process in velocity space.

One can look at the problem from a different point of view. Let $P(I,v,s)dIdv$ be the joint probability of finding at s the intensity I in dI and the velocity v in dv. Then with the assumption of a velocity field governed by two-point correlations only, the transition probability for the velocity depends only on v at s and on no other data. Since the transfer equation is a first order differential equation, the change of I depends also only on the values of I and v at s, so that one can consider I and v as being subjected to a Markovian process in s. In this case the smooth change of $P(I,v,s)$ with s can be described by an equation of the Einstein-Smoluchowski type:

$$P(I,v,s)= \int_{v'} dv' \int_{I'} dI' \delta(I-I'+\kappa(v')(I'-S)\Delta s)\cdot W(v|v',\Delta s) P(I',v',s') . \quad (26)$$

In order to establish the relation to eq. (25) one has to make use of the fact that $Q(v,s)$ is the first moment of $P(I,v,s)$ with respect to I

$$Q(v,s)= \int_{I} P(I,v,s) I \, dI. \quad (27)$$

The relations of this mesoturbulence approach to the micro- and macroturbulence limits and to hydrodynamic turbulence can best be illustrated by means of the corresponding two-point velocity correlations

$$\rho(s)= \langle v(s')v(s'+s)\rangle / \langle v^2 \rangle \; . \quad (28)$$

Qualitatively the graphs of $\rho(s)$ are:

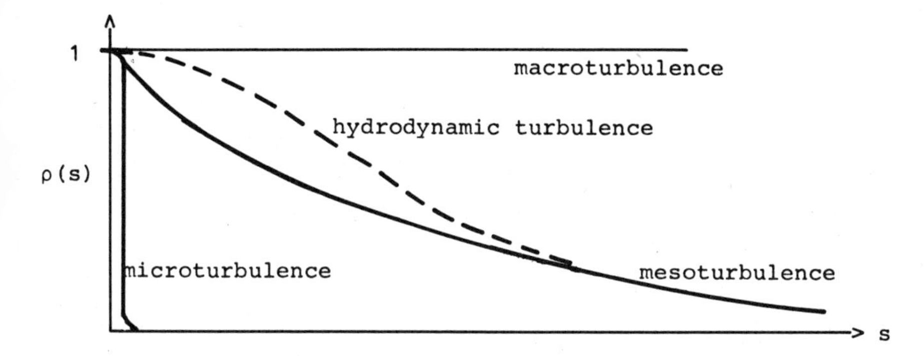

In order to work out eq. (25) in more detail the transition probability for the velocities has to be specified. This is the point where different approaches have been followed.

Auvergne et al. (1973) conceived a process (the Kubo-Anderson process) in which the two-point velocity distribution is a linear combination of a) a completely uncorrelated part consisting of the product of two one-point distribution functions $\mathbb{P}_1(v,1)\cdot \mathbb{P}_1(v',s')$ and b) a part which is completely correlated $\mathbb{P}_1(v',s')\cdot\delta(v-v')$. Then the transition probability is

$$W(v|v',\Delta s)=(1-\rho(\Delta s))\,\mathbb{P}_1(v')+\rho(\Delta s)\,\delta(v-v'). \quad (29)$$

Then with

$$\rho(ds) = 1 - \frac{ds}{l} \quad (30)$$

eq. (25) turns into

$$\frac{\partial Q}{\partial s}=-\kappa(Q-\mathbb{P}_1 S)-\frac{1}{l}\{Q-\mathbb{P}_1\int_{-\infty}^{+\infty} Q(v')dv'\}=-(\kappa+\frac{1}{l})Q+(\kappa S+\frac{1}{l}\langle I\rangle)\mathbb{P}_1 \quad . \quad (31)$$

The sink and source terms in the curly bracket can easily be interpreted as being due to "complete redistribution in velocity space"

Gail et al. (1974), on the other hand, preferred Gaussian one- and two-point velocity distributions with the consequence that

$$\mathbb{P}_1 = (2\pi\sigma^2)^{-1/2} \exp\left(-\frac{v^2}{2\sigma^2}\right) \tag{32}$$

and

$$W(v|v',\Delta s) = (2\pi\sigma^2(1-\rho^2))^{-1/2} \exp\left(-\frac{(\rho v'-v)^2}{2\sigma^2(1-\rho^2)}\right) . \tag{33}$$

With these assumptions which define an Uhlenbeck-Ornstein process (Wang and Uhlenbeck, 1945) eq. (25) can be written as

$$\frac{\partial Q}{\partial s} = -\kappa(Q - \mathbb{P}_1 S) + \frac{1}{l}\frac{\partial}{\partial v}\left(v + \sigma^2 \frac{\partial}{\partial v}\right) Q . \tag{34}$$

The differential operator on the rhs. of eq. (34) is consistent with the assumption of a continuous velocity field, hence only infinitesimal changes of v within ds have non zero probabilities; the"scattering"is almost "coherent". Eq. (34) is a partial differential equation of parabolic type which is easily solved by standard numerical techniques.

The central eq. (25) and hence also eq. (31) and eq. (34) have been derived starting from the macroturbulence limit ($l \to \infty$). We use eq. (34) in order to show that they contain also the microturbulence limit ($l \to \infty$).

Noting that the Hermite functions

$$\phi_n\left(\frac{v}{\sigma}\right) = \phi_n(\hat{v}) \tag{35}$$

are the eigenfunctions of the differential operator occurring in eq.(34)

$$\frac{\partial}{\partial \hat{v}}\left(\hat{v} + \frac{\partial}{\partial \hat{v}}\right) \phi_n = -n\,\phi_n \tag{36}$$

we use the expansion

$$Q(s,\hat{v}) = \sum_{n=0}^{\infty} T_n(s)\,\phi_n(\hat{v}) \tag{37}$$

and obtain due to the orthonormality relations of the Hermite functions the following system of ordinary differential equations

$$\frac{dT_m}{ds} = -\sum_{n=0}^{\infty} \kappa_{mn} T_n + \kappa_{mo} S - \frac{m}{l} T_m \, , \quad m = 0,1,2.... \tag{38}$$

with

$$\kappa_{mn} = \int_{-\infty}^{+\infty} \phi_o^{-1} \phi_m \phi_n \kappa(\hat{v}) d\hat{v} . \tag{39}$$

It is obvious that in the limit $l \to 0$ all modes with $m \neq 0$ will have zero amplitude due to the last term on the rhs. of eq. (38). Hence in this limit eq. (38) reduces to

$$\frac{dT_o}{ds} = - \kappa_{oo} (T_o - S) \tag{40}$$

This is the microturbulence equation since

$$\phi_o(\hat{v}) = \mathbb{P}_1 (\hat{v}) \tag{41}$$

and

$$\langle I(s) \rangle = \int_{-\infty}^{+\infty} Q(s,\hat{v}) d\hat{v} = \sum_{n=0}^{\infty} T_n(s) \int_{-\infty}^{+\infty} \phi_o^{-1} \phi_o \phi_n d\hat{v} = T_o . \tag{42}$$

Conclusion

1) The word "turbulence" denotes a velocity field which can be described only in statistical terms. However, the reasons for such a description may be different. For a hydrodynamicist it is the very nature of the flow, for a spectroscopist, however, it is lack of information concerning the flow pattern.

2) The basic parameter of such a velocity field is the mean square velocity $\langle v^2 \rangle$. The scale length l is a further relevant information. Micro- and macroturbulence are limiting cases $l \to 0$ and $l \to \infty$ respectively.

3) If the limit $l \to 0$ is taken literally the spectroscopists concept of turbulence would be in conflict with basic principles of hydrodynamics. Hence an approach with finite l -Mesoturbulence- is needed.

4) Whereas in principle the radiative transfer depends on all correlations of the velocity field, the two-point correlation, which contains most of the coherence properties of the field, seems to provide a sensible first order approximation.

5) The formalism of the mesoturbulent radiative transfer in this approximation is simple. It may be interpreted correctly in terms of scattering in velocity space.

6) Efficient numerical methods exist for the solution of the resulting equations.

7) The possibility to extend the mesoturbulence formalism to non-LTE problems has not been touched upon (see Gail et al. 1975, Frisch and Frisch 1976, Traving 1976).

Stochastic Approach

H.-P. Gail
Institut für Theoretische Astrophysik
der Universität Heidelberg
Heidelberg, FRG

Abstract

A general formalism for describing the radiation transfer in a medium with arbitrary velocity fields is presented. It is demonstrated that classical microturbulence and mesoturbulent models based on Markov processes can be considered as the two lowest order members within a hierarchy of model equations with an increasing degree of approximation to reality. Some preliminary results concerning the relevance of low order model equations are presented.

1. Introduction

In interpreting line profiles originating from stellar atmospheres with internal motions, one encounters the problem that the observed flux within a spectral line is composed of contributions from regions with quite different velocities. In order to obtain the line profile which actually is observed, an averaging process has to be applied with respect to the ensemble of flow situations which occur along all lines of sight which contribute to the measured flux.

If the velocity distribution $\vec{v}(\vec{x},t)$ within the atmosphere is known, the calculation of the radiation intensity exhibits no special problem (except for numerical difficulties) and the required average is simply calculated as an integral of the emergent intensity over all directions of interest. In this case, there is no need for a statistical treatment of the line transfer problem.

If, however, the flow is turbulent and thus can only be described in terms of its statistical properties or if the information on the state of motion of the matter is incomplete, it is not possible to assign in an unambiguous way to each line of sight a velocity profile v(r) along this ray. On the other hand, a definitive knowledge of this function v(r) is a pre-requisite for solving the ordinary equation of radiative transfer in a moving medium. In this case, one has to take recourse to statistical methods.

Even if the statistical properties of the flow are completely known (for instance, if the complete hierarchy of probability densities defined in chapter 2 is known), it is not possible to specify uniquely the quantity v(r) for each line of sight. However, for a large ensemble of equivalent rays it is possible to determine for each possible distribution of velocities along a ray the probability, that this specific v(r) is realized. Since, for a given v(r), one is able to solve the radiative transfer problem, one obtains a certain intensity distribution I(r) along the ray. The probability of realization of this distribution I(r) equals the probability of realization of the specific v(r), on which this solution I(r) is based. Thus it is natural to describe the radiation field in terms of probability densities and to reformulate the radiative transfer equation in terms of these quantities. A general theory of this kind has been developped for the case of LTE and negligable scattering (see chapter 3 and 4).

If no or incomplete information about the velocity field is available, it is nevertheless useful to apply statistical methods. The best procedure in this case would be (i) to isolate the basic parameters of the velocity field which are relevant for the line transfer problem and (ii) to substitute the equations of the original problem by model equations, which depend on the relevant parameters only.

One method to proceed in this direction is, to derive a hierarchy of statistical model equations, which incorporate an increasing degree of information on the structure of the velocity field. By a study of the properties of the members of such a hierarchy it will be possible to find out the relevant parameters and the adequate model equations. The classical microturbulence model and the Markov-process models of Auvergne et al.(1973) and Gail et al.(1974) can be considered as the zeroth order and the first order models within such a hierarchy (see chapter 5). Model equations of higher order have not been derived up to now. Thus, it is not possible at present to decide, whether first order model equations already are sufficient to treat the line transfer problem or not.

A second method is to start with a statistical theory, valid for general velocity fields, and to derive from this the model equations. Some preliminary results in this direction are presented in chapter 6.

2. Description of the velocity field

The ensemble of different flow situations, which one encounters along different rays in a stellar atmosphere is most conveniently described by the hierarchy of n-point probability densities

$$\mathbb{P}_n(\vec{x}_1,\vec{v}_1;\vec{x}_2,\vec{v}_2;\ldots;\vec{x}_n,\vec{v}_n) = \mathbb{P}_n(1,\ldots,n) \quad . \tag{2.1}$$

$\mathbb{P}_n$ is the probability of finding at $\vec{x}_1$ the velocity $\vec{v}_1$ <u>and</u> at $\vec{x}_2$ the velocity $\vec{v}_2$ ... <u>and</u> at $\vec{x}_n$ the velocity $\vec{v}_n$. The following properties of the $\mathbb{P}_n$ are self evident:

$$\mathbb{P}_n(1,\ldots,n) \geq 0 \quad , \tag{2.2}$$

$$\int d^3\vec{v}_i \mathbb{P}_n(1,\ldots,i-1,i,i+1,\ldots,n) = \mathbb{P}_{n-1}(1,\ldots,i-1,i+1,\ldots,n) \quad , \tag{2.3}$$

$$\int d^3v_1 \mathbb{P}_1(1) = 1 \quad . \tag{2.4}$$

Since only the component of the velocity parallel to the ray under consideration enters into the radiative transfer problem, we define a new probability density by

$$\mathbb{P}_n(x_1,v_{1\parallel};x_2,v_{2\parallel};\ldots;x_n,v_{n\parallel};\vec{k}) = \int d^2v_{1\perp}\ldots\int d^2v_{n\perp}\mathbb{P}_n(1,\ldots,n) \tag{2.5}$$

which gives the corresponding probabilities for the $\parallel$ component of the velocity. For anisotropic $\mathbb{P}_n(\vec{x}_1,\vec{v}_1;\ldots)$ they depend explicitly on the direction $\vec{k}$ of the ray. The x_i $(i=1,\ldots,n)$ are the coordinates of the points to which $\mathbb{P}_n$ refers along the ray. We assume these points to form an ordered sequence with $x_1 \leq x_2 \leq \ldots \leq x_n$, since in the application to the radiative transfer problem the $\mathbb{P}_n$ occur only in this special form. In the following we simply write v_i instead of $v_{i\parallel}$ and omit the $\vec{k}$ from our notation.

Since every hydrodynamic flow has the property of being continuous for distances between the points x_i, x_{i+1} smaller than a certain length, the probability densities $\mathbb{P}_n$ have to satisfy the condition:

$$\lim_{x_{i+1}\to x_i} \mathbb{P}_n(1,\ldots,i,i+1,i+2,\ldots,n) = \mathbb{P}_{n-1}(1,\ldots,i,i+2,\ldots,n)\cdot\delta(v_{i-1}-v_i) \tag{2.6}$$

since, if x_{i+1} equals x_i then v_{i+1} necessarely equals v_i due to the continuity of the flow. Especially, it follows

$$\lim_{x_n \to x_1} \mathbb{P}_n(1,\ldots,n) = \mathbb{P}_1(1) \prod_{i=2}^{n} \delta(v_{i-1}-v_i) \quad . \tag{2.7}$$

A wide class of flows has the property, that there exists a finite correlation length l such that the velocities v_{i+1} and v_i become statistically independent, if the distance between x_{i+1} and x_i becomes large compared to l. If the flow has this special property, then the $\mathbb{P}_n$'s have to satisfy the condition

$$\mathbb{P}_n(1,\ldots,i,i+1,\ldots,n) = \mathbb{P}_i(1,\ldots,i)\, \mathbb{P}_{n-i}(i+1,\ldots,n) \quad \text{if} \quad x_{i+1}-x_i \gg 1 \quad . \tag{2.8}$$

This condition is valid for instance for turbulent flows. It is not valid for instance for harmonic waves.

The concept of a description of the velocity field by means of the hierarchy of the $\mathbb{P}_n$'s is flexible enough to allow a unified description of such different cases like (for instance) a purely deterministic velocity field v(x):

$$\mathbb{P}_n(1,\ldots,n) = \delta(v_1-v(x_1))\cdot\delta(v_2-v(x_2)) \cdots \delta(v_n-v(x_n)) \tag{2.9}$$

or a pure noise

$$\mathbb{P}_n(1,\ldots,n) = \mathbb{P}_1(x_1,v_1) \cdots \mathbb{P}_1(x_n,v_n) \quad . \tag{2.1o}$$

3. Description of the Radiation Field

In analogy to the description of the velocity field by means of the $\mathbb{P}_n$'s we describe the joint process (I,v) by means of the hierarchy of n-point probability densities (Gail et al, 1979, henceforth called paper I):

$$P_n(x_1,v_1,I_1;x_2,v_2,I_2;\ldots;x_n,v_n,I_n;\vec{k},\nu) = P_n(1,\ldots,n) \quad . \tag{3.1}$$

P_n is the probability of finding at x_1 the velocity v_1 and the intensity of radiation I_1 <u>and</u> at x_2 the velocity v_2 and the intensity I_2 ... <u>and</u> at x_n the velocity v_n and the intensity I_n. From the physical properties of radiative transfer it follows, that we are interested only in probability densities, where the x_i (i=1,...,n) form an ordered sequence along the ray under consideration (with direction $\vec{k}$).

The probability densities P_n have to satisfy conditions analogue to (2.2), (2.3) and (2.4). In paper I it is shown, that the general P_n can be written as

$$P_n(1,\ldots,n) = P_1(1)\prod_{i=2}^{n-1} P_2(i|i+1)\, \mathbb{P}_n(1,\ldots,n)/\Big(\mathbb{P}_1(1)\prod_{i=2}^{n-1} \mathbb{P}_2(i|i+1)\Big) \; . \tag{3.2}$$

Here we have introduced the conditional probabilities

$$P_2(1|2) = P_2(1,2)/P_1(1) \tag{3.3}$$

$$\mathbb{P}_2(1|2) = \mathbb{P}_2(1,2)/\mathbb{P}_1(1) \quad . \tag{3.4}$$

Hence , the complete information on the radiation field is already contained in P_2.

The conditional probability $P_2(1|2)$ has a simple interpretation. The analogue of (2.3) may be written as

$$\int dI_1 \int dv_1 P_1(1) P_2(1|2) = P_1(2) \tag{3.5}$$

and from this we infer, that $P_2(1|2)$ is the kernel function of an evolution operator, which solves the transfer problem for P_1.

In order to construct $P_2(1|2)$ we approximate $\kappa(x,v(x))$, $S(x,v(x))$ and $v(x)$ by step-functions. Then we consider one arbitrary realization of the velocity field between x_1 and x_2 with fixed velocities v_1 at x_1 and v_2 at x_2. If the final intensity at x_2 is just equal to I_2, the initial intensity I_1 at x_1 is given by

$$I_1 = I_2 \exp\Big(\sum_{i=1}^{n} \kappa_i \Delta x_i\Big) - \sum_{l=1}^{n} \kappa_l S_l \exp\Big(\sum_{i=1}^{l} \kappa_i \Delta x_i\Big) \Delta x_l = \tilde{I} \tag{3.6}$$

with $\kappa_i = \kappa(x_i, v(x_1))$, $S_i = S(x_i, v(x_i))$ and Δx_i beeing the length's of the intervals of the step-functions. This is simply a discretized version of the solution of the ordinary equation of radiative transfer. Then $P_2(1|2)$ is given by

$$P_2(1|2) = \int dv_{\alpha_1} \ldots \int dv_{\alpha_{n-1}} \left(\mathbb{P}_{n+1}(1,\alpha_1,\ldots,\alpha_{n-1},2)/\mathbb{P}_1(1)\right) \exp\left(\sum_{i=1}^{n} \kappa_i \Delta x_i\right) \cdot \delta(I_1 - \tilde{I}) \quad . \tag{3.7}$$

The delta function assures, that only those realizations of the velocity field contribute, which have the correct final intensity I_2 for fixed I_1. The factor $\mathbb{P}_{n+1}/\mathbb{P}_1$ is the probability of realization of the considered step-function approximation of the velocity field. The exponential function takes care of the contraction of the interval dI_1 to dI_2 in going from x_1 to x_2. Finally, we integrate over all possible velocities at the division points of the interval x_1, x_2. More details with respect to the derivation of Eq. (3.7) can be found in paper I.

4. The Mean Intensity and the Conditional Intensity

In many cases, interest is concentrated on the mean intensity $\langle I \rangle$. This quantity can be calculated from P_1 as follows:

$$\langle I \rangle = \int dv \int dI \cdot I \cdot P_1 \quad . \tag{4.1}$$

The direct calculation of $\langle I \rangle$ or P_1 may become quite tedious. However, for the quantity

$$Q = \int dI \cdot I \cdot P_1 \tag{4.2}$$

one derives from (3.5) and (3.6) an equation, which can often be solved much easier than the equation for $\langle I \rangle$ or P_1. The mean intensity $\langle I \rangle$ is obtained from Q by a simple integration.

In order to derive the equation for Q, one multiplies (3.7) with I_2 and integrates with respect to I_2 with the result

$$Q(2) = \int dv_1 \int dv_{\alpha_1} \ldots \int dv_{\alpha_{n-1}} \left(\mathbb{P}_{n+1}(1,\alpha_1,\ldots,\alpha_{n-1},2)/\mathbb{P}_1(1)\right) \exp\left(-\sum_{i=1}^{n} \kappa_i \Delta x_i\right) \cdot \left\{Q(1) + \sum_{l=1}^{n} \kappa_l S_l \exp\left(\sum_{i=1}^{l} \kappa_i \Delta x_i\right) \mathbb{P}_1(1) \Delta x_l\right\} \quad . \tag{4.3}$$

Since one easily shows (of paper I):

$$\lim_{n\to\infty} \exp\{\pm\sum_{i=1}^{n} \kappa_i \Delta x_i\} = 1 + \sum_{\mu=1}^{\infty} (-1)^{\mu} \int_{x_1}^{x_2} ds_1 \int_{s_1}^{x_2} ds_2 \ldots \int_{s_{n-1}}^{x_2} ds_n \prod_{\nu=1}^{\mu} \kappa(\nu) \quad , \tag{4.4}$$

one arrives at the final equation

$$Q(2) = \int dv_1 Q(1) \mathbb{E}(1,2) + \int_{x_1}^{x_2} ds_\alpha \int dv_\alpha \mathbb{P}_1(\alpha) \kappa(\alpha) S(\alpha) \mathbb{E}(\alpha,2) \tag{4.5}$$

with

$$\mathbb{E}(1,2) = \mathbb{P}_2(1|2) + \sum_{n=1}^{\infty} (-1)^n \int_{x_1}^{x_2} ds_{\alpha_1} \ldots \int_{s_{\alpha_{n-1}}}^{x_2} ds_{\alpha_n} \int dv_{\alpha_1} \ldots \int dv_{\alpha_n} \cdot$$

$$\left(\mathbb{P}_{n+2}(1,\alpha_1,\ldots,\alpha_n,2) / \mathbb{P}_1(1)\right) \sum_{\mu=1}^{n} \kappa(\alpha_\mu) \quad . \tag{4.6}$$

Eqs. (4.5) and (4.6) are the basic equations, which serve to calculate the mean intensity for arbitrary velocity fields, described by their n-point probability densities.

5. Stochastic Models

Different stochastic models have been used to treat the line transfer problem in presence of velocity fields. These models are discussed in some detail in the preceding contribution of Traving. Thus we limit ourselves at this place to show, how they fit into our general formalism.

a) The classical microturbulence-macroturbulence approach

Pure microturbulence is described by n-point probability densities of the type (2.1o). Pure macroturbulence on the other hand is described by $\mathbb{P}_n$'s of the type (2.6). The superposition of both yields n-point probability densities of the type

$$\mathbb{P}_n((v_1,\ldots,v_n) = \int dw_1 \mathbb{P}_1^{mac}(w_1) \prod_{i=1}^{n} \mathbb{P}_1^{mic}(v_i - w_1) \quad . \tag{5.1}$$

Then, a simple calculation shows that the mean value <I> is just

$$\langle I \rangle = \int dw_1 \mathbb{P}_1^{mac}(w_1)\Big[I_o \exp\Big(-\int_{x_1}^{x_2} ds\, \kappa_{mic}(w_1)\Big) + \int_{x_1}^{x_2} ds\, \exp\Big(-\int_{s}^{x_2} ds' \kappa_{mic}(w_1)\Big) S(s')\kappa_{mic}(w_1)\Big] \qquad (5.2)$$

where

$$\kappa_{mic}(w_1) = \int dv\, \mathbb{P}_1^{mic}(v)\kappa(v-w_1) \qquad (5.3)$$

and we have used the obvious initial condition

$$Q(1) = I_o \mathbb{P}_1(1) \quad . \qquad ((5.4)$$

Eq. (5.2) is just the classical microturbulence-macroturbulence result, as was to be expected.

b) Markov-processes

Markov-processes can be defined by the property

$$\mathbb{P}_n(1,\ldots,n)/\mathbb{P}_{n-1}(1,\ldots,n-1) = \mathbb{P}_2(n-1|n) \quad . \qquad (5.5)$$

Then the general $\mathbb{P}_n$ can be expressed by $\mathbb{P}_2(i-1|i)$ as follows:

$$\mathbb{P}_n(1,\ldots,n) = \mathbb{P}_1(1)\prod_{i=2}^{n} \mathbb{P}_2(i-1|i) \quad . \qquad (5.6)$$

The conditional probability $\mathbb{P}_2(i-1|i)$ is due to condition (2.3) subject to the restriction to be a solution of

$$\int dv_2\, \mathbb{P}_2(1|2)\mathbb{P}_2(2|3) = \mathbb{P}_2(1|3) \quad . \qquad (5.7)$$

Examples of $\mathbb{P}_2$ are given in the contribution of Traving. For other examples see for instance Brissaud and Frisch (1974).

While pure microturbulence can be interpreted as a stochastic process without memory on the velocities encountered at x_i if we go from x_i to x_{i+1}, the Markov-process is a stochastic process with "short" memory. The velocities encountered at x_{i+1} are not independent of the velocity, which we have found at x_i, but are completely uncorrelated with all previous velocities at x_j with $j<i$.

In the case of Markov-processes, a simple equation for Q can be derived. By multiplying Eq. (4.6) with $\kappa(0)\ \mathbb{P}_2(0|1)$ and integrating with respect to v_o and x_o, one derives the following integral equation for $\mathbb{E}$ (see paper I):

$$\int_{x_1}^{x_2} ds_1 \int dv_1 \kappa(1)\mathbb{P}_2(1|2)\mathbb{E}(0,1) = -\mathbb{E}(0,2) + \mathbb{P}_2(0|2) \quad . \tag{5.8}$$

Then, by multiplying (4.5) by $\kappa(\alpha)\ \mathbb{P}_2(\alpha|2)$ and integrating with respect to v_α, x_α one derives by using (5.8):

$$Q(2) = \int dv_o \mathbb{P}_2(0|2)Q(0) - \int_{x_o}^{x_2} dx_1 \int_{-\infty}^{+\infty} dv_1 \kappa(1)\mathbb{P}_2(1|2)\{Q(1) - \mathbb{P}_1(1)S(1)\} \quad . \tag{5.9}$$

Differentiating this with respect to x_2, we obtain

$$\frac{\partial Q(0)}{\partial x_o} = \int dv_o [\lim_{x_1 \to x_o} \frac{\partial}{\partial x_1} \mathbb{P}_2(0|1)]Q(0) - \kappa(0)\{Q(0) - \mathbb{P}_1(0)S(0)\} \tag{5.1o}$$

which is equivalent with equation (25) of the preceding contribution of Traving. For a discussion of the special model of Auvergne et al (1973) and Gail et al (1974) see that contribution.

c) Higher order models

The microturbulence model and the Markov-process model may be considered as the two lowest order members of a hierarchy of model equations in the following sense:

(i) The microturbulence model assumes, that the general $\mathbb{P}_n$ can be factorized into a product of one-point probability densities $\mathbb{P}_1(v_i)$.

(ii) The Markov-process model assumes, that the general $\mathbb{P}_n$ can be factorized into a product of two-point conditional probability densities $\mathbb{P}_2(i|i+1)$ (cf Eq. (5.6)).

(iii) The next step would be to assume, that the general $\mathbb{P}_n$ can be factorized into a product of three-point conditional probability densities $\mathbb{P}_3(i,i+1|i+2)$ and to derive a model equation based on this special form of the $\mathbb{P}_n$.

In this way, one would obtain a hierarchy of model equations which allow to incorporate an increasing degree of information on the structure of the velocity field into the theory. However, higher order models have not been studied up to now.

6. Some comments on the relevance of low-order model equations

In this chapter, we consider velocity fields with finite correlation length. The starting point are Eqs. (4.5) and (4.6). From these one derives

$$\langle I(x_2)\rangle = S(x_2) - \int_{x_1}^{x_2} dt' \langle E(x_2,t')\rangle \frac{dS(t')}{dt'} + (I(x_1)-S(x_1))\langle E(x_2,x_1)\rangle \tag{6.1}$$

with

$$\langle E(x_2,x_1)\rangle = \exp\{-\int_{x_1}^{x_2} dt' \kappa_2(t')\}\{\ 1 + \sum_{n=1}^{\infty} (-1)^n\ e_n(x_2,x_1)\} \quad , \tag{6.2}$$

where

$$e_n(x_2,x_1) = \int_{x_1}^{x_2} dt_1 \int_{x_1}^{t_1} dt_2 \ldots \int_{x_1}^{t_{n-1}} dt_n \int_{-\infty}^{+\infty} dv_1 \ldots \int_{-\infty}^{+\infty} dv_n \mathbb{P}_n(1,\ldots,n) \sum_{\mu=1}^{n} \kappa(\mu) \quad . \tag{6.3}$$

Here we have assumed, that the absorption coefficient consists of two parts

$$\kappa = \kappa_1(v) + \kappa_2 \quad , \tag{6.4}$$

one of which, κ_2, is independent of the velocity. The actual choice of κ_1 and κ_2 will be specified later.

a) The case $x_2 - x_1 << 1$

At this place we choose

$$\kappa_1 = \kappa_{line} \quad , \quad \kappa_2 = \kappa_{continuum} \quad . \tag{6.5}$$

We introduce the new integration variable $s_i = (t_i - x_i)/l$. By assumption we have

$$\varepsilon = (x_2 - x_1)/l << 1 \quad . \tag{6.6}$$

It is natural, to expand the integrand in (6.3) into a Taylor series with respect to the small quantities $s_1,\ldots,s_n$. Then all s-integrations are easily done and one obtains to first order in the small quantity ε

$$e_n = l^n\int dv_1\ldots\int dv_n\left[\mathbb{P}_1(1)\prod_{\mu=2}^{n}\delta(v_{\mu-1}-v_\mu) + \frac{\varepsilon}{n+1}\sum_{i=1}^{n}\left.\frac{\partial\mathbb{P}_n}{\partial s_i}\right|_o (n+1-i)\right]\cdot$$

$$\int_0^{\varepsilon} ds_1\ldots\int_0^{s_{n-1}} \prod_{\nu=1}^{n}\kappa(\nu) \quad . \tag{6.7}$$

The dominating contribution corresponds to pure macroturbulence, as was to be expected. The first order correction depends only on $\mathbb{P}_3$, since due to (2.6) and assuming $\mathbb{P}_n$ to be uniform continuous at the origin, we have

$$\left.\frac{\partial\mathbb{P}_n}{\partial s_i}\right|_o = \lim_{s_{i+1}\to 0}\ \lim_{s_n\to s_{i+1}}\ \lim_{s_{i-1}\to s_1}\ \frac{\partial\mathbb{P}_n}{\partial s_i} =$$

$$= \lim_{s_{i+1}\to 0}\frac{\partial\mathbb{P}_3(0,i,i+1)}{\partial s_i}\prod_{\mu=2}^{i-1}\delta(v_{\mu-1}-v_\mu)\prod_{\nu=i+2}^{n}(v_{\nu-1}-v_\nu) \quad . \tag{6.8}$$

Since preliminary results indicate, that l is of the order of the skale height of the atmosphere (see the subsequent contribution of Sedlmayr), the present case applies to strong lines. Thus, in strong lines no information on the structure of the velocity field is contained, which extends beyond the three-point probability density $\mathbb{P}_3$.

b) The case $x_2-x_1 >> l$

At this place we choose

$$\kappa_{mic} = \int dv_1\mathbb{P}_1(1)\kappa_{line}(1) \tag{6.9}$$

$$\kappa_1 = \kappa_{line} - \kappa_{mic} \quad , \quad \kappa_2 = \kappa_{continuum} + \kappa_{mic} \quad . \tag{6.1o}$$

The integration in (6.3) is extentended over a n-dimensional simplex. Within this volume, the quantity

$$\langle\kappa_1(1)\cdots\kappa_1(n)\rangle = \int dv_1\ldots\int dv_n\ \mathbb{P}_n(1,\ldots,n)\kappa_1(1)\cdots\kappa_1(n) \tag{6.11}$$

is different from zero only in regions of the integration volume, where all points s_i form clusters of at least two points with mutual distances between the members of a cluster of at most $\sim$1 correlation length l. If at least one point is isolated, then due to (2.8) there will occur a factor $\langle\kappa_1\rangle$ in (6.11) which is zero according to the definition of κ_1. By analyzing the various possible clusters, one can show that, provided the condition

$$\max_{\forall v} \kappa_1(v)\cdot l \ll 1 \qquad (6.12)$$

is satisfied (cf Brissaud and Frisch, 1974, and Frisch, 1968), the dominating contribution is provided by two cases: (i) only clusters of two points occur and (ii) besides one or at most 2 clusters of three points only clusters of pairs occur. Since $\max(\kappa_1(v))$ is of the order of the line absorption coefficient in the centre of the line, this case corresponds to weak lines, since l itself is probably of the order of the scale height. Thus, weak lines, just as strong lines, do not contain any significant information on the structure of the velocity field extending beyond $\mathbb{P}_3$.

These results suggest, that model equations for the radiative transfer problem in moving media based on $\mathbb{P}_2$ or $\mathbb{P}_3$ are sufficient, at least for strong and weak lines.

Acknowledgement

This work has been performed as part of the program of the Sonderforschungsbereich 132 "Theoretische und praktische Stellarastronomie" which is sponsored by the Deutsche Forschungsgemeinschaft.

The Application of Mesoturbulence to Stellar Atmospheres

E. Sedlmayr
Institut für Theoretische Astrophysik
der Universität Heidelberg
Heidelberg, FRG

Abstract

For realistic stellar atmospheres the equations describing mesoturbulent line formation are solved numerically. The general dependence of theoretical line profiles and equivalent widths on the correlation length l and the mean square turbulent velocity σ is demonstrated. Also empirical relations between the basic parameters of the micro-macroturbulence description (v_{mic}, v_{mac}) and the fundamental mesoturbulence parameters (l,σ) are derived.

Introduction

It has been shown (see Traving's contribution) that turbulence in stellar atmospheres is necessarily of finite scale length. So mesoturbulence is a very common phenomenon which should be taken into account in line formation studies.

In this contribution we shall apply the mesoturbulence formalism (e.g. Auvergne et al. (1973), Gail et al. (1974), Gail and Sedlmayr (1974), Frisch (1975)) to line formation in realistic atmospheres. Our aimes are

(i) to demonstrate the basic effects of the correlation length l and the mean square turbulent velocity σ on the line profiles and equivalent widths, and

(ii) to derive empirical relations between the classical parameters (microturbulent velocity v_{mic} and macroturbulent velocity v_{mac}) and the fundamental parameters of the mesoturbulence description l and σ.

1. The Numerical Method

Two particular stochastic models have been adopted for discussing mesoturbulent line formation in stellar atmospheres:

a) The Uhlenbeck-Ornstein Process (UOP)

The UOP is a stationary Markovian process continuous in space with both Gaussian one-point and two-point velocity distribution functions. The resulting transfer equation for the monochromatic local conditional intensity q(s,v) is therefore a parabolic partial differential equation of Fokker-Planck type (Gail et al. (1974), Gail and Sedlmayr (1974):

$$\frac{\partial q}{\partial s} = \frac{1}{l}\left(-v \frac{\partial q}{\partial v} + \sigma^2 \frac{\partial^2 q}{\partial v^2}\right) - (\kappa_o \phi + \kappa_c)\ (q - S), \tag{1}$$

with s being the distance along the ray, v the actual velocity, $\kappa_o\phi$ and κ_c the line and continuum absorption coefficients. $\phi(s,v,\sigma,\Delta\lambda)$ is the profile function and S the monochromatic local source function. κ_o,κ_c and S are assumed to be independent of the velocity v.

The expectation value for the monochromatic local intensity $\langle I(s)\rangle$ is obtained by multiplication of q by the one-point velocity distribution $\mathbb{P}_1(v)$ and integration with respect to the velocity variable:

$$\langle I(s)\rangle = \int_{-\infty}^{+\infty} dv\ \mathbb{P}_1(v) q(s,v). \tag{2}$$

Given a model atmosphere which provides κ_o, κ_c and S, and given appropriate initial and boundary conditions the above transfer equation can be solved numerically straight-forward by the Crank-Nicholson method (e.g. Richtmyer and Morton,(1967)).

Without going into any technical details I shall point to the only problem which may arise and which results from an inconsistent incorporation of the boundary conditions: The natural boundary conditions for our problem are given by the asymptotic limits $v \to \pm\infty$, where the line absorption vanishes and the radiative transfer is controlled by continuous absorption only. It turns out that in the case of strong lines or large turbulent velocities one has to take into account rather large $|v|$ to approach these limits with sufficient accuracy and to avoid slight instabilities near the boundaries.

b) The Kubo-Anderson Process (KAP)

The KAP is a stationary Markovian process discontinuous in space with a Gaussian one-point velocity distribution function (like UOP) but a non-Gaussian two-point velocity distribution function which bridges the gap between the two limiting cases of complete correlation and complete noncorrelation by a linear ansatz (Auvergne et al. (1973), Frisch and Frisch (1975); see also Traving's contribution).

Using the KAP one arrives at an integro-differential equation for the monochromatic local conditional intensity $q(s,v)$:

$$\frac{\partial q(s,v)}{\partial s} = \frac{1}{l}\left\{\int_{-\infty}^{+\infty} dv' \mathbb{P}_1(v') q(s,v') - q(s,v)\right\} - (\kappa_o \phi + \kappa_c)(q(s,v) - S). \tag{3}$$

From this equation the expectation value of the emergent intensity can be calculated

(i) by means of the semianalytical methods used by Auvergne et al. (1973), and Frisch (1975) or

(ii) by a direct numerical solution of (3) and a subsequent integration according to (2). A proper discretisation in the velocity coordinate reduces this equation to a system of coupled linear ordinary differential equations which can be solved by standard methods (Gail et al. (1976).

However, both procedures have inherent specific difficulties either connected with instabilities in the region of very small l (microturbulent limit) or problems of large matrix size respectively.

Thus for numerical reasons we consider the straigthforward solution of the Fokker-Planck equation described in a) to be more efficient.

2. Effect of a Correlated Velocity Field on Line Profiles and Equivalent Widths

Adopting the UOP line profiles, equivalent widths and curves of growth have been calculated for selected lines of particular elements formed in the solar photosphere. In order to account for the effects of model atmospheres different from the sun we have extended these computations to atmospheres of earlier and later spectral type (AOV (Wega) and K2III (Arcturus)) also.

All calculated profiles show a similar monotonic dependence on l and σ respectively. Thus for demonstration of principle effects we may restrict ourselves to the discussion of an arbitrary line in the solar photosphere.

For a given mean square turbulent velocity σ and a non-negative correlation function microturbulence always yields the maximum absorption for a spectral line. Hence line profiles which originate in a velocity field determined by a small scale length are always deeper than the corresponding profiles formed in a velocity field with equal turbulent velocity but larger correlation length.

This is clearly seen in Fig. 1a, where for the line FeI λ6200 theoretical profiles are plotted for σ = 2 km/s and different values for the correlation length l. l = 1 km and l = 10^4km refer to the microturbulent and macroturbulent limit respectively.

If the correlation of the velocity field is increased for fixed σ both the line depth and the width of the corresponding profiles decrease monotonically until for very large l ($\gtrsim$ 3000 km) saturation is achieved (macroturbulent limit).

This general behaviour is also demonstrated in Fig. 2 which for different values of σ - but for the same line - shows the effect of the correlation length l on the central depth r_o and the equivalent width W_λ of the emergent profile.

For $\sigma \lesssim$ 1 km/s thermal broadening exerts the essential influence on the absorption coefficient. Hence for small mean square turbulent velocities both the central depths and equivalent widths depend only slightly on l. With increasing σ the thermal contribution becomes more and more negligible and the influence of the correlation of the turbulence field on r_o and W_λ is considerably increased.

For large σ both r_o and W_λ become monotonically strongly decreasing functions with increasing l approaching their minimum value in the macroturbulent limit. In this case the equivalent widths become independent on σ; a fact which causes all W_λ - curves in Fig. 2 to converge asymptotically.

The theory of mesoturbulent line formation provides a simple criterion which allows a quantitative estimate of a profile's deviation from the corresponding microturbulent result (see Traving's contribution). According to this criterion microturbulence should be a good approximation if

Fig. 1: All calculations have been performed on the basis of the empirical solar model atmosphere of Holweger (1967) assuming a mean square turbulent velocity σ = 2 km/s and a normal iron abundance.

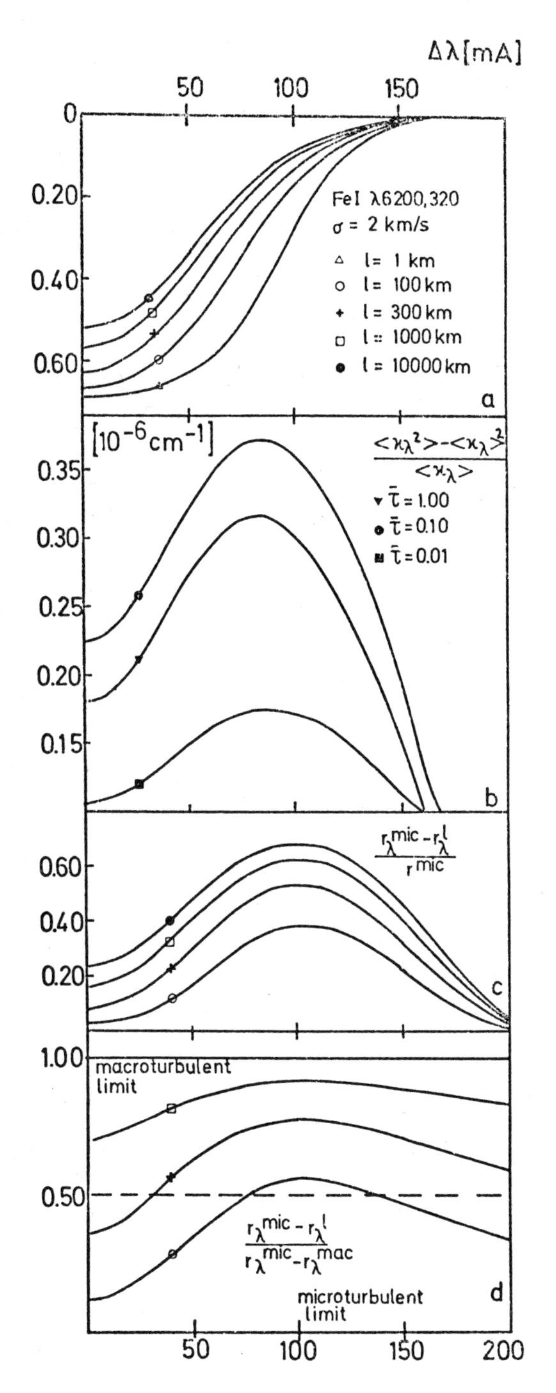

a) Theoretical profiles of the line FeI λ6200 for different values of the correlation length l. l $\lesssim$ 1 km and l $\gtrsim$ 3000 km correspond to the micro- and macroturbulent limit respectively.

b) Mean square root of the line absorption coefficient $\kappa_\lambda = \kappa_o(s)\phi(s,v,\sigma,\Delta\lambda)$ versus λ at different optical depths (Rosseland depth scale).

c) Deviation of a profile (calculated with a finite correlation length) from the microturbulent profile normated to the microturbulent value.

d) Deviation of a profile (calculated with a finite correlation length) from the microturbulent profile normated to the difference between the micro- and macroturbulent profile. This quantity indicates quantitatively which part of a profile is mainly determined by micro-turbulent or macroturbulent conditions respectively.

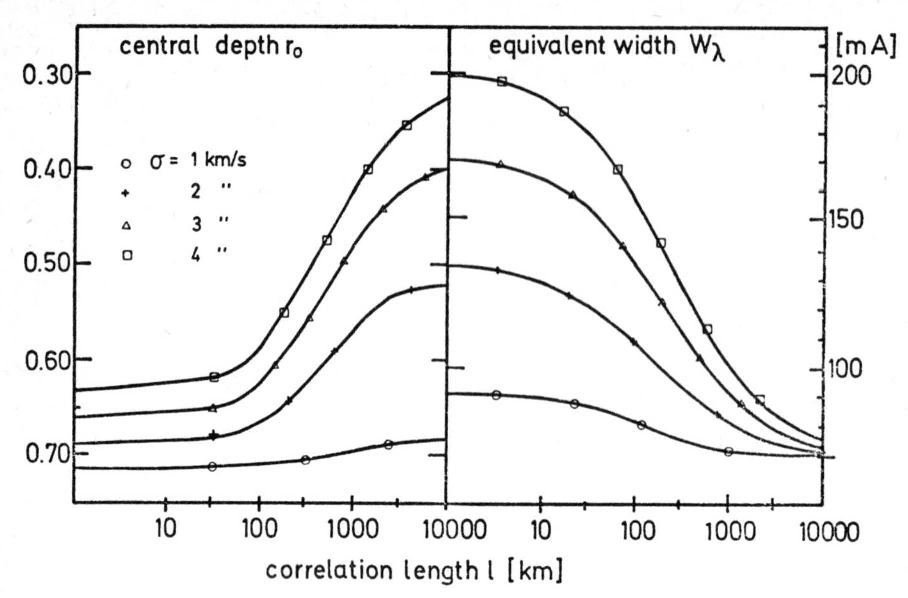

Fig. 2: Dependence of the central depth r_o and the equivalent width W_λ of the line FeI λ62oo on the correlation length l for different values of the mean square turbulent velocity σ.

the condition

$$\frac{1}{2}\,\tau_1 \cdot \tau_s \frac{\langle\kappa^2\rangle - \langle\kappa\rangle^2}{\langle\kappa\rangle^2} \ll 1 \tag{4}$$

holds. $\langle\ldots\rangle$ indicates the average values calculated by means of the one-point velocity distribution function, and $\tau_1 = 1\langle\kappa\rangle$ and $\tau_s = s\langle\kappa\rangle$ are characteristic optical depths corresponding to the correlation length and the relevant geometrical depth of line formation respectively. Thus by definition τ_s is usually of the order of unity. With $\tau_s = 1$ and $\tau_1 = 1\langle\kappa\rangle$ the proper wavelength dependence of (4) is essentially given by the quantity $(\langle\kappa^2\rangle - \langle\kappa\rangle^2)/\langle\kappa\rangle$, which is plotted in Fig. 1b for relevant optical depths $\bar{\tau}$.

From Fig. 1b we expect the largest deviations for a line profile calculated with fixed values 1,σ from the corresponding microturbulent result to occur at the transition from the core to the wing of the line where the curves $(\langle\kappa^2\rangle - \langle\kappa\rangle^2)/\langle\kappa\rangle$ show a pronounced maximum.

However, condition (4) is primarily a mathematical condition which for each wavelength measures the magnitude of the difference between a profile calculated with finite correlation length and the same profile calculated under the assumption of microturbulence. Obviously Fig. 1b shows that for the line centre and the outer wings the deviations from the microturbulent profile are considerably smaller than for the transition region even for large l. This is confirmed by the results in Fig. 1c where for different values of l the normated deviations from the microturbulent profile are plotted versus λ. However, this does not indicate that for large l these parts of the line do form under microturbulent conditions, but only that for these parts of the profile the microturbulent result is a reliable approximation for the true profile.

In order to decide whether micro- or macroturbulent conditions prevail we have plotted in Fig. 1d the quantity

$$d_\lambda^{l,\sigma} = (r_\lambda^{mic,\sigma} - r_\lambda^{l,\sigma})/(r_\lambda^{mic,\sigma} - r_\lambda^{mac,\sigma}) \qquad (5)$$

for σ = 2 km/s and several values of l. Essentially two regions can be distinguished:

1) $o \lesssim d_\lambda^{l,\sigma} \lesssim \frac{1}{2}$: Microturbulence governs the radiative transfer. For $l \simeq$ 1oo...5oo km this case applies only to the line cores and the outer wings. We infer that for smaller l the entire line is formed under microturbulent conditions.

2) $\frac{1}{2} \lesssim d_\lambda^{l,\sigma} \lesssim 1$: Macroturbulence governs the radiative transfer. This effect is most pronounced at those parts of the profile where the curves of Fig. 1b have their maximum. For correlation lengths $l \gtrsim$ 1ooo km the entire line is formed under macroturbulent conditions.

Throughout this discussion the UOP has been adopted. Using the KAP would yield quantitatively very similar results. However, due to the underlying cell structure of this model, KAP results always show a higher degree of correlation than the corresponding UOP results obtained for the same line with an identical correlation length and mean square turbulent velocity, (Frisch (1975), Gail et al. (1976)).

3. Empirical Relations between (v_{mic}, v_{mac}) and (l,σ)

In this section we want to present preliminary results obtained by a cooperation with H. Holweger (Kiel).

There are essentially two reasons which provide the motivation for the study of the relations between the classical micro-macroturbulence parameters (v_{mic}, v_{mac}) and the fundamental parameters of the mesoturbulence description (l,σ).

1) In classical theory v_{mic} and v_{mac} have been introduced as free parameters to fit observed and calculated profiles and curves of growth. If it is possible to relate these quantities in a simple way to l and σ the corresponding parameters of the turbulent velocity field, we arrive at a physically more justified interpretation of v_{mic} and v_{mac}.

2) In the past analyses of stellar atmospheres on the basis of classical micro- macroturbulence theory have been performed providing empirical values for v_{mic} and v_{mac}. By means of relations between these quantities and l and σ one is in the position to determine the mean turbulent velocity and the correlation length of the turbulence field for such atmospheres without reanalizing them by means of the more complex mesoturbulence formalism.

In order to find out the relations between these two sets of parameters the following procedure has been used:

For a given model atmosphere and artificial data for FeI- and FeII-lines curve of growth have been computed for a l- σ- grid. On the other hand the conventional microturbulence approach has been used to calculate an independent set of curves of growth for the same transitions. By varying the parameter v_{mic} an optimum fit between these two sets of data has been found using as criterion the minimum value of

$$\phi = \sum_{\text{lines}} \left[(W_\lambda^{mic} - W_\lambda^{l,\sigma})/W_\lambda^{l,\sigma}\right]^2 \quad . \tag{6}$$

Having thus determined a relation between v_{mic} and l,σ we use the calculated profiles $r_{\Delta\lambda}^{mic,mac}$ and $r_{\Delta\lambda}^{l,\sigma}$ to derive an optimum macroturbulent velocity v_{mac} by minimizing

$$\psi = \sum_{\text{lines}} \sum_{\Delta\lambda_j} (r_{\Delta\lambda_j}^{mic,mac} - r_{\Delta\lambda_j}^{l,\sigma})^2 \quad . \tag{7}$$

In this way for each pair (l,σ) a corresponding pair (v_{mic}, v_{mac}) has been determined empirically.

A typical result for the sun is shown in Fig. 3 where for three different artificial iron lines the optimum values for v_{mic} and v_{mac} are plotted versus l for a given mean square turbulent velocity σ. We see the expected strong correlation between the "decrease" of microturbulence and the corresponding "increase" of macroturbulence with growing correlation of the turbulence field.

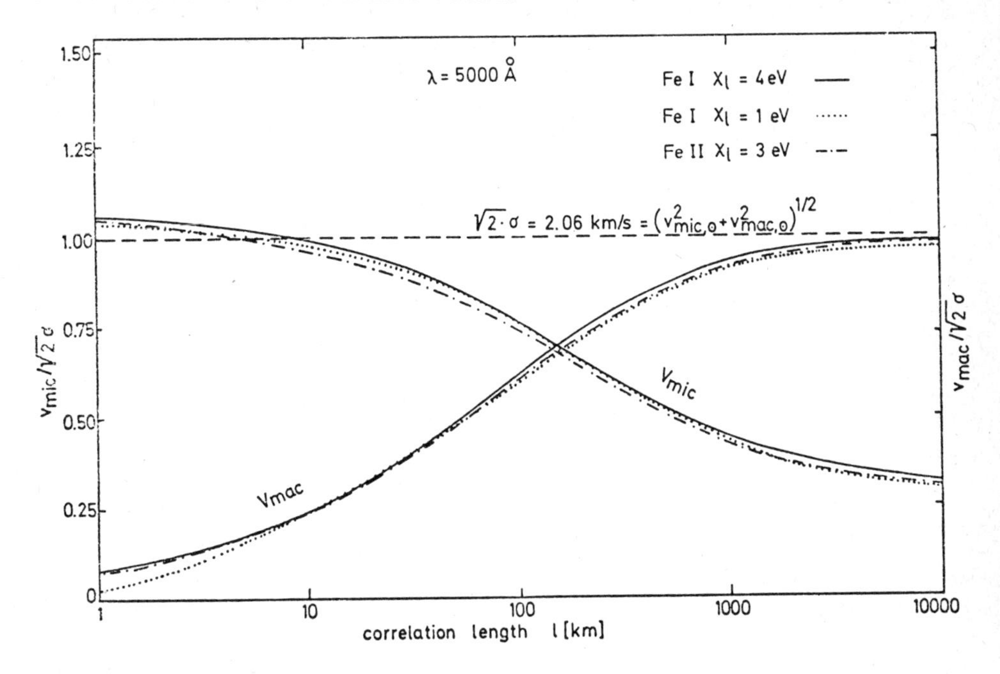

Fig. 3: Empirical dependence of the classical micro- and macroturbulent velocity on the correlation length l, derived on the basis of three artificial iron lines with wavelength λ = 5000 A and an excitation energy of the lower level χ_l. (A mean square turbulent velocity $\sqrt{2}\sigma$ = 2.06 km/s has been chosen, because this value - according to relation (8) - corresponds exactly to the empirical values $v_{mic,\odot}$ = 1 km/s and $v_{mac,\odot}$ = 1.8 km/s. (Empirical model photosphere of Holweger (1967)).

The curves of Fig. 3 resemble to some extent the filter functions introduced by de Jager (1972, 1979), de Jager and Vermue (1977) and Vermue and de Jager (1979) in order to describe the fraction of energy contained in the microturbulent and macroturbulent mode respectively. However, there are significant conceptual differences:

(i) In the description of de Jager and Vermue the adopted definition of microturbulence is based solely on the investigation of weak lines, whereas in our approach the microturbulence velocity is derived by a classical curve of growth analysis with no restriction concerning the line strength.

(ii) The filter function method is based on the consideration of one single mode u(k) of the Fourier spectrum of the turbulent field, whereas our statistical approach by means of probability distributions and correlation functions describes the mean values of the turbulent field to which all modes contribute.

The fact, that in the filter function presentation the transition from the micro- to the macroturbulent regime is much steeper than in our approach may be due to this difference.

In order to demonstrate the empirical relations between (v_{mic}, v_{mac}) and (l,σ) more clearly for an arbitrary line of the solar photosphere the quantity $(v_{mic}^2 + v_{mac}^2)^{1/2}$ is plotted in Fig. 4a versus l for a given σ. This plot provides strong evidence for a relation of the form

$$2\sigma^2 = v_{mic}^2 + v_{mac}^2 \tag{8}$$

which seems to hold with considerable accuracy at least within the region of relevant l values.

This relation, which has been confirmed by computing a large number of lines, also holds for different stellar atmospheres (AOV, K2III). In no case deviations from the relation (8) of more than few percents have been found.

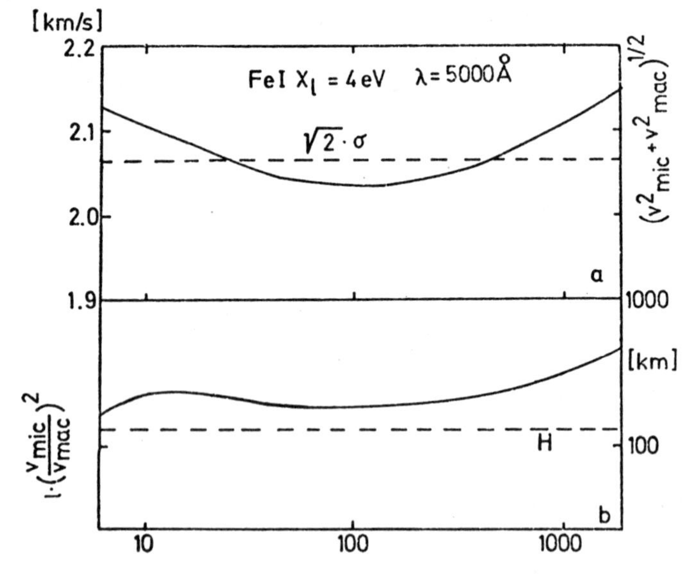

Fig. 4: The quantities $(v_{mic}^2 + v_{mac}^2)^{1/2}$ and $l(\frac{v_{mic}}{v_{mac}})^2$ versus λ for the same line as Fig. 3. H is the pressure scale height at the depth of line formation.

To relate the correlation length l to the velocity parameters an additional typical scale length H characterizing the atmosphere has to be introduced. This scale length turns out to be the pressure scale height taken at an optical depth which corresponds to the region of line formation. From Fig. 4b, where the quantity $l(\frac{v_{mic}}{v_{mac}})^2$ is plotted versus l, we conclude that for l values which can be determined reliably a satisfactory fit is obtained by the relation

$$\frac{H}{l} \approx (\frac{v_{mic}}{v_{mac}})^2 . \tag{9}$$

Relation (9), like (8), has been confirmed for the considered atmospheres by calculating a large number of lines. In no case a deviation of more than 1o percent has been found.

Such empirical relations should reflect similar relations connecting the basic physics of these approaches. In order to give an at least heuristic explanation for these basic relations we require that the corresponding one-point and two-point velocity distribution functions describing classical micro-macroturbulence and mesoturbulence respectively are identical:

$$\mathbb{P}_1^{mic,mac}(v) \equiv \mathbb{P}_1^{l,\sigma}(v) \tag{1o}$$

and

$$\mathbb{P}_2^{mic,mac}(v_1,v_2) \equiv \mathbb{P}_2^{l,\sigma}(v_1,v_2,\rho) . \tag{11}$$

With

$$\mathbb{P}_1^{mic,mac}(v) = \left[\pi(v_{mic}^2 + v_{mac}^2)\right]^{-1/2} \exp\left[-\frac{v^2}{v_{mic}^2 + v_{mac}^2}\right] \tag{12}$$

and

$$\mathbb{P}_1^{l,\sigma}(v) = \left[\pi 2\sigma\right]^{-1/2} \exp\left[-\frac{v^2}{2\sigma^2}\right] \tag{13}$$

one immediately obtains relation

$$(8) \quad 2\sigma^2 = v_{mic}^2 + v_{mac}^2 .$$

From

$$\mathbb{P}_2^{mic,mac}(v_1,v_2) = \left[\pi^2 v_{mic}^2 (v_{mic}^2 + 2v_{mac}^2)\right]^{-1/2} \cdot \tag{14}$$

$$\exp\left[-\frac{(v_{mic}^2 + v_{mac}^2)(v_1^2 + v_2^2) - 2v_{mac}^2 v_1 v_2}{v_{mic}^2(v_{mic}^2 + 2v_{mac}^2)}\right]$$

and

$$\mathbb{P}_2^{1,\sigma}(v_1,v_2,\rho) = \left[\pi 2\sigma^2(1-\rho^2)^{1/2}\right]^{-1} \exp\left[-\frac{v_1^2 + v_2^2 - 2\rho v_1 v_2}{2\sigma^2(1-\rho^2)}\right] \tag{15}$$

with

$$\rho(s) = \exp(-\frac{s}{l}) \cdot \tag{16}$$

One sees that eq. (11) can hold only for one particular value of the correlation function $\rho(s)$.

Using eq. (8) one obtains for this particular value

$$\rho(s) = \frac{v_{mac}^2}{v_{mic}^2 + v_{mac}^2} \tag{17}$$

which for $s \ll l$ immediately yields the approximate relation

$$\frac{s}{l} \simeq \left(\frac{v_{mic}}{v_{mac}}\right)^2 \cdot \tag{18}$$

So at least the analytical form of the empirical result (9) is recovered. As s is a typical geometrical dimension of the line forming region, it should be closely connected with the scale height H introduced in (9).

In order to justify the assumption $s \ll l$ we plot in Fig. 3 two horizontal lines indicating the empirically determined values for $v_{mic,\odot}$ and $v_{mac,\odot}$ respectively (Holweger (1967)). This plot is shown in Fig. 3a. The abscissas of the intersection points of each line with the corresponding v_{mic}- or v_{mac}- curves determine the region of best fitting l values. We see that the optimum l values are confined to a rather narrow interval of 53o km $\lesssim$ l $\lesssim$ 64o km. These l values are considerably larger than the scale height H shown in Fig. 4b, which justifies our above assumption.

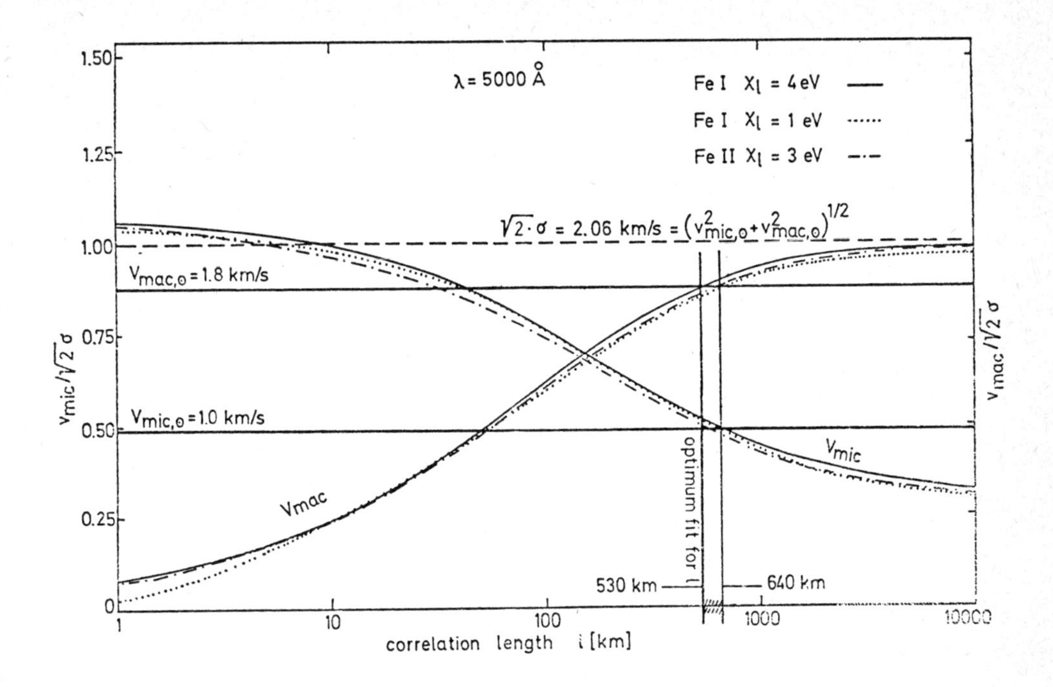

Fig. 3a: See Fig. 3. The horizontal lines indicate the empirically determined photospheric values for $v_{mic,\odot}$ and $v_{mac,\odot}$ (Holweger (1967)). The vertical lines determine the interval of best fitting l values (strongly crosshatched on the abcissa).

Our result for the optimum correlation of $l \simeq 600$ km for the solar photosphere is in close agreement with the average value $\frac{2\pi}{k} \simeq 600$ km for the inverse wavenumber of the velocity spectrum derived by de Jager and Vermue (1977) by means of the filter function method. The fact that these very different methods yield approximately the same results provides confidence for this value of the scale length.

Conclusion

Our analysis of the radiative transfer of selected lines in three particular atmospheres (AOV (Wega), G2V (Sun), K2III (Arcturus)) demonstrates that a finite correlation length of the turbulent velocity field exerts a strong influence on the line formation. The general dependence of this effect on the basic parameters (l,σ) is similar in all atmospheres and for all profiles considered.

A comparison between profiles and curves of growth computed by means of the mesoturbulence formalism at one hand and by the classical micro-macroturbulence method at the other, provides strong evidence for very close relations between (l,σ) and (v_{mic},v_{mac}). These relations allow a reinterpretation of the classical micro-macroturbulence dichotomy which is closer to physics.

Acknowledgements. This work has been performed as part of the program of the Sonderforschungsbereich 132 "Theoretische und praktische Stellarastronomie", which is sponsored by the Deutsche Forschungsgemeinschaft.

References

Auvergne, M.; Frisch, H.; Frisch, U.; Froeschlé, Ch.; Pouquet, A.;1973, Astron. & Astrophys. 29, 93

Brissaud, A.; Frisch, U.; 1974, J. Math. Phys., 15, 524

De Jager, C.; 1972, Solar Phys. 25, 71

De Jager, C.; 1978, Astrophys. Space Sci. 59, 165

De Jager, C.; Vermue, J.; 1977, Solar Phys. 54, 313

Frisch, U.; 1968, in Probabilistic Methods in Applied Mathematics, edited by A.T. Bharucha-Reid, Academic Press, New York and London

Frisch, H.; Frisch, U., 1976 in Physique des mouvements dans les atmosphères stellaires, Colloques internationaux du centre national de la recherche scientifique, Paris p.113

Frisch, H.; Frisch, U., 1976, Monthly Notices Roy.Astron. Soc. 175, 157

Gail, H.-P.; Hundt, F.; Kegel, W.H.; Schmid-Burgk, J.; Traving, G., 1974, Astron. & Astrophys. 32, 65

Gail, H.-P.; Sedlmayr, E., 1974, Astron. & Astrophys. 36, 17

Gail, H.-P.; Kegel, W.H.; Sedlmayr, E., 1975, Astron. & Astrophys.42,81

Gail, H.-P.; Sedlmayr, E.; Traving, G., 1975, Astron. & Astrophys.44,421

Gail, H.-P.; Sedlmayr, E.; Traving, G., 1976, Astron. & Astrophys.46,441

Gail, H.-P.; Sedlmayr, E.; Traving, G., 1979, JQSRT, in press

Gray, D.F., 1977, Astrophys. J. 218, 53o

Gray, D.F., 1978, Solar Phys. 59, 193

Hearn, A.G., 1974, Astron. & Astrophys. 31, 415

Holweger, H., 1967, Z. Astrophys. 65, 365

Hundt, E., 1973, Astron. & Astrophys. 29, 17

Magnan, C., 1976, JQSRT 16, 281

Magnan, C., 1976, in Physique des mouvements dans les atmosphères stellaires, Colloque internationaux du CNRS, Paris p. 179

Rybicki, G., 1976, in Physique des mouvements dans les atmosphères stellaires, Colloque internationaux du CNRS, Paris p. 189

Schmid-Burgk, J., 1974, Astron. & Astrophys. 32, 73

Sedlmayr, E., 1976, in Physique des mouvements dans les atmosphères stellaires, Colloque internationaux du CNRS, Paris p. 157

Struve, O.; Elvey, C.T., 1934, Astrophys. J. 79, 4o9

Traving, G., 1964, Z. Astrophys. 6o, 167

Traving, G., 1975, in Problems in Stellar Atmospheres and Envelopes eds. B. Baschek, W.H. Kegel and G. Traving, Springer-Verlag, Berlin-Heidelberg-New York p. 325

Traving, G., 1976, in Physique des mouvements dans les atmosphères stellaires, Colloques internationaux du CNRS, Paris p. 145

Vermue, J.; de Jager, C., 1979, Astrophys. Space Sci. 61, 129

Wang, M.CH.; Uhlenbeck, G.E., 1945, Rev. Mod. Phys. 17, 323

EFFECTS OF ACOUSTIC WAVES ON SPECTRAL LINE PROFILES

Lawrence E. Cram
Sacramento Peak Observatory*
Sunspot, NM 88349, U.S.A.

Abstract

Almost all studies of spectral line formation in the presence of non-thermal velocity fields have been made assuming that the only effect of the velocity field is to produce a Doppler shift of the absorption and emission coefficients. However, a non-thermal velocity field will entail velocity-correlated fluctuations in temperature, pressure, level populations, and other parameters of the line formation problem. Using a time-dependent dynamical calculation describing the propagation of non-linear, radiatively-damped short period (P = 30s) acoustic waves in the solar photosphere, Cram, Keil and Ulmschneider (1980) have shown that velocity-correlated fluctuations in state variables (particularly the temperature) may lead to important effects in line broadening, line shifts and asymmetries, and in line-shift oscillations. Upwardly propagating waves generally produce significant redshifts in the cores of medium-strong Fe I lines, and the increased ratio of observed line shift to wave velocity amplitude would significantly modify the results of kinematic studies of high frequency line shifts such as those of Deubner (1976) and Keil (1980).

Cram (1980) has further explored dynamical effects in the formation of Fe I and Fe II lines by using the "microturbulence" limit, wherein an average is made over the phase of the wave before the transfer equation is solved. Except for weak, high EP Fe II lines, the predicted solar lines are redshifted and show a "red" asymmetry. For a model of Arcturus the lines are often shifted to the blue, but it does not appear that this model can account for the observed differences between solar and Arcturan line asymmetries (Gray 1980).

References

Cram, L.E., 1980 (in preparation).
Cram, L.E., Keil, S.L., Ulmschneider, P., 1980, Ap.J. (in press).
Deubner, F.-L., 1976, Astr. Ap. 51, 189.
Gray, D., 1980, Ap.J. (in press).
Keil, S.L., 1980, Ap.J. (submitted).

* Operated by the Association of Universities for Research in Astronomy, Inc. under contract AST-78-17292 with the National Science Foundation.

SOME EFFECTS OF STRONG ACOUSTIC WAVES ON STRONG SPECTRAL LINES

Pierre Gouttebroze
Laboratoire de Physique Stellaire et Planetaire
Verriéres-le-Buisson, France
John Leibacher
Lockheed Palo Alto Research Laboratory
Palo Alto, CA 94304, U.S.A.

Abstract

We have studied the formation of optically thick lines in time dependent, non-linear hydrodynamic model of the solar chromosphere. Models of the 200 second, chromospheric oscillation indicate that the emission peaks of self-reversed profiles such as those of Mg II and Ca II are formed at very different depths depending on the phase of the oscillation, while the central absorption feature is emitted at a very nearly constant mass depth. The figure shows the Mg II k line emitted by the mean atmosphere, including "microturbulence" (triangles) and the mean profile (squares). In addition to the substantial intensity increase interior to the emission peaks, one should note that the peaks are broadened only towards line center; i.e. the intensity fluctuations are symmetric outside of the peaks and strengthen more than they weaken within the peaks. A more detailed version has been submitted to the Astrophysical Journal.

J.L. wishes to acknowledge support by NASA contracts NASA-3053 and 5-23758.

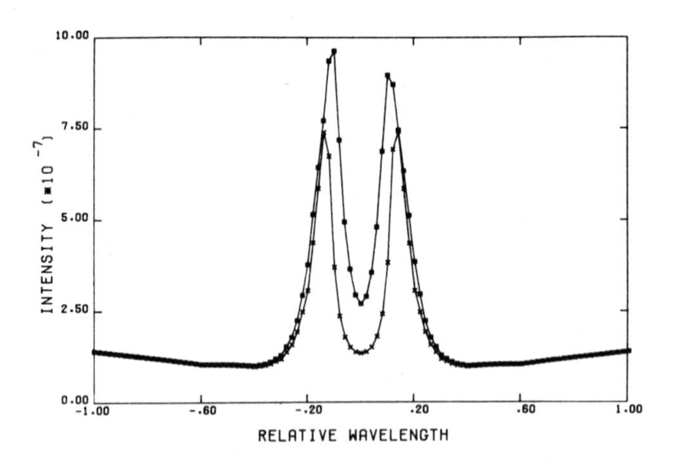

NUMERICAL SIMULATION OF GRANULAR CONVECTION: EFFECTS ON PHOTOSPHERIC SPECTRAL LINE PROFILES

Åke Nordlund
Nordita, Blegdamsvej 17
Dk-2100, København Ø, Denmark

ABSTRACT

The results of numerical simulations of the solar granulation are used to investigate the effects on photospheric spectral lines of the correlated velocity and temperature fluctuations of the convective granular motions. It is verified that the granular velocity field is the main cause for the observed broadening and strengthening of photospheric spectral lines relative to values expected from pure thermal and pressure broadening. These effects are normally referred to as being due to "macro-turbulence" and "micro-turbulence", respectively. It is also shown that the correlated temperature and velocity fluctuations produce a "convective blue shift" in agreement with the observed blue shift of photospheric spectral lines. Reasons are given for the characteristic shapes of spectral line bisectors, and the dependence of these shapes on line strength, excitation potential, and center to limb distance are discussed.

1. INTRODUCTION

Stellar as well as solar spectral lines are observed to be broader than expected from purely thermal and pressure broadening. A weak Fe line at $\lambda \simeq 500$ nm, observed at solar center disc, has a full width at half maximum (fwhm) of $\simeq 7$ pm or, expressed in velocity units $c\Delta\lambda/\lambda$, $\simeq 4.2$ kms^{-1}, whereas the value expected from pure thermal broadening is $2(\ln 2)^{\frac{1}{2}}(2kT/m)^{\frac{1}{2}} \simeq 2.0$ kms^{-1}. Close to the limb, at an angular inclination cosine, $\mu \equiv \cos\theta = 0.16$, weak Fe lines have a width of $\simeq 6$ kms^{-1}. Classically, this broadening is attributed to a "large-scale" velocity field, with assumed Gaussian distribution, $\exp(-v^2/v^2_{macro})$, of line of sight velocity. Typical required values of the "macro-turbulence" parameter are: $v_{macro} = (v^2_{fwhm}/(4\ln 2)-2kT/m)^{\frac{1}{2}} \simeq 2.2$ kms^{-1} at $\mu = 1$, $\simeq 3.4$ kms^{-1} at $\mu = 0.16$.

It is also well known that spectral lines are stronger than expected from purely thermal and pressure broadening of the line absorption profile. The concept of "micro-turbulence" was introduced to enforce a fit to the observed line strengths (Struve & Elvey 1934). Recently, accurate measurements of FeI oscillator strengths (Blackwell et al. 1975, 1976a, 1979a, 1979b) have made possible careful comparisons of observed and expected line strengths for a number of solar FeI lines, at several values of μ. If interpreted in terms of a "micro-turbulence" parameter, typical values required range from $v_{micro} = 0.6 - 0.9$ kms^{-1} at disc center, to 1.4 - 1.7 kms^{-1} close to the limb (Blackwell et al. 1976b, 1979c).

Taken together, the broadening and strengthening data show that the velocity field responsible for the broadening cannot be a small scale velocity field; if so, it would give rise to more strengthening than is actually observed. As discussed previously (Nordlund 1976b, 1978), the convective granular velocity field is the likely candidate for this 2 á 3 kms^{-1} velocity field, being on a bit too small scale to be fully spatially resolved observationally, yet on a large enough scale to show up mainly as broadening and only to a lesser extent as strengthening.

In a previous paper (Nordlund 1978) it was shown that the velocity field obtained from the equations of motion using an assumed, time-independent driving force, could be made to match both the broadening and strengthening of photospheric spectral lines if an appropriate horizontal scale and amplitude were chosen. The necessary size agrees well with typical granular sizes, and the necessary amplitude of the driving force is consistent with temperature fluctuations obtained in a simple, two-component model of granular convection (Nordlund 1976a).

Recently (Nordlund 1979) it has been possible to numerically solve the full set of hydrodynamical equations describing granular convection. In this treatment, the driving force is determined consistently from the energy equation, which governs the time-development of the temperature. Motions are allowed on a range of scales. The result is a realistic simulation of the granular convection. Results of these simulations are used in this paper to verify that the granular convection is the main cause of the broadening and strengthening of photospheric spectral lines (Section 3). The physical and numerical limitations of the simulations relevant to the spectral line formation problem are discussed in Section 2. Spectral line bisectors are discussed in Section 4 as a diagnostic tool to analyse the finer details of the granular convection. Section 5 summarizes the discussion.

2. THE NUMERICAL SIMULATION OF THE GRANULAR CONVECTION

A description of the numerical simulations is given elsewhere (Nordlund 1979). However, some details relevant to the synthetic spectral line calculations should be mentioned here:

The heavy computer storage and time requirements of three-dimensional hydrodynamic calculations naturally enforce strong numerical restrictions on the simulations. Allowing for 16×16 Fourier components in 16 layers, a compromise has to be made between spatial extent and resolution in the model. To cover the observed range of granular sizes (cf. Bray and Loughhead 1967, Fig. 2.1), the horizontal period was chosen = 3600 km. A vertical grid spacing of 100 km, with a vertical extent from z = 1100 km (depth relative to optical depth unity at λ = 500 nm) up to z = -400 km is a compromise between the vertical resolution rquired by typical scale heights ≃ 150 km in the photosphere; a large enough depth for the stratification to be almost adiabatic at the lower boundary, with small temperature fluctuations; and an upper boundary close to the top of the photosphere. This compromise was aimed primarily at a simulation of continuum brightness fluctuations for a computer movie. For the purpose of spectral line calculations, the upper boundary should preferredly have been placed at a somewhat higher level (z ≃ -600 km) to avoid influences from the necessarily imperfect boundary conditions.

Another important limitation concerns the energy balance in the line formation layers. The radiative part of the energy equation is necessarily handled with only a few wavelength points (typically three). However, the radiative part of the energy balance in optically thin regions is dominated by transfer in thousands of spectral lines. Energy is most efficiently exchanged between the gas and the radiation field at wavelengths where the optical depth is close to unity. This leads to a substantial cooling of the upper photosphere, relative to radiative equilibrium without spectral lines (spectral line "blanketing", cf. discussion in Gustafsson et al. 1975, Gustafsson 1979). Furthermore, radiative relaxation times in the line formation layers are typically underestimated by a factor of 5 - 10 when radiative transfer in the continuum alone is treated.

3. BROADENING AND STRENGTHENING OF PHOTOSPHERIC SPECTRAL LINES

For weak spectral lines, atoms see approximately the same radiation intensity regardless of local fluid + thermal velocity along the line of sight. Therefore, the strength of a weak spectral line averaged over horizontal area (and/or time) does not depend on the scale of the velocity field. The emergent average line profile simply reflects the distribution of line of sight velocities, convolved with the thermal velocity distribution. Thus, from the broadening data, one would estimate a ratio of horizontal to vertical velocities of approximately $3.4/2.2 \simeq 1.6$.

This velocity ratio is consistent with what would be expected of the velocity field of granules of typical dimensions. To see this, consider the condition of continuity applied to a simplified, one term representation of the granular velocity field,

$$\rho u_z = a(z)\cos(kx)\cos(ky) \quad . \tag{1}$$

(In the numerical simulation a Fourier sum is used, but for the present illustration this single term suffices). The "anelastic" approximation to the continuity equation,

$$\mathrm{div}(\rho \underline{u}) \simeq \mathrm{div}(\rho \underline{u}) + \partial\rho/\partial t = 0 \quad , \tag{2}$$

plus x-y symmetry, requires

$$\rho u_x = -a(z)/(2kH_a)\sin(kx)\cos(ky) \quad , \tag{3}$$

$$\rho u_y = -a(z)/(2kH_a)\cos(kx)\sin(ky) \quad , \tag{4}$$

where H_a is the scale height of the vertical mass flux amplitude $a(z)$, and $2^{\frac{1}{2}}\pi/k$ corresponds to the intergranular distance. Estimating $H_a \lesssim$ a density scale height $\simeq 140$ km, and $k \simeq 2^{\frac{1}{2}}\pi/1800\ \mathrm{km}^{-1}$ (cf. Bray & Loughhead 1967, Tab. 2.1), one obtains $u_x^{rms}/u_z^{rms} \gtrsim 1.4$, in good agreement with the broadening of weak spectral lines.

The results of the numerical simulations of the solar granulations (Nordlund 1979) have been used to calculate synthetic spectral lines as averages over the simulation sequence (2 hours solar time) and area (3600×3600 ≃ 5×5 arc sec). Local thermodynamic equilibrium (LTE) was assumed, and pressure broadening was calculated according to Unsöld (1955) (with no enhancement factor). The time averaging was performed as a sampling with a 6 minute sampling interval. The horizontal averaging was performed using bundles of 256 parallel rays, through all the grid points at z = 0.

An example of the granular velocity and temperature field is given in Fig. 1. In Fig. 2, observed full widths at half maxima of photospheric spectral lines are compared with widths of synthetic spectral lines. The good agreement in Fig. 2 shows that the granular velocity field (which is required to transport the bulk of the solar energy output to the surface) does have the average horizontal and vertical velocity amplitudes required to fit the broadening data.Note, however, that the broadening at center disc is slightly too small and that the broadening close to the limb is slightly too large. These discrepancies are probably due to the neglect of the spectral line blanketing in the numerical simulations. In the upper photosphere, temperature fluctuations are induced by convective motions overshooting into a stable region. These temperature fluctuations tend to retard the motion. If spectral line blanketing were correctly allowed for, the temperature fluctuations would be reduced in magnitude (cf. the discussion in Section 2), and the retardation would decrease. Vertical velocities would then decrease less rapidly with height, and the horizontal velocities would be reduced, as required by the condition of continuity (cf. discussion above, Eqs (1) - (4)). Attempts are presently being made to include spectral line blanketing in a schematic way with a very small number of wavelength points.

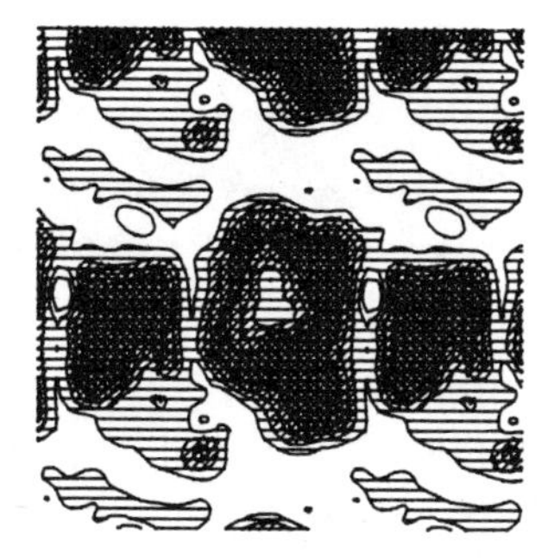

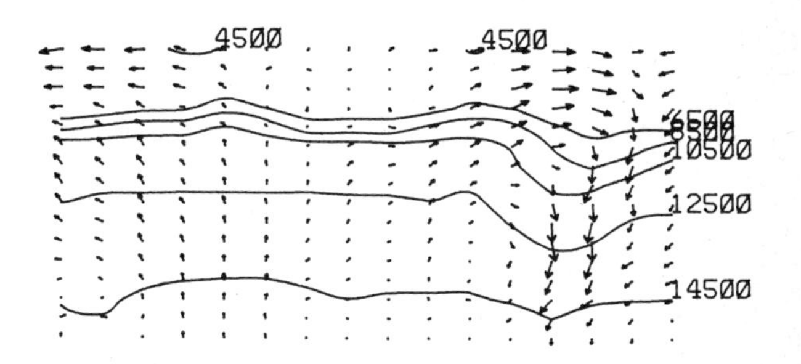

Fig. 1. The granular temperature field at z = 0 and the granular velocity field in an yz-plane through the center of the center granule, at a time when this granule has expanded to form an "exploding" granule. The temperature plot (left) is shaded above 6500 K, with the shading increasing in steps of 1000 K. The velocity (right) is shown over the 3600×1500 km vertical plane, with arrows showing the distance covered in 15 seconds. Temperature contours are labelled with the temperature in K.

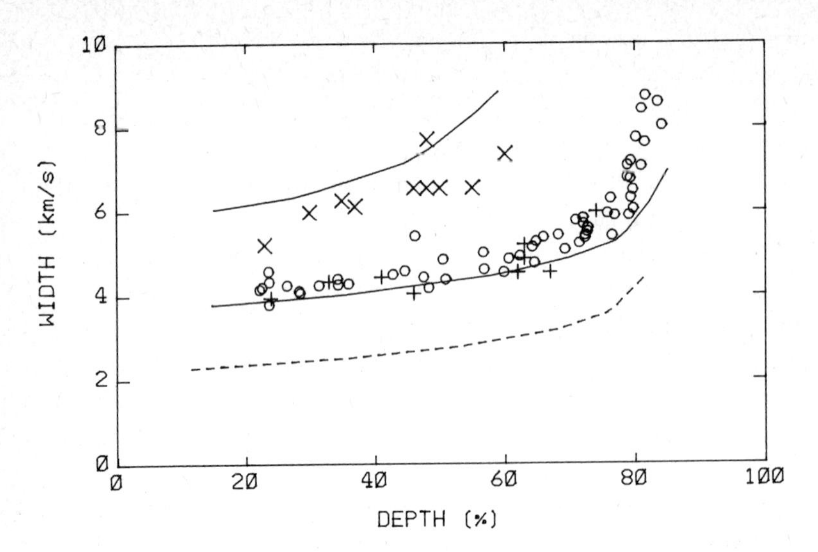

Fig. 2. Observed and calculated widths (fwhm) of solar FeI lines, as a function of line center depth, at $\mu = 1$ and $\mu = 0.16$. Observations are from Evans et al. (1975), for $\mu = 1$ (+) and for $\mu = 0.16$ (x); and from Stenflo & Lindegren (1977), for $\mu = 1$ (o). Synthetic spectral lines were calculated using $\lambda = 500$ nm and $\chi_{exc} = 3$ eV (which is typical for the observed lines). Full drawn lines show the results obtained (at $\mu = 1$ and 0.16) from an average (see text) of a 2 solar hour simulation of the solar granular convection. The dashed line shows widths expected at $\mu = 1$ (these are similar at $\mu = 0.16$ for small depths) with only thermal and pressure broadening.

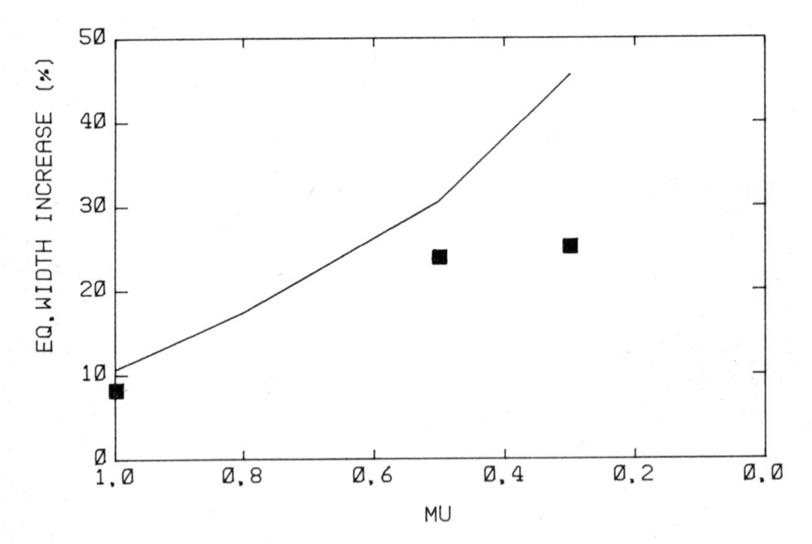

Fig. 3. The strengthening of a saturated FeI line, with $\lambda = 500$ nm, $\chi_{exc} = 1$ eV, and equivalent width 5 pm, as a function of μ. The full drawn line shows the equivalent width increase of the spectral line in the simulated granular velocity field, relative to equivalent width obtained with the velocity set to zero. The squares show the equivalent width increases corresponding to the "micro-turbulence" parameter values derived by Blackwell et al. (1979c) for similar lines.

In Fig. 3, the spectral line strengthening effects of the granular velocity field are illustrated. Traditionally, the observed and calculated equivalent widths of spectral lines are reconciled using a "micro-turbulence" parameter. Thus, Blackwell et al. (1976b, 1979c) interpret their accurate data on line strengthening in terms of an angle-dependent "micro-turbulence", $v_{micro}(\mu)$. In Fig. 3, the line strengthening due to the simulated granular velocity field is compared with that of the "micro-turbulence" derived by Blackwell et al. It is obvious that the simulated granular velocity field has a line strengthening effect similar to the one required by the observations. The effects of the simulated granular velocity field are too large towards the limb, just as was the case for the broadening data.

The line strengthening effect of the granular velocity field is due to gradients of the velocity along the line of sight. In the approximately exponential photosphere, the optical length scale

$$\ell_{\tau} = ds/d\ln\tau = (dz/d\ln\tau)/\mu = H_{\tau}/\mu \tag{5}$$

is approximately inversely proportional to μ. Thus, in terms of optical depth along the line of sight, velocity fluctuations appear to occur on a smaller scale for inclined lines of sight. In itself, the granular velocity field is not isotropic; as discussed above, the horizontal velocities are generally larger than the vertical ones (cf. also Fig. 1). Together, these circumstances contribute to the the increased line strengthening ("micro-turbulence") towards the limb. In fact, as with the line broadening, the line strengthening at small μ may be reproduced by a simple, stationary, one-mode model of the granular velocity field (Nordlund 1976b, 1978). However, at disc center, line strengthening occurs mainly as a result of the velocity gradients associated with the time dependence of the granular motions.

4. CONVECTIVE BLUE SHIFTS AND SPECTRAL LINE BISECTORS

Photospheric spectral line profiles show a net blue-shift because of the larger contributions to the emergent intensity of the bright granules, with their locally blue shifted spectral lines. Observationally, this blue shift is known to be of the order 300 - 400 ms^{-1} for weak FeI lines, decreasing with increasing line strength, and with a weak dependence on excitation potential (e.g. Beckers & de Vegvar, 1978). Moreover, the blue-shifted spectral line profiles are asymmetric (Adam et al., 1976). A concise and powerful way to present the blue-shift and asymmetry of spectral lines is to use the spectral line bisectors; i.e., the loci of points midway between equal intensity points on either side of the line profile (Adam et al. 1976, Dravins 1979, Dravins et al. 1979). The shape of the spectral line bisector reflects, in a complicated way, an average of the thermal and velocity fluctuations of the photosphere. With the wealth of spectral lines available, spanning a range of different elements, strength, excitation potential and wavelength, the ensemble of spectral line bisectors form - in a way - a "fingerprint" of the temperature and velocity fluctuations of the photosphere.

Fig. 4 shows bisectors of weak FeI lines, synthesized as above, using the results of the numerical simulations of the solar granulation. The order of magnitude of the shifts, the shapes of the bisectors, and the weak dependence on excitation potential are all consistent with the properties of observed photospheric spectral lines (Dravins 1979, Dravins et al. 1979). For stronger spectral lines, the agreement is less satisfactory. Again, this is probably a consequence of the numerical inadequacies of the upper photospheric region of the simulation model (cf. section 2).

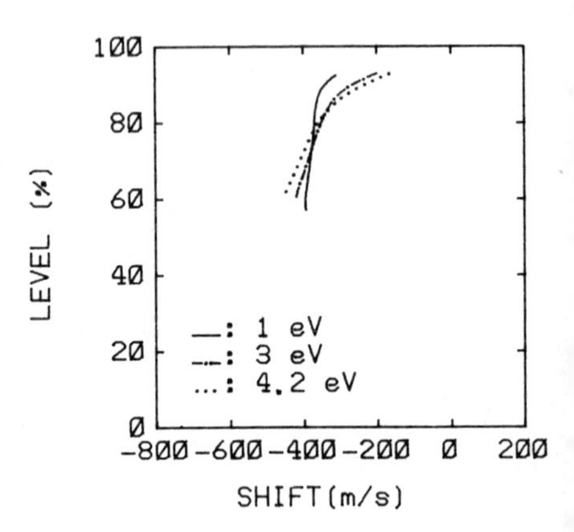

Fig. 4. Spectral line profile bisectors, for spectral lines with λ = 500 nm, line center depths ≃ 40 %, χ_{exc} = 1 eV (full drawn), 3 eV (dashed), and 4.2 eV (dotted). Synthesized line profiles as in Figs 2 and 3.

To investigate some of the mechanisms influencing the shape of the bisectors, some calculations were performed with simpler velocity and temperature models. One of the important qualitative properties of the granular velocity field is its distinct asymmetry with respect to up and down: Granules with upward velocities are separated by narrow intergranular lanes with relatively larger downward velocities. This influences the shape of the bisectors in a characteristic way which is illustrated in Fig. 5. Due to the large red-shifts in the intergranular lanes, as compared to relatively smaller blue-shifts in the granules, the red wing of the average line profile is strengthened relative to the blue wing. An alternative way to see this is to consider the idealised case of a very narrow (δ-function) local line profile. The average spectral line profile then has the shape of the distribution function for the vertical velocities. The granular/intergranular asymmetry mentioned above corresponds to a distribution function with an extended "red" tail. This is the major cause of the upper redward bend of the spectral line bisectors. Fig. 5 illustrates how the characteristic C-shape of the bisector vanishes when the granular velocity field is replaced by a velocity field of similar $v_{rms}(z)$, but with a purely sinusoidal horizontal variation (and thus symmetrical with respect to up and down). A depth-independent velocity amplitude, chosen consistent with the line broadening data results in approximately the same shift as with the depth-dependent velocity amplitude, but with a different slope of the bisector. A stronger penetration of the temperature fluctuations up into the photosphere results in a smaller shift of weak FeI lines. This is due to the strong decrease in the number of FeI atoms with increasing temperature, which results in a weaker (in terms of equivalent width) blue-shifted contribution.

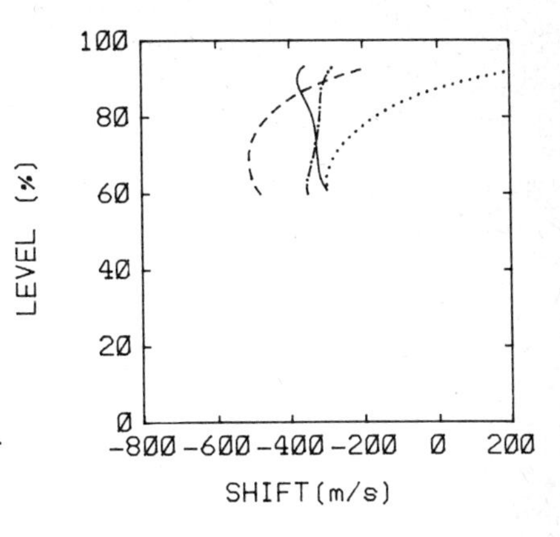

Fig. 5. Bisectors of spectral lines with λ = 500 nm, line depths $\simeq$ 40 %, and χ_{exc} = 3 eV. The dashed line shows the bisector obtained using a quasi-stationary granular velocity field (cf. Nordlund 1978, Fig. 6), and a parameterised temperature fluctuation, $\Delta T = \Delta T_o (z-z_o)^2/z_o^2 \cos(kx)\cos(ky)$ for $z > z_o$, $\Delta T = 0$ for $z < z_o$, ΔT_o = 1100 K, z_o = -100 km. The dashed-dotted line shows the bisector obtained with a velocity field with similar $v_{rms}(z)$, but with a purely sinusoidal horizontal variation, $v = -2v_{rms}(z)\cos(kx)\cos(ky)$. The full drawn line shows the bisector obtained with a depth-independent velocity amplitude = 2.5 kms^{-1} (v_{rms} = 1.25 kms^{-1}). The dotted line shows the bisector obtained with the granular velocity field, but with ΔT_o = 1500 K, z_o = -300 km.

Fig. 6 illustrates the excitation potential dependence of bisectors of strong spectral lines. The lower portions of these bisectors show a reversed excitation potential dependence; with higher excitation potential resulting in a smaller blue shift (in agreement with the lower portions of observed bisectors of strong lines). These parts of the bisectors are most influenced by the red flanks of the most blue-shifted (granular) contributions to the average line profile. The radiation intensities in these red flanks decrease strongly with excitation potential, both because of the Boltzman factor and because of the increased pressure broadening.

Fig. 7 shows the center to limb behaviour of the bisector of a spectral line selected to be similar to one of the three lines observed by Adam et al. (cf. their Fig. 8). The C-shape of the bisector at center disc disappears as one approaches the solar limb. As discussed in connection with

Fig. 5, the C-shape at $\mu = 1$ is caused by the up/down asymmetry of the line of sight velocity. For small μ, the line of sight is almost horizontal, and therefore the line of sight velocities are nearly symmetrical with respect to blue- and red-shift.

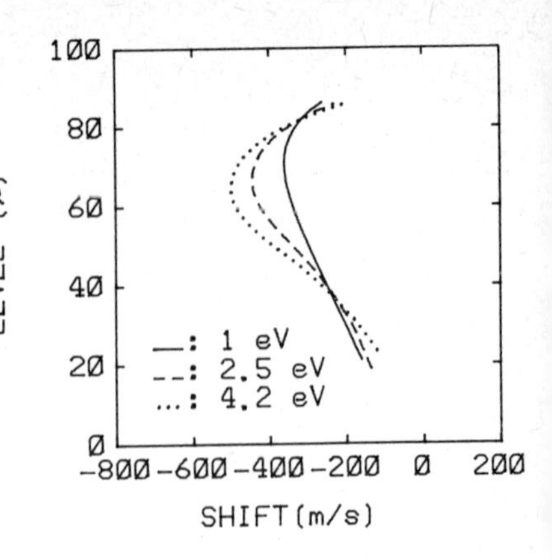

Fig. 6. Bisectors of strong spectral lines with λ = 500 nm and line center depths ≃ 80 %, for χ_{exc} = 1 eV (full drawn), 2.5 eV (dashed), and 4.2 eV (dotted). Synthesized with a granular velocity field and a parameterized temperature fluctuation, as in Fig. 5. ΔT_o = 1100 K, z_o = -100 km.

CONCLUSIONS

Numerical simulations of the solar granular convection show that the convective velocities in the solar photosphere are on the order of 1.5 kms^{-1} (typical rms vertical velocity) to 2.5 kms^{-1} (typical rms horizontal velocity). The broadening caused by this velocity field is consistent with the broadening of weak photospheric spectral lines. The velocities are also consistent with observed granular velocities corrected for limited spatial resolution (cf. the review by Wittman, 1979). Gradients of the granular velocity field are sufficient to explain the strengthening of photospheric spectral lines classically attributed to "micro-turbulence". The correlation of temperature and velocity in the granular convection causes a "convective blue--shift" and asymmetry of photospheric spectral lines. This is potentially, with sufficient accurate observations of stellar spectra, a powerful tool to investigate granular convection in stars other than the sun (cf. further discussion in Dravins et al., 1979).

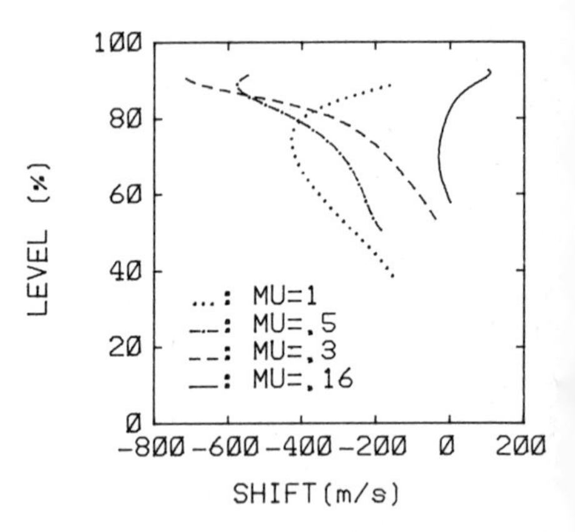

Fig. 7. Center to limb behaviour of the bisector of a spectral line with λ = 630 nm, line center depth ≃ 60 %, and χ_{exc} = 3.6 eV. Synthesized as in Figs 5 and 6. Compare Fig. 8 of Adam et al. (1976).

ACKNOWLEDGEMENTS

Support by the Danish Natural Science Research Council, the Danish Space Research Committee, and the Swedish Natural Science Research Council is gratefully acknowledged.

REFERENCES

Adam, M.G., Ibbetson, P.A., Petford, A.D.,1976, Monthly Notices Roy. Ast. Soc. 177, 687

Beckers, J.M., de Vegvar, P. 1978, Solar Phys. 58, 7.

Blackwell, D.E., Ibbetson, P.A., Petford, A.D., 1975, Monthly Notices Roy. Ast. Soc. 171, 195

Blackwell, D.E., Ibbetson, P.A., Petford, A.D., Willis, R.B., 1976a, Monthly Notices Roy. Ast. Soc. 177, 219

Blackwell, D.E., Ibbetson, P.A., Petford, A.D., Willis, R.B., 1976b, Monthly Notices Roy. Ast. Soc. 177, 227

Blackwell, D.E., Ibbetson, P.A., Petford, A.D., Shallis, M.J., 1979a, Monthly Notices Roy. Ast. Soc. 186, 633

Blackwell, D.E., Petford, A.D., Shallis, M.J., 1979b, Monthly Notices Roy. Ast. Soc. 186, 657

Blackwell, D.E., Shallis, M.J., 1979c, Monthly Notices Roy. Ast. Soc. 186, 673

Bray, R.J., Loughhead, R.E., 1967, The Solar Granulation, Chapman & Hall Ltd., London

Dravins, D., 1979, this colloquium

Dravins, D., Lindegren, L., Nordlund, Å., 1979, in preparation

Evans, J.C., Ramsey, L.W., Testerman, L., 1975, Astron. Astrophys. 42, 237

Gustafsson, B., Olander, N., 1979, Physica Scripta, in press

Gustafsson, B., Bell., R.A., Eriksson, K., Nordlund, Å., 1975, Astron. Astrophys. 42, 407

Nordlund, Å., 1976a, Astron. Astrophys. 50, 23

Nordlund, Å., 1976b, in "Problems of Stellar Convection", Lecture Notes in Physics no. 71, ed. Spiegel, E.A., Zahn, J.P.

Nordlund, Å., 1978, in "Astronomical Papers dedicated to Bengt Strömgren", ed. A. Reiz, T. Andersen; Cop. Univ. Obs.

Nordlund, Å., 1979, this colloquium

Stenflo, J.O., Lindegren, L., 1977, Astron. Astrophys. 59, 367
Struve, O., Elvey, C.T., 1934, Astrophys. J. 79, 409
Unsöld, A., 1955, Physik der Sterneatmosphären, Berlin, Springer
Wittman, A., 1979, in "Small Scale Motions on the Sun", Mitteilungen aus dem Kiepenheuer-Institut 179

MECHANICAL ENERGY TRANSPORT

Robert F. Stein
Department of Astronomy and Astrophysics
Michigan State University
East Lansing, MI U.S.A.

and

John W. Leibacher
Space Astronomy Group
Lockheed Palo Alto Research Laboratory
Palo Alto, CA, U.S.A.

I. INTRODUCTION

Ladies and Gentlemen, we now reveal to you the secrets of how to create chaos out of order. The existence of a chromosphere or corona requires the existence of motions. A chromosphere or corona requires some non-radiative heat input. There has to be some kind of motion, either oscillatory or quasi-static, to transport the energy up to the chromosphere or corona. This ordered motion may be observed as chaos: microturbulence, macroturbulence, line asymmetries or shifts. Of course, it is necessary to actually compute the effects of motions on line profiles in order to see what will really happen.

We review the properties, generation and dissipation mechanisms of three kinds of waves: acoustic, gravity and Alfven waves. These are not the only kinds that can exist, but they will give you some idea of most of the range of wave properties, at least for the low frequency waves for which plasma effects are unimportant. They are pure cases. These different wave modes are distinguished by their different restoring force--pressure for acoustic waves, buoyancy for gravity waves, and magnetic tension for the Alfven waves. Their properties are summarized in Table I. From an observational viewpoint, the most important properties are the relation between temperature and density variations (which change the intensity) and fluid velocity (which shifts the line). Acoustic waves are compressive: In propagating waves the temperature and density vary in phase with the velocity, which is parallel to the energy flux. However, for standing or evanescent waves the temperature and density are 90° out of phase with the velocity. Gravity waves are slightly compressive: The temperature and density vary oppositely to each other and 90° out of phase with the velocity, which is parallel to the energy flux. Alfven waves are not compressive: The temperature, density, pressure and the total magnetic field strength remain constant and the motion is transverse to the energy flux. We will come back to these properties in more detail later.

There are basically two kinds of generation mechanisms: One is direct coupling from the convective motions to the wave motions, either inside the convection zone or penetrating into the photosphere; and the other is thermal overstability. There are only a few basic dissipatiom mechanisms also: Radiation can destroy the restoring force and damp a wave. The other major dissipation processes occur by collisions which diffuse momentum and energy, and produce viscosity, thermal conductivity, and resistivity. Sometimes, instabilities can clump the particles so that collisions occur with a large collective "particle" rather than an individual one. This increases the effective collision rate and enhances the diffusion of momentum and energy. These diffusive transport processes dissipate acoustic waves in shocks, gravity waves in shear layers, and Alfven waves by viscous or Joule heating.

II. ACOUSTIC WAVES

A. PROPERTIES

This material is well-known, so let us quickly run through the basics. As we said, the restoring force for acoustic waves is the pressure. The energy flux is

$$F = p\underset{\sim}{u} = \rho u^2 V_g ,$$

and the group velocity is

$$V_g \simeq s(1-N_{ac}^2/\omega^2)^{1/2},$$

where s is the sound speed. The acoustic cutoff frequency is

$$N_{ac} = s/2H = \gamma g/2s.$$

Acoustic energy propagation can occur only for $\omega > N_{ac}$ ($2\pi/200s$ for the sun). In the absence of dissipation or refraction the flux must be constant, and the sound speed is roughly constant throughout the photosphere and chromosphere and increases by a factor of 10 going up to the corona, hence the velocity amplitude will scale roughly as

$$u \propto \rho^{-1/2}.$$

Acoustic waves can propagate or be evanescent or standing. The essential difference is that propagating waves transport energy, but evanescent or standing waves don't. In propagating waves with vertical wavelength small compared to the scale height, the pressure, temperature and density vary in phase with the velocity:

$$\frac{\delta\rho}{\rho} \simeq \frac{u}{s}, \ \frac{\delta T}{T} \simeq (\gamma -1)\frac{u}{s}, \ \frac{\delta p}{p} \simeq \gamma \frac{u}{s}.$$

In standing or evanescent waves although the pressure, temperature and density fluctuations are of the same order as in propagating waves, they vary 90 degrees out of phase with the velocity

$$\frac{\delta\rho}{\rho} \simeq -i\frac{u}{s}, \ \frac{\delta T}{T} \simeq i(\gamma-1)\frac{u}{s}, \ \frac{\delta p}{p} \simeq -i(2-\gamma)\frac{u}{s}.$$

Since the energy flux is the average over a period of the pressure times the velocity, the average flux will be zero. Also the vertical phase velocity of evanescent waves will be infinite, so that the motions will be in phase all the way up and down through the atmosphere. Figure 1 shows a portion of the diagnostic diagram that Jacques Beckers showed yesterday, to remind you that the acoustic waves occur in the high frequency region. Later we will come to gravity waves, which occur in the low frequency region. (For more details, see Lighthill, 1978.)

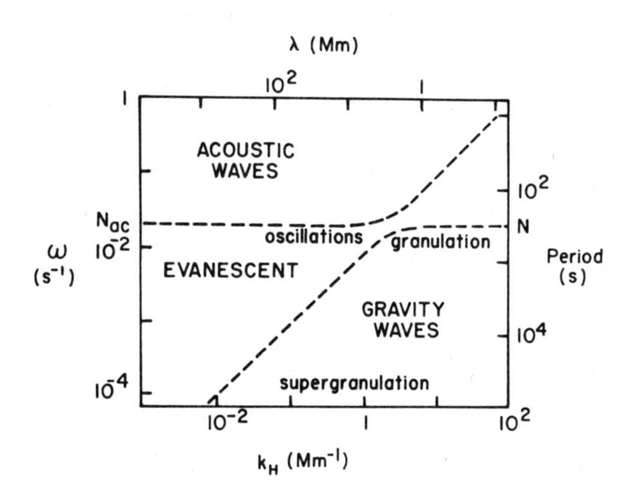

Fig. 1. The k-ω diagnostic diagram.

B. GENERATION

1. Direct Generation by Turbulent Motions

Acoustic waves may be generated directly by the turbulent convective motions. This is what is usually called the Lighthill mechanism. The radiated power is roughly the energy density in the turbulent motions divided by the time scale for the turbulent motions, times some efficiency factor. The turbulent energy density is ρu^2, the time scale is the eddy to turn over time, which is the length scale of the eddies divided by their velocity, $\tau \simeq \ell/u$, and the efficiency factor is the wave number of the wave times the size of the eddy to some power, $(k\ell)^{2n+1}$.

Thus, the radiated power is

$$P \simeq \frac{\rho u^3}{\ell} (k\ell)^{2n+1}.$$

The exponent n depends on the kind of emmission: If there is monopole emmission, which corresponds to a mass source, then n = 0; there is no mass source in the convection zone. If there is dipole emission, which corresponds to a momentum source, which corresponds to an external force, then n = 1; in a uniform medium there would be no external force and no dipole emission, but in stars there is an external gravitational field so there is some dipole emission. Finally, for quadruple emission, which corresponds to the action of the Reynolds stresses, n = 2, and that is the dominant process (Stein, 1967). For acoustic waves:

$$k \simeq \omega/s \text{ and } \omega \simeq \tau^{-1} \simeq u/\ell,$$

so

$$k\ell \simeq u/s = M,$$

the Mach number of the turbulent motions. What you get is the familiar result that the radiated power is proportional to the eighth power of the turbulent velocity:

$$P \simeq \frac{\rho u^3}{\ell}\left(\frac{u}{s}\right)^5 \propto u^8.$$

The turbulent velocity that one chooses is very sensitive to the model that one takes for the turbulence, and therefore the emission is very uncertain. But if one makes some crude estimates for the sun,

$$\rho \sim 10^{-6},\ s \sim 10^6,\ u/s \sim 1/4,$$

then

$$F \sim P\ell \sim 10^7 \text{ ergs/cm}^2\text{s}.$$

One can also use mixing length theory to see how the flux will depend on stellar properties. From mixing length theory

$$u \sim \left(\frac{g}{T}\,\beta\right)^{\frac{1}{2}} \ell,$$

and

$$\Delta T \sim \beta\ell,$$

where

$$\beta = -\left[dT/dz - (dT/dz)_{AD}\right]$$

is the superadiabatic temperature gradient. Hence

$$F \sim \rho c_p\, \Delta T u \sim \rho c_p\, (g/T)^{\frac{1}{2}}\, \beta^{3/2}\, \ell^2,$$

so

$$u \sim \left(\frac{gF\ell}{\rho c_p T}\right)^{1/3}.$$

From hydrostatic equilibrium

$$\rho \propto \frac{P}{T} \simeq \frac{g}{\kappa T}.$$

Assume $$\kappa \propto P^{0.7}\, T^{10} \propto g^{0.41}\, T^{5.88},$$

then $$F \sim \frac{\rho u^8}{s^5} \propto g^{-1}\, T_{eff}^{17}.$$

This means that the flux decreases very rapidly as you come down the main sequence, and increases rapidly as you go up to the giants and the supergiants. There are some problems with this. Linsky and co-workers find that the MgII flux is a good measure of the chromospheric emission and they claim that the ratio of the MgII flux to the total flux of the star is independent of g, which is contrary to what the Lighthill mechanism predicts (Basri and Linsky, 1979). Also if you look at the cool main sequence stars you find that the predicted flux is much less than the scaled chromospheric losses. The predicted wave flux may be increased by including effects of molecular hydrogen on the specific heats and the adiabatic gradient. Just how much is not known. Ulmschneider and Bohn are working on that now. But as of the moment there is still over an order of magnitude discrepancy in those results (Schmitz and Ulmschneider, 1979).

The turbulent motions may also directly excite the "five-minute" oscillation. In a steady state the amplitude of a given mode will be determined by the balance between turbulent generation by the Lighthill mechanism and dissipation by turbulent viscosity (Goldreich and Keeley, 1977):

$$\frac{\rho u_\lambda^3}{\lambda}(k\lambda)^5 = \nu k^2\, \varepsilon_k,$$

where

$$\nu = u_\lambda \lambda.$$

Hence, the energy density of an oscillation mode k will be

$$\varepsilon_k \simeq \rho\lambda^3\, u_\lambda^2\, k^3,$$

where λ is the size of the eddy whose turnover time equals the oscillation period, $u_\lambda/\lambda = \omega_k$. For a Kolmogorov turbulence spectrum, where

$$u_\lambda = u_H\left(\frac{\lambda}{H}\right)^{1/3},$$

and H is the scale height, which is assumed to be the size of the largest turbulent eddies which contain most of the turbulent energy, the oscillation mode energy density is

$$\varepsilon_k \simeq \rho u_H^2 \left(\frac{u_H}{s}\right)^{11/2}\left(\frac{\omega}{N_{ac}}\right)^{-5/2}.$$

2. Thermal Overstability

Overstability is an oscillating, thermal instability. There are several kinds of thermal instabilities. The one that works for acoustic waves is the κ- mechanism or the Eddington Valve. If you have an opacity which increases with temperature, then when the gas is compressed, it gets hotter, the opacity goes up, it blocks the flow of radiation, so heat accumulates which raises the gas pressure, which means there is more pressure expanding the gas than would be obtained from just compressing the gas, and so it will have a stronger expansion than its compression and the amplitude will increase. When you actually calculate the growth rates as Ando and Osaki (1975) did, you find that they are very slow. The time scale for a mode to grow is about a thousand periods, and that is so long that the turbulent viscosity has a chance to destroy the overstability. On the other hand, as we have seen, the turbulent motion may also directly excite the modes. And since we certainly see them in the sun, we know something is exciting them. It ought to be pointed out that the calculations show that the fundamental mode is stable. It is, however, seen on the sun, although at a somewhat smaller amplitude than the higher modes. If the calculations are right, at least the fundamental mode must be excited by some other mechanism besides thermal overstability. So thermal overstability may or may not work for the five minute oscillation. Some people have proposed a mechanism of Doppler shifted line opacities as a generation mechanism for sound waves in stellar winds.

C. DISSIPATION

What about the dissipation of acoustic waves?

1. Radiation

In the first place photons can transfer energy from the hotter to the cooler regions of a wave which will reduce the restoring force and damp the wave. Calculations show that about 90% of the wave energy of the acoustic waves is removed in the photosphere. Radiative damping also alters the phase of the temperature and density relative to the velocity (Noyes and Leighton, 1963).

2. Shocks

The other damping mechanism, of course, is shocks. As the wave propagates its front steepens, and when the thickness of the wave front becomes comparable to the mean free path of the particles, one gets a shock. It should be remarked that a shock has to do with the steepness of the gradient, not with the size of the velocity. You can have a shock where the velocity amplitude is small compared to the sound speed. The distance a wave must travel for a crest to overtake a trough and a shock develop is

$$\Delta Z = 2H \ln \left(1 + \frac{\lambda}{2H} \frac{s}{u} \frac{1}{\gamma+1}\right).$$

Short period acoustic waves will dissipate near the temperature minimum, but longer period waves with periods around the acoustic cutoff period (200 sec) will dissipate higher up. And the five minute oscillation which is evanescent doesn' steepen at all until it gets high enough to become nonlinear. The dissipation length is

$$L \simeq \begin{cases} \lambda/(M-1) & \text{weak shocks} \\ \lambda & \text{strong shocks,} \end{cases}$$

and the strength of weak shocks varies as

$$M = V_{shock}/s \propto \rho^{-1/4}$$

(Stein and Schwartz, 1972).

III. GRAVITY WAVES

A. PROPERTIES

In gravity waves the restoring force is buoyancy, which is similar to convection. The different thing about gravity waves is that, while for acoustic waves there is a natural speed, the sound speed, for gravity waves there is a natural frequency, the buoyancy or Brunt-Vaisala frequency at which a blob will oscillate if displaced:

$$N = \left(-\frac{g}{T}\beta\right)^{\frac{1}{2}},$$

where β is the superadiabatic temperature gradient. In an isothermal atmosphere

$$N \to (\gamma - 1)^{\frac{1}{2}} g/s.$$

Gravity waves only propagate at frequencies less than this natural buoyancy frequency, and the buoyancy frequency is only real and nonzero in convectively stable regions. You cannot have gravity waves propagating in a convectively unstable region. They cannot propagate or be generated inside the convection zone, only by motions in the stable photosphere. Gravity waves propagate energy in a particular direction, which depends on frequency. The cosine of the angle between the flux and the vertical is

$$\cos\theta = \omega/N.$$

For the sun, for values that are appropriate for the granulation,

$$\tau_{granulation} \simeq 10\text{-}20 \text{ min and } N_{Tmin} \simeq 0.03 = 2\pi/3\text{min},$$

the direction of gravity wave energy propagation is

$$\cos\theta \simeq 1/5, \quad \theta \simeq 75^{o}.$$

Figure 2 shows what you would see if you looked at a schlieran photograph of gravity waves produced by an oscillating source. The solid lines are the wave crests and the dashed lines are the wave troughs. The lines intersect at the source. Energy is propagating radially outward at the angle theta and the velocity of the fluid is also radial, parallel to the energy flux, but the phase propagates perpendicular to the energy,

$$\underset{\sim}{u} \parallel \underset{\sim}{F} \perp \underset{\sim}{k}.$$

You see different waves moving across the fan, while the fan extends out further and further with time as the energy gets out further and further. The group velocity of gravity waves is

$$V_g = \frac{N}{k} \sin\theta.$$

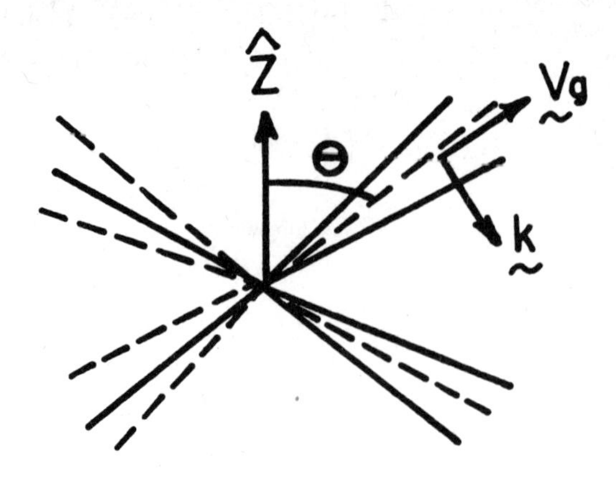

Figure 2: Gravity Wave Crests (——) and troughs (---).

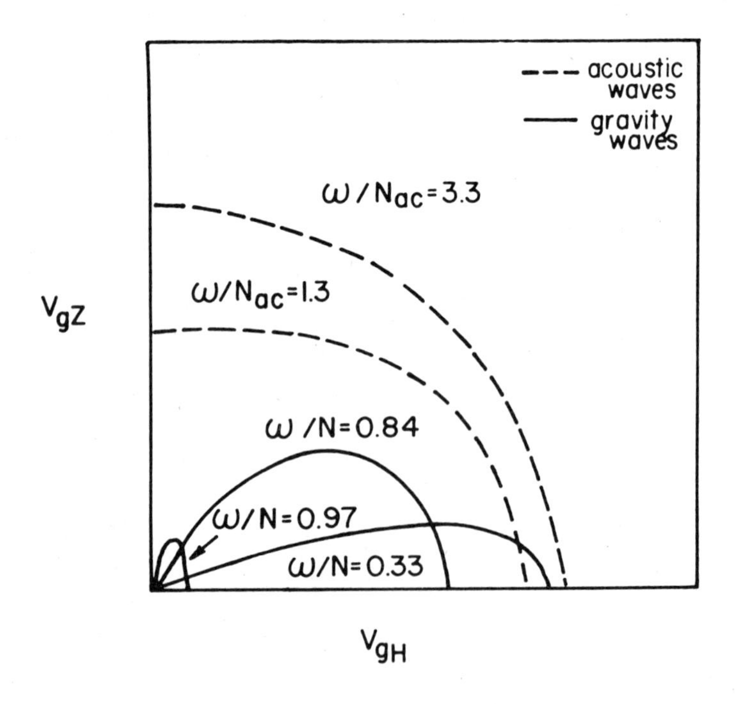

Figure 3: Group Velocity for gravity and acoustic waves.

For solar granulation $N/k \sim 1$ km/s. Figure 3 shows the group velocity for gravity (and acoustic) waves as a function of frequency and direction. Notice that one would observe low frequency, long wavelength waves first. The phase relations between temperature, density, pressure and velocity for low frequency gravity waves are similar to evanescent acoustic waves:

$$\frac{\delta\rho}{\rho} = i\,\frac{u}{s}, \quad \frac{\delta T}{T} = -i\,\frac{u}{s}, \quad \frac{\delta p}{p} = \frac{\omega}{sk_H}\,\frac{u}{s} << \frac{u}{s}.$$

Pressure fluctuations are in phase with velocity but very small, while temperature and density are 90° out of phase with velocity (Pittway and Hines, 1965; Lighthill, 1978, Chapter 4).

B. GENERATION

Gravity waves are ubiquitous in the atmosphere of the earth; they are produced by any slow motion. Therefore, they should also be present in the Sun. There are two common generation mechanisms.

1. Penetrative Convective Motions

One of the main ways gravity waves will likely be produced is by the penetrative convective motions. One can think of the penetrative convection as blobs pushing on the boundary of the stably stratified layer. The amplitude of the wave produced will be comparable to the amplitude of the penetrative motion,

$$|\underset{\sim}{u}_{wave}| = |\underset{\sim}{u}_{penetration}|.$$

This has been verified in laboratory experiments (Townsend, 1966). However, since only frequencies that are less than the buoyancy (Brunt-Väisälä) frequency can propagate, only that part of the penetrative convective power that satisfies $\omega = \underset{\sim}{k}.\,\underset{\sim}{V} < N$ will contribute to the production of gravity waves. This mechanism is similar to the Lighthill mechanism, but with an efficiency near one.

If you make a rough estimate for the solar granulation, taking a velocity of 1 km/sec and the appropriate length scales, then

$$F \sim \rho u^2 V_{gz} \sim 3\times10^{-7} \times 10^{10} \times \frac{1}{3}\,10^5 \sim 10^8 \text{ erg/cm}^2\text{s}.$$

This flux will, however, be greatly reduced by the strong radiative dissipation of gravity waves, which we discuss below.

2. Shear

Gravity waves can also be produced by the shear that will arise from the supergranule motions. Supergranule flows have a cellular structure. Conservation of mass requires that a gradient of the vertical momentum flux produces a horizontal momentum flux. Braking is large in the photosphere and produces a large horizontal flow there. Even though the horizontal momentum flux is small in the chromosphere, the chromospheric horizontal velocity is large, because of the small density. The horizontal supergranule flow is observed to decrease from ~ 0.8 km/s in the low photosphere to ~ 0.4 km/s in the low chromosphere and then increase to ~ 3 km/s in the mid-chromosphere (November, et al. 1979). Where the size of shear becomes comparable to the buoyancy frequency,

$$\frac{d\,U_H\,(z)}{dz} \geq \min\left[\, N, N^2\,\tau_{cool} \,\right]$$

(where τ_{cool} is the radiative cooling time) the shear layer becomes unstable and radiates gravity waves. Most of the energy is radiated near

$$k \simeq N/\sqrt{2}\;U_H,$$

and the growth times are of order

$$\gamma \simeq 10^{-1}\,dU_H/dz$$

(Lindzen, 1974). This mechanism will operate in the low photosphere where the cooling time is short and in the high chromosphere where the shear is large.

C. DISSIPATION

How do gravity waves dissipate?

1. Wave Breaking

Gravity waves steepen, but instead of forming a shock front they form a thin shear layer, where the fluid velocity changes direction over a very short distance. When that shear becomes comparable to the buoyancy frequency, $dU/dz \simeq N$, turbulence will develop along that wave front. Small scale motions are produced which dissipate the wave motion and damp the wave. To find the condition on the wave amplitude for breaking to occur, we need to calculate du/dz. Let $\underset{\sim}{u} = (u, 0, w)$ where u is the horizontal and w the vertical component of the velocity. For gravity waves the Boussinesq approximation holds, so

$$\nabla \cdot \underset{\sim}{u} = 0,$$

which implies that

$$\frac{dw}{dz} = -\,i\,k_H\,u.$$

The wave equation for gravity waves is

$$\frac{d^2w}{dz^2} + \left(\frac{N^2}{\omega^2} - 1\right)k_H^{\,2}\,w = 0,$$

so

$$\frac{du}{dz} = \frac{i}{k_H}\,\frac{d^2w}{dz^2} = -i\left(\frac{N^2}{\omega^2} - 1\right)k_H w.$$

Hence, the condition for gravity wave breaking is

$$N^{-1}\,\frac{du}{dz} = \left(\frac{N^2}{\omega^2} - 1\right)\frac{k_H w}{N} > 1$$

or

$$w > \frac{N}{k_H}\left(\frac{N^2}{\omega^2} - 1\right)^{-1}$$

2. Radiative Damping

As we mentioned, there is very severe radiative damping of the gravity waves. Radiation tends to make a wave isothermal, which destroys the buoyancy restoring force. The radiative cooling rate is

$$\gamma_R = \tau_R^{-1} = \frac{16\kappa\sigma T^3}{\rho c_v}\left(1 - \frac{\kappa}{k} \cot^{-1} \frac{\kappa}{k}\right)$$

$$\rightarrow \frac{16\kappa\sigma T^3}{\rho c_v} \times \begin{cases} 1 - \frac{\pi}{2}\frac{\kappa}{k} & \text{optically thin} \\ \frac{1}{3}\left(\frac{k}{\kappa}\right)^2 & \text{optically thick.} \end{cases}$$

where κ is the inverse of the photon mean free path (Spiegel, 1957). The optically thin damping time $\rho c/16\kappa\sigma T^3$, increases rapidly with height and exceeds 10 min above

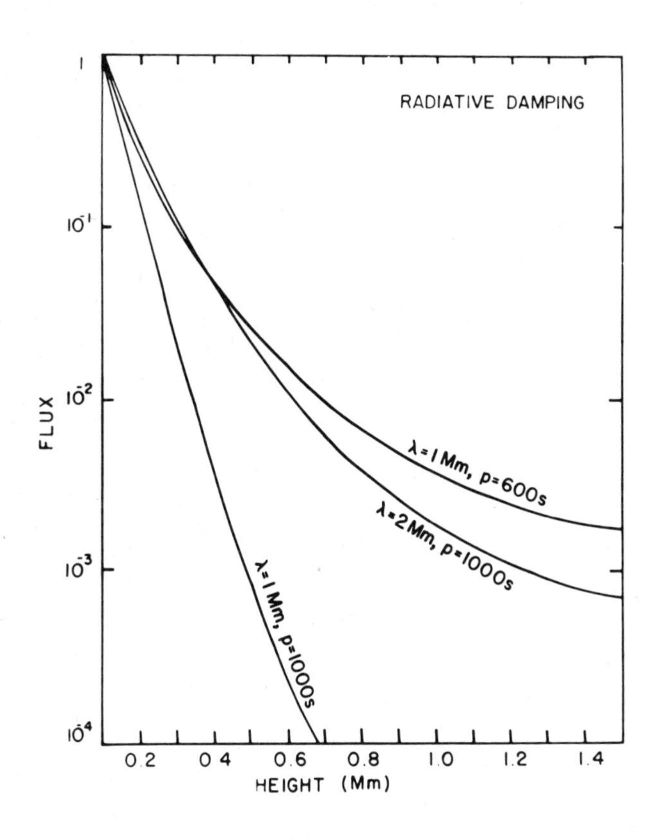

Figure 4. The fraction of flux remaining as a function of height.

700 km above $\tau_{5000} = 1$. Damping is severe when the radiative cooling time is shorter than the wave period. The long period, short wavelength waves are the most highly damped. Figure 4, from Barbara Mihalas' thesis (1979), shows the fraction of flux remaining at each height for several wavelengths and periods. In order of magnitude, the flux decreases by 10^3 between the bottom of the photosphere and the mid-chromosphere, for parameters appropriate to granule produced gravity waves. On the other hand, gravity waves generated by penetrative motions a few hundred kilometers above the bottom of the photosphere suffer reduced radiative dissipation and can transmit 2% of their flux to the chromosphere. Since the radiation which produces the damping of gravity waves also allows for easier penetration of the convection, the wave flux reaching the chromosphere from penetrative convective motions high up in the photosphere is only slightly less than that from the bottom of the photosphere with its severe radiative damping.

3. Critical Layers

Gravity waves have one other property which is very different from sound waves: they can interact very strongly with the mean horizontal fluid flow. In particular, they can give the fluid a horizontal acceleration. A layer where the horizontal fluid flow velocity is equal to the horizontal phase speed of the wave, is a "critical layer." Here in the fluid frame, the doppler shifted frequency goes to zero, $\omega - \underset{\sim}{k}.\underset{\sim}{u} \to 0$, so the waves propagate along the critical layer, $\cos\theta = (\omega - \underset{\sim}{k}.\underset{\sim}{u})/N \to 0$, and their energy is absorbed. Such a critical layer can arise from the horizontal supergranulation flow or may be produced by absorption of horizontal gravity wave momentum. For the sun, the horizontal gravity wave phase speed is about 1.5 km/s, which will match the horizontal supergranulation flow somewhere in the mid-chromosphere. The transmission through a critical layer is

$$T = \frac{F_{trans.}}{F_{inc.}} = \exp\left\{ - 2\pi \left(Ri - \frac{1}{4} \right)^{\frac{1}{2}} \right\},$$

where the Richardson number,

$$Ri = N^2/(dU/dz)^2 ,$$

is the order of 50. Hence, the absorption at critical layers is very large (Booker and Bretherton, 1967; Acheson, 1976). Such critical layers will occur in the chromosphere and will produce localized dissipation of gravity waves there.

The important thing to emphasize about gravity waves is that because they propagate energy mostly horizontally and because of radiative damping, one does not expect too much heating from them. But they are a source of chaos and may contribute substantially to any microturbulence, because they are certain to be there. Their horizontal wavelengths are comparable to granules, $\lesssim$ 1Mm, and their vertical wavelengths are only 1/4 as large, which is comparable to the scale height or smaller.

IV. ALFVEN WAVES

A. PROPERTIES

Moving on apace, we now come to Alfven waves. Here the restoring force is magnetic tension. The group velocity is the Alfven velocity,

$$V_g = a = B/\sqrt{4\pi\rho}.$$

Unlike the sound speed, the Alfven speed increases substantially between the photosphere and the corona, by a factor of 10^3. The flux is

$$F = \rho u^2 a = \frac{\delta B^2}{4\pi} a,$$

and is parallel to the magnetic field. In the absence of dissipation or refraction, the flux remains constant, so the velocity amplitude scales as

$$u \propto \rho^{-1/4} B^{-1/2}.$$

The velocity is perpendicular to the magnetic field, i.e. transverse to the direction of energy propagation. Hence Alfven waves act like a vibrating string. There is no compression,

$$\delta T = \delta p = \delta \rho = 0,$$

so there are no opacity changes. Only Doppler shifts affect the line profiles. The total magnetic field strength, $|\underset{\sim}{B}_o + \delta B|$, is constant, so the wave is polarized in part of the arc of a circle, along which the magnetic vector swings back and forth (see e.g., Bazer and Fleischman, 1959; Barnes and Hollweg, 1974).

Most people up until recently have considered uniform magnetic fields. But we know that in the corona the magnetic field is very inhomogeneous. Luckily it turns out that waves in inhomogeneous fields are rather similar to the internal waves in a uniform field. If you consider a thin flux tube surrounded by plasma in a weaker field, several different modes occur (Roberts and Webb, 1978; Wilson, 1979; Wentzel, 1979a). First, there is an axisymmetic mode which is just the torsional Alfven wave, propagating along the flux tube at the Alfven speed inside the tube, $c = a_i$. Second, there is another axisymmetic mode which is like the slow mode, a sound wave propagating along the flux tube with

$$c^2 = a_i^2 s_i^2/(a_i^2 + s_i^2).$$

It has a short wavelength and its amplitude is concentrated at the surface of the tube. For these waves to excite waves outside the tube, the phase velocity inside the tube would have to be greater than the sound or the Alfven speed outside. In general that is only possible for these modes when the inside density is less than the outside density. Third, there are other modes which are not axisymmetric, and act like vibrating string modes. Their phase velocities are essentially an average of the Alfven speed inside and outside the tube,

$$c^2 = (B_i^{\,2} + B_e^{\,2})/\ 4\pi\ (\rho_i + \rho_e).$$

These modes can have a resonance where the phase velocity of the tube mode equals the local Alfven speed. At that resonant point the amplitude of the perpendicular (to the magnetic field) velocity and electric field becomes large. Other perturbed quantities are unaffected by the resonance. Thus, even in an inhomogeneous corona the waves are similar, although not identical, to those in a homogeneous corona. The main difference is that for these tube or surface wave modes there exists a resonance at places in the flux tube where $c = a$.

B. GENERATION

What about Alfven wave generation?

1. Convective Motions

Alfven waves can be generated by the convective motions, similar to the Lighthill mechanism for the sound waves. But, because the magnetic field channels the motions, monopole rather than quadrupole emission occurs. The radiated power is

$$P = \frac{\rho u^3}{\ell}\ (k\ell).$$

For Alven waves,

$$k = \frac{\omega}{a}\ ,\ \omega = \frac{u}{\ell}\ ,$$

so

$$k\ell = \frac{u}{a} = M_B.$$

Thus

$$P = \frac{\rho u^3}{\ell}\left(\frac{u}{a}\right)$$

(Kulsrud, 1965: Kato 1968). Rough estimates for the sun $\rho \sim 10^{-6}$, $u \sim 10^5$, $a \sim 10^6$ predict an Alfven wave flux from strong field regions of

$$F \sim P\ell \sim 10^8\ \mathrm{erg/cm^2 s}.$$

However, anything which jiggles a magnetic field line will also generage Alfven waves, so granules will generate Alfven waves and supergranules will generate Alfven waves. Granules have larger velocities and so are more important. They produce a flux

$$F = \rho u^2 a.$$

For typical granule velocities of 1 km/s,

$$F \simeq 3\mathrm{x}10^{-7}\ \mathrm{x}\ 10^{10}\ \mathrm{x}\ 10^6 = 3\mathrm{x}10^9\ \mathrm{ergs/cm^2 s}.$$

(See, however, Hollweg, 1979.)

What we are really interested in is the average flux, so we have to include the fact that the flux tubes will spread with height and that the Alfven speed increases with height. The waves produced by granular motions have fairly long wavelength so they see the change in Alfven speed roughly as a discontinuity. In this case the transmission coefficient is

$$T = \frac{4a_1a_2}{\left(a_1 + a_2\right)^2} \simeq 4 \frac{a_1}{a_2} \simeq 10^{-3}.$$

The average flux is

$$F = F_o \frac{A_{photo}}{A_{corona}} \; 4 \frac{a_{photo}}{a_{corona}},$$

where A is the flux tube area. Since the magnetic flux is constant along a flux tube, BA = constant, so $Aa \propto \rho^{-1/2}$. Thus, the average flux will be

$$F = F_o \left(\rho_{corona}/\rho_{photosphere}\right)^{1/2}$$

$$\simeq 10^{-4} F_o$$

$$\simeq 3x10^{5} \text{ erg/cm}^2\text{s},$$

which is fairly substantial, enough to heat the corona.

I have not made a distinction between Alfven waves and magnetohydordynamic waves. In a strong field the Alfven and the fast mode are similar and the slow mode is an acoustic wave propagating along the flux tube.

2. Thermal Overstability

Alfven waves may also be generated by thermal overstability. This is not the κ mechanism, but the Cowling-Spiegel mechanism (Cowling, 1957; Moore and Spiegel, 1966). Here buoyancy acts as a driving force which tends to destabilize the system and make it depart from equilibrium. Magnetic tension acts as a restoring force which tends to bring it back to equilibrium and radiative transfer decreases the destabilizing effect of the buoyancy force. Since there will be less destabilizing effect on the way back than there was on the way out from equilibrium, the magnetic tension will return the system to equilibrium faster than it departed, and the wave amplitude will grow. Because the buoyancy must be destabilizing this mechanism works only in convectively unstable regions. One can calculate the growth rate by equating the rate of working of this buoyancy force with the kinetic energy of the waves (Parker, 1979). The buoyant force is

$$F_B = \Delta\rho g = \rho g \frac{\Delta T}{T}.$$

The temperature fluctuation is the temperature difference between the adiabatically displaced fluid in the wave and mean temperature at its displaced level, reduced by diffusive radiation cooling:

$$\Delta T = A\lambda\beta \frac{t}{\tau_{cool}},$$

Where A is the wave amplitude, λ is the wave length, β is the superadiabatic temperature gradient, t is the period, and the diffusive radiation cooling time is

$$\tau_{cool} \approx \frac{\lambda^2}{c\ell_{mfp}} \frac{E_{gas}}{E_{rad}} = \frac{\kappa\lambda^2}{c} \frac{\rho c_p T}{aT^4} \approx \frac{\rho c_p}{\kappa\sigma T^3} (\kappa\lambda)^2,$$

which is assumed to be much greater than the period. The growth time γ^{-1}, is the time it takes the buoyant work, vF_B, to supply the wave energy $\frac{1}{2} \rho v^2$, where $v = \Lambda a = A\lambda/t$:

$$\gamma^{-1} v F_B = \frac{1}{2} \rho v^2.$$

Thus

$$\gamma \approx \frac{g\beta}{T\omega^2 \tau_{cool}} \approx \frac{t_{osc}^2}{t_{eddy}^2 \tau_{cool}}.$$

How does this growth rate depend on stellar properties? For an opacity

$$\kappa = \ell_{mfp}^{-1} \propto P^{1.7} T^9,$$

hydrostatic equilibrium gives

$$P \simeq \frac{g\rho}{\kappa} \propto g^{.6} T^{-6},$$

so

$$\kappa \propto g/T.$$

The wave frequency is

$$\omega^2 \sim a^2/\lambda^2 \sim B^2/\rho\lambda^2 \propto B^2 g^{-.6} T^{+7} \lambda^{-2},$$

and the cooling rate is

$$\tau_{cool}^{-1} \approx \frac{4F}{\rho c_p T\kappa\lambda^2} \propto g^{-1.6} T^{11} \lambda^{-2},$$

From mixing length theory,

$$\tau_{eddy}^{-2} = \frac{g\beta}{T} \approx \left(\frac{gF}{\rho c_p T\ell^2}\right)^{2/3}$$

$$\propto g^{1.6} T^{5.25},$$

where $\ell = H \propto T/g$. Hence the growth rate varies with stellar gravity, surface temperature, and magnetic field as

$$\gamma \propto g^{3/5} T^{9.25} B^{-2}.$$

C. DISSIPATION

1. Viscous and Joule Heating

How do Alfven waves dissipate? Alfven waves don't steepen and form shocks, because the Alfven speed is independent of wave amplitude, since the magnetic field strength is constant and they are not compressive. They can still dissipate by particle collisions which produce Joule or viscous heating. However, the damping lengths are large, although they may be comparable to coronal loop dimensions. The Joule and viscous heating rates are

$$Q_J = \eta J^2 = \eta c^2 k^2 (\delta B)^2/(4\pi)^2$$

and

$$Q_V = \mu k^2 u^2 .$$

The damping rate is

$$\gamma = Q/2E,$$

where

$$E = \rho u^2 = \delta B^2/4\pi ,$$

and the damping length is

$$L = a/\gamma .$$

For Joule heating

$$L_J \simeq 8\pi a^3/\eta c^2 \omega^2$$

$$\simeq 10^{20} n^{-3/2} T^{3/2} B^3 P^2 \text{ cm},$$

where the resistivity is

$$\eta = \frac{m_e \nu_{coll}}{ne^2} = \frac{\pi e^2 m^{1/2}}{(kT)^{3/2}} \ln \Lambda \simeq 10^{-7} T^{-3/2} .$$

For viscous heating

$$L_V \simeq 2\rho a^3/\mu\omega^2$$

$$\simeq 10^{23} n^{-1/2} T^{-5/2} B^3 P^2 \text{cm},$$

where the viscosity is

$$\mu \simeq \frac{1}{3} n m_p v_e^2/\nu_{coll} \simeq 10^{-15} T^{5/2} .$$

Viscous damping is more important than Joule heating in the corona and visaversa in the photosphere and chromosphere. If the fields are weak, then the Joule dissipation will be significant in the photosphere. Hence Alfven waves can only get through from the photosphere up into the corona in strong field regions. That used to be a serious problem before we knew that fields come in little patches of high field strength. Now it is not. For typical coronal loop parameters ($n \sim 10^{10} \text{cm}^{-3}$, $T \sim 2 \times 10^{6\,o}$K, $B \sim 100$G, $L \sim 10^{10}$cm) the viscous damping length is

$$L_V \sim 10^8 P^2 \text{cm},$$

which is comparable to the loop length for short period waves.

Because the corona is inhomogeneous, the Alfven waves are really tube modes, and have a resonance where the tube phase speed is equal to the local Alfven speed. In this resonant region the wave amplitude is large. Hence large currents and large Joule heating will occur in the narrow resonant layer. The rate of heating is controlled by the rate at which energy can flow into that resonant region, and has been calculated by Ionson (1978) and Wentzel (1979b):

$$\gamma = \pi\omega k_r \Delta r \quad (\Delta a^2/a^2) \quad (2 + \rho_i/\rho_e + \rho_e/\rho_i)^{-1}.$$

They thought this was the rate at which waves were radiating energy away. It isn't. It is the rate of energy flow into the resonant region (Hollweg, 1979). There is a problem of how the energy released in this very small resonant region volume is transferred to the rest of the large coronal volume where it is needed. Nobody has figured out how that is done. This is a problem for just about every type of Alfven wave dissipation, except for the Joule and the viscous dissipation. The reason is that in order to rapidly dissipate the Alfven wave energy the currents must be clumped, and so the dissipation occurs in a small region.

2. Mode Coupling

In the presence of inhomogenieties the Alfven wave will couple to other wave modes. Coupling will be large between wave modes whose wave vectors' difference is comparable to the inverse of the inhomogeniety scale length. The coupling ratio between two modes is roughly

$$|\Delta kL|^{-1},$$

where Δk is the difference in k between the two modes and L is the length scale of the inhomogeniety (Melrose, 1977). In a strong field the major coupling occurs between Alfven and fast mode waves that are propagating in the direction of the magnetic field, because both of them are propagating nearly at the Alfven speed, so they will stay together. The difference in wave vector is

$$\Delta k = \Delta\left(\frac{\omega}{c}\right) = \frac{\omega}{c}\frac{\Delta c}{c},$$

where c is the phase speed of the mode. For propagation close to the magnetic field direction

$$c_+^2 = a^2 + s^2\theta^2,$$

$$c_-^2 = s^2\left(1 - \frac{s^2}{a^2}\theta^2\right).$$

$$c_A^2 = a^2(1 - \theta^2).$$

So

$$\frac{\Delta c}{c} = \frac{c_+ - c_A}{a} = \left(1 + \frac{1}{2}\,\frac{s^2}{a^2}\,\theta^2\right) - \left(1 - \theta^2/2\right)$$

$$\approx \theta^2/2.$$

and

$$\Delta k/k \simeq \theta^2/2.$$

Thus the coupling ratio is

$$\simeq 2\,(kL)^{-1}\,\theta^{-2}.$$

When $\theta < (\omega/\Omega_i)^{\frac{1}{2}}$, where Ω_i is the ion-cyclotron frequency, ion cyclotron effects increase Δk. So the maximum coupling occurs at this critical angle and is

$$|\Delta kL|^{-1} \simeq (kL)^{-1}\,(\Omega_i/\omega)$$

(Melrose, 1977). There will not be much coupling if the waves are not propagating along the field direction, nor will there be much coupling between the Alfven and slow modes. The fast mode, if it gets any energy, will form shocks and dissipate, so that this is a round about way in which the Alfven waves can dissipate their energy.

3. Alfven Wave Decay

The Alfven waves can also decay. If a set of waves satisfy the resonance condition,

$$\omega_o = \omega_1 + \omega_2,$$

$$\underset{\sim}{k}_o = \underset{\sim}{k}_1 + \underset{\sim}{k}_2,$$

which is essentially energy and momentum conservation, then one wave can decay into two. In this case a forward moving Alfven wave can decay into a backward moving Alfven wave and a slow mode pressure wave, at the rate

$$\gamma \simeq \omega \left(\frac{a}{s}\right) \frac{k}{\Delta k} \left(\frac{\delta B}{B}\right)^2.$$

(Kaburaki and Uchida, 1971). For a broad spectrum of incident waves, only those that stay in resonance for a decay time, i.e. have $\Delta k/k \sim \Delta\omega/\omega \lesssim \gamma/\omega$, can decay. So the decay rate is

$$\gamma \simeq \omega \left(\frac{a}{s}\right)^{\frac{1}{2}} \left(\frac{\delta B}{B}\right).$$

(Sagdeev and Galeev, 1969). There seems to be some disagreement between the calculated decay rates and the fact that one sees the Alfven waves in the solar wind at the earth.

4. Current Dissipation

Finally, I want to talk a little about current dissipation. Currents are produced not only by Alfven waves, but by any twisting motion of the magnetic flux tubes, for instance a quasi-static twisting motion. Current or magnetic field dissipation is a diffusive process due to single or collective particle collisions. The characteristic resistive diffusion time scale is

$$\tau_R = \frac{2E}{Q_J} = \frac{2B^2}{8\pi\eta J^2} = 4\pi L^2/\eta c^2.$$

For typical coronal parameters $\tau_R \sim 10^4$ yrs, too long to be significant. This Joule dissipation time can be reduced either by reducing the width L of the region through which the currents flow or by increasing the resitivity by increasing the effective collision rate. If some instability or resonance filaments the current so the current density is high in a small region, then there can be significant dissipation of the currents. If that occurs and if the current density $J = n_e ev_{drift}$ becomes large enough so that the drift velocity approaches the electron thermal velocity then substantial numbers of electrons will tend to run away and generate several different types of electrostatic waves. These waves bunch the ions, so that the electrons collide with the electric field of a large collective charge rather than that of a single ion. This scattering of electrons by the waves increases the effective collision rate, the rate of momentum transfer and hence the resistivity. The enhanced resistivity due to electron scattering by plasma waves is called "anomalous resistivity," and since it occurs in conjunction with current filamentation will shorten the resistive diffusion time tremendously (Papadopolous, 1977; Rosner et al, 1978; Hollweg, 1979). Also if the current density becomes large it will develop large shears or gradients in the magnetic field, which will lead to tearing mode instabilities. Parallel currents attract one another and tend to clump. The clumping of current produces a fluid flow that forces the sheared magnetic field into X-type neutral points. Filamented currents are produced with small enough length scales so that the classical, Coulomb collision, resistive diffusion time becomes small and the magnetic field can tear and reconnect and the currents can dissipate (Drake and Lee, 1977). This mechanism has been invoked for the violent energy release in flares (see Spicer & Brown, 1980). However, shorter wavelength tearing modes distort the field lines more, which produces a greater restoring force, and also have a smaller volume of magnetic energy they can release, so they may produce a more tranquil quasi-static heating appropriate for coronal flux tubes. The tearing instability has a lower threshold than the current driven instabilities which lead to anomalous resistivity. To get significant current dissipation by any mechanism, the dissipation must occur in such small volumes that the transfer of the resulting heat to the rest of the corona is a serious problem.

V. CONCLUSION

In conclusion, there are many kinds of different wave motions in the sun; acoustic, gravity, Alfven waves, and other kinds of magneto-acoustic-gravity wave modes; maybe even higher frequency waves like whistlers. All of these ordered motions may contribute to the chaos observed in stellar atmospheres. In order to develop diagnostics, somebody has to take self-consistent calculations of these waves with the right velocities, temperatures, densities and pressures and calculate the effects on the line profiles of each wave mode.

To summarize the major roles of the waves we have discussed today: Acoustic waves can heat the low chromosphere but not the corona. The evidence is partly observational: the observed nonthermal line widths due to waves in the upper chromosphere are too small, and also theoretical: increasing the driving amplitude at the bottom of the atmosphere only increases the dissipation of the acoustic waves in the chromosphere, but doesn't increase the flux through the transition region to the corona.

For gravity waves the motion is mainly horizontal. Their amplitudes will be comparable to the amplitudes of the penetrative convection (granulation). They may contribute to the observed microturbulent velocities, but they are unlikely to be important in the heating.

Alfven waves will dissipate primarily by highly clumped currents in very small regions, so there is the problem of how to get that energy from the small volume where the dissipation occurs to the larger volume of the corona. The nice thing about Alfven waves is that they are observed in the solar wind, and they seem to be important in providing an energy and momentum input to the wind. Someplace between the photosphere and the earth, where the wind is observed, those Alfven waves must be produced.

TABLE I.

WAVE MODE	ACOUSTIC	GRAVITY	ALFVEN
PROPERTIES	Pressure $\underset{\sim}{u}$ // $\underset{\sim}{F}$ // $\underset{\sim}{k}$ $V_g \sim s$ Propagating: $\frac{\delta T}{T} \sim \frac{\delta\rho}{\rho} \sim \frac{\delta p}{p} \sim u/s$ Evanescent $\frac{\delta T}{T} \sim \frac{\delta\rho}{\rho} \sim \frac{\delta p}{p} \sim i\,\frac{u}{s}$	Buoyancy $\underset{\sim}{u}$// $\underset{\sim}{F} \perp \underset{\sim}{k}$ $\cos\theta = \omega/N$ $\frac{\delta p}{p} \sim \frac{\omega}{k_H s}\,\frac{u}{s} \ll u/s$ $\frac{\delta\rho}{\rho} \sim \frac{\delta T}{T} \sim i\,u/s$	Magnetic Tension $\underset{\sim}{u} \perp \underset{\sim}{F}$, $\underset{\sim}{F}$//$\underset{\sim}{B}$ $V_g = a$ $\delta T = \delta p = \delta p = 0$ $\lvert \underset{\sim}{B}_o + \delta\underset{\sim}{B} \rvert$ = const.
GENERATION	Convective Motions Thermal Overstability	Penetrative Convection	Convective Motions Thermal Overstability
DISSIPATION by single or collective partical collisions	Shocks radiation	Shear Critical Layers Radiation	Viscous and Joule Heating Plasma Instabilities Mode Coupling

R.F.S. is grateful for support from N.S.F. grant AST-76-22479, NASA grant NSG 7293, and Air Force contract F19678-77-C-0068. J.W.L. is grateful for support from NASA contract NASw-3053 and NAS-5-23758, and the Lockheed Independent Research Fund.

REFERENCES

Acheson, D. J., 1976, J. Fluid Mech. 77, 433.
Ando, H., and Osaki, Y., 1975, Publ. Astron. Soc. Jap. 27, 581.
Barnes, A., and Hollweg, J. V., 1974, J. Geophys, Res. 79, 2302.
Basri, G. S., and Linsky, J. L., 1979, Astrophys. J. (in press).
Booker, J. R., and Bretherton, F. P., 1967, J. Fluid Mech. 27, 513.
Cowling, T. G., 1957, Magnetohydrodynamics, Interscience, New York.
Drake, J. F., and Lee, Y. C., 1977, Phys. Fluids 20, 1341.
Goldreich, P., and Keeley, D. A., 1977, Astrophys. J., 212, 243.
Hollweg, J. V., 1979, Proc. Skylab Active Region Workshop, ed. F. Q. Orrall, NASA. (in press)
Ionson, J.A., 1978, Astrophys, J., 226, 650.
Kaburaki, O., and Uchida, Y., 1971, Publ. Astron., Soc. Jap. 23, 405.
Kato, S., 1968, Publ. Astron. Soc. Jap. 20, 59.
Kulsrud, R., 1965, Astrophys. J., 121, 461.
Lighthill, J., 1978, Waves in Fluids, Cambridge University Press, Cambridge, England.
Lindzen, R. S., 1974, J. Atm. Sci. 31, 1507.
Melrose, D. B., 1977, Aust. J. Phys. 30, 495.
Mihalas, B., 1979, Thesis, University of Colorado.
Moore, D. W., and Spiegel, E.A., 1966, Astrophys. J., 143, 871.
Noyes, R. W., and Leighton, R. B., 1963, Astrophys. J., 138, 631.
November, L. J., Toomre, J., Gebbie, K. B., and Simon, G. W., 1979, Astrophys, J., 227, 600.
Papadopoulos, K., 1977, Rev. Geophy. Space Phys. 15, 113.
Parker, E. N., 1979, private communication.
Phillips, O.M., 1966, The Dynamics of the Upper Ocean, Cambridge University Press, Cambridge, England.
Pitteway, M. L. V., and Hines, C. O., 1965, Can. J. Phys. 43, 2222.
Roberts, B., and Webb, A. R., 1978, Solar Phys. 56, 5.
Rosner, R., Golub, L., Coppi, B., and Vaiana, G. S., 1978, Astrophys. J., 222, 317.
Sagdeev, R. Z., and Galeev, A. A., 1969, Plasma Physics, Benjamin, N. Y.
Schmitz, F., and Ulmschneider, P., 1979, Astron. and Astrophys. (in press).
Spicer, D. S., and Brown J. C., 1980, The Sun as a Star, NASA/CNRS (in press).
Spiegel, E. A., 1957, Astrophys. J., 126, 202.
Stein, R. F., 1967, Solar Phys. 2, 385.
Stein, R. F. and Schwartz, R. A., 1977, Astrophys. J., 177, 807.
Townsend, A. A., 1966, J. Fluid Mech. 24, 307.
Wentzel, D. G., 1979, Astron. and Astrophys. 76, 20.
____ 1979b, Astrophys. J. (in press).
Wilson, P. R., 1979, Astron. and Astrophys. 71, 9.

STELLAR CHROMOSPHERES

Jeffrey L. Linsky*
Joint Institute for Laboratory Astrophysics
National Bureau of Standards and University of Colorado
Boulder, Colorado 80309 U.S.A.

I. INTRODUCTION

Important progress in our understanding of stellar chromospheres has occurred in the past few years as a result of new observations, developments in spectral line formation theory, and the application of that theory to the construction of detailed model chromospheres. Significant trends are beginning to emerge from such analyses, and we are on the threshold of a meaningful confrontation between purely theoretical models and the data. The range of stars thought to possess chromospheres may be widening, and we now have a better understanding of the enigmatic problem of why the Wilson-Bappu relation between Ca II emission core widths and stellar absolute visual magnitudes actually works.

An important element in this progress has been the realization that much can be learned by studying the outer atmospheres of the Sun and a wide range of stars in the same context, and that such an approach is a two-way street. Not only are the theoretical techniques for analyzing spectra, modeling atmospheric structures, and computing the consequences of different heating processes the same, but also the wide variety of structures and phenomena seen on the Sun with high spatial and spectral resolution may be useful prototypes for stellar atmospheric structures and phenomena that we cannot hope to resolve but whose existance is implied by indirect evidence. Needless to say, our understanding of the Sun can be strengthened by studying phenomena in stars that have values of gravity, rotational velocity, chemical composition, and luminosity very much different from those of the Sun.

Despite the importance of solar-stellar cross fertilization, it is unwise to pursue solar analogies too far. At some point most and perhaps all of the solar analogies will fail to explain observables for certain stellar chromospheres. When we run into such situations, as I think we have in several cases, we are in a position to make important advances in our understanding of underlying physical processes operating in stars.

---

*Staff Member, Quantum Physics Division, National Bureau of Standards.

The general topic of stellar chromospheres has been reviewed recently by Linsky (1977, 1979, 1980), Praderie (1977), Ulmschneider (1979), and Snow and Linsky (1979). Important earlier reviews include those of Praderie (1973), Doherty (1973) and Kippenhahn (1973) in the proceedings of the IAU Colloquium on Stellar Chromospheres held in 1972 (Jordan and Avrett 1973), which the reader is encouraged to pursue. Recent reviews and monographs on the solar chromosphere include Athay (1976) and Withbroe and Noyes (1977). Because of the extensive review literature in this field and the rapid advances made most recently and presently under way, I will adopt a nonstandard approach here. I will not specifically discuss plasma diagnostics since this topic has been reviewed by Praderie (1973), Linsky (1977), and Ayres and Linsky (1979). However, several interesting developments in our understanding and use of chromospheric diagnostics are described below under the relevant headings.

II. WHAT TRENDS ARE EMERGING FROM SEMIEMPIRICAL CHROMOSPHERIC MODELS OF SINGLE STARS?

During the past several years two important developments have greatly facilitated the computation of semiempirical models of stellar chromospheres. The first is the acquisition of absolute flux profiles of important diagnostics such as Ca II H and K, the Ca II infrared triplet, He I λ10830, and Hα from the ground; and Mg II h and k, Lα, and the resonance lines of C II, Si II, and Si III from space experiments, particularly IUE. The second development is the refinement in our understanding of optically thick resonance line formation. We now have increased confidence in the use of these diagnostics to probe can provide insight into the gross properties of stellar chromospheres.

The various diagnostics used in building semiempirical model chromospheres have been reviewed by Linsky (1977), Ulmschneider (1979), Linsky (1979), and Ayres and Linsky (1979). The usefulness of different spectral features for identifying chromospheres has been reviewed by Praderie (1973, 1977). Here I will describe some recent developments in the field, identify interesting trends that are emerging, and call attention to uncertainties in the conventional analyses.

First, I will mention important observational programs that are producing chromospheric line profiles, calibrated in absolute flux units, which are the backbone of model chromosphere studies. Linsky _et al_. (1979a) have obtained 120 mÅ spectra of the Ca II H and K lines in a wide variety of stars later than spectral type F0, using the Kitt Peak 4 m echelle spectrograph. These data were calibrated in absolute flux units at the stellar surface based on Willstrop's (1972) narrow band photometry and the Barnes and Evans (1976) relation for deriving stellar angular diameters. Giampapa (1979) is extending the program to dMe stars. Blanco _et al_. (1974, 1976) have published absolute K line fluxes for F5-G8 dwarfs and G2-M5 giants, and Blanco et al. (1978) have presented absolute fluxes and brightness temperatures for the H_1 and K_1 features in 21 late-type stars. Echelle spectra of λ And obtained with the Mt. Hopkins Kron camera system have been discussed by Baliunas and Dupree (1979). Anderson (1974) and Linsky et al. (1979b) have obtained spectra of the Ca II infrared triplet lines, and Young (1979) is obtaining spectra of these features in RS CVn-type systems and other active chromosphere stars using the KPNO CID system. The Balmer lines in dMe flare stars have been studied by many observers, most recently by Worden et al. (1979). Zirin (1976) has obtained He I λ10830 equivalent widths for some 200 stars later than F5, and in several stars such as R Aqr, T Tau, and 12 Peg that have circumstellar envelopes. O'Brien and Lambert (1979) are studying λ10830 with an echelle-reticon system at McDonald Observatory. They find λ10830 emission in α Boo and α^1 Her and are presently studying nearly 60 F-M stars, with monitoring programs on several particularly interesting objects.

A number of important observing programs are under way in the ultraviolet. Surveys of Mg II emission fluxes include the *Copernicus* observations of 49 stars by Weiler and Oegerle (1979), BUSS observations with 0.1 Å resolution (e.g. Kondo et al. 1979; de Jager et al. 1979; van der Hucht et al. 1979), and IUE observations by Pagel and Wilkins (1979), Basri and Linsky (1979), Carpenter and Wing (1979), Stencel and Mullan (1979), and several other groups. The 1175-2000 Å short wavelength spectral region of IUE permits the study of prominent chromospheric resonance lines of H I, O I, C I, Si II, C II, and Si III, and transition region lines of C III, Si IV, C IV, N V, and O V. Initial observations of cool stars in the short-wavelength region include Linsky et al. (1978), Linsky and Haisch (1979), Ayres and Linsky (1979b,c), Dupree et al. (1979a), Hartmann et al. (1979), Brown et al. (1979), Carpenter and Wing (1979), and Bohm-Vitense and Dettmann (1979).

Table 1 summarizes semiempirical chromospheric models that have been constructed to match various diagnostic features observed in quiet and active regions on the Sun and in stars cooler than spectral type F0 V. For the most part these models were constructed to fit the Ca II and Mg II emission cores and damping wings using partial redistribution (PRD) radiative transfer codes, although prior to 1975 only the less accurate complete redistribution (CRD) codes were available. Because the Ca II and Mg II lines are formed in chromospheric layers cooler than 8000 K, these models may have validity only below that temperature. The Lyman and millimeter continua are useful for extending the models to 10,000 K, but suitable data are available only for the Sun. Other diagnostics of the 6500-8000 K temperature range include Si II λλ1808, 1817, 1265, and 1553 and the damping wings of Lα. Tripp et al. (1978) have described the formation of the Si II lines in the quiet Sun, while Basri et al. (1979) have used Lα wing observations to construct quiet and active solar models. However, the accuracy of the Lα diagnostic is compromised by the uncertain amount of frequency redistribution beyond the Doppler core. Because the Si II lines and Lα wings can be observed in stars by IUE, these diagnostics are

TABLE 1. SEMIEMPIRICAL CHROMOSPHERIC MODELS

Stars or Solar Features	Diagnostics Used	Approximations Used	Chromospheric Temperature Range	References
	Solar Models			
Quiet Sun (one comp.)	H, H^-, C I, Si I, IR continua	nonLTE	4100-25,000	Vernazza *et al.* (1973)
Quiet Sun (one comp.)	Mg II h + k, Ca II H + K	PRD	4450-5320	Ayres and Linsky (1976)
Quiet Sun (one comp.)	UV and IR continua	nonLTE	4150-5360	Vernazza *et al.* (1976)
Quiet Sun (one comp.)	C II $\lambda\lambda$1334, 1335; Lyman lines Lyman continuum	CRD	6884-57,000	Lites *et al.* (1978)
	Si II λ1816, 1265, 1533; Si III 1206	CRD	VAL (1973)	Tripp *et al.* (1978)
Plage	Ca II H + K, λ8498, λ8542, λ8662	CRD	4200-8000	Shine and Linsky (1974)
Plage	Mg II h + k, Ca II H + K	PRD	4600-8000	Kelch and Linsky (1978)
4 Component Sun	Lα, Lyman and mm continua	PRD	4460-25,000	Basri *et al.* (1979)
6 Component Sun	UV and IR continua	PRD	4150-25,000	Avrett (1979)
Flares	Ca II H + K	CRD	5000-8400	Machado and Linsky (1975)
Flares	Ca II K, UV continua	PRD	4890-6100	Machado *et al.* (1978)
	Main Sequence and Subgiant Stars			
α CMi (F5 IV-V)	Ca II K, λ8542	CRD	4750-8000	Ayres *et al.* (1974)
α Cen A (G2 V), α Cen B (K1 IV)	Ca II K	PRD	3650-8000	Ayres *et al.* (1976)
70 Oph A (K0 V), ϵ Eri (K2 V)	Ca II K, Mg II h + k	PRD	3850-8000	Kelch (1978)
γ Vir N (F0 V), θ Boo (F7 IV-V) 59 Vir (F8 V), HD 76151 (G4 V) 61 UMa (G8 V), ξ Boo A (G8 V) EQ Vir (dK 7e), 61 Cyg B (dM0)	Ca II H + K	PRD	3000-8000	Kelch *et al.* (1979)
dMe stars	Hα, Hβ, Hγ	CRD	3000-15,000	Cram and Mullan (1979)
ϵ Eri (K2 V)	Mg II h + k, C II, Si II, Si III	PRD, CRD	3850-30,000	Simon *et al.* (1979)

Table 1. (continued)

Stars or Solar Features	Diagnostics Used	Approximations Used	Chromospheric Temperature Range	References
	Giants			
α Boo (K2 III)	Ca II H + K, λ8542; Mg II h + k	CRD, PRD wings	3200–8000	Ayres and Linsky (1975)
β Gem (K0 III), α Tau (K5 III)	Ca II K, Mg II h + k	PRD	2700–8000	Kelch *et al.* (1978)
	Supergiants			
β Dra (G2 II)	Ca II K, Mg II k, C II	PRD	4800–16,000	Basri (1979)
ε Gem (G8 Ib), α Ori (M2 Iab)	Ca II K, Mg II k	PRD	2730–7000	Basri (1979)
	RS CVn-Type Systems			
α Aur (G6 III + F9 III)	Ca II K, Mg II h + k	PRD	4700–8000	Kelch *et al.* (1978)
λ And (G8 III–IV) α Aur (G6 III + F9 III)	Ca II K, Mg II k, Hα	PRD	3800–10,000	Baliunas *et al.* (1979)
HR 1099 (G5 V + K1 IV), UX Ari (G5 V + K0 IV)	Mg II k, C II, Si II, Si III	PRD, CRD		

potentially very important. In dMe stars, Hα and other Balmer series members go into emission and are useful diagnostics of the 8000–15,000 K temperature range (Fosbury 1974; Cram and Mullan 1979).

Vernazza et al. (1973) have proposed that the solar upper chromosphere has a plateau near 20,000 K with small temperature gradients, presumably produced when Lα becomes optically thin and an efficient radiator. Such a plateau can be studied by analyzing the Lα core. However, interstellar H I absorption prevents the observation of the Lα core in any star other than the Sun, or perhaps a few short period binary systems for which the orbital motions are large enough to unmask the stellar Lα core from the saturated interstellar absorption feature. Instead, one can study the C II λλ1334, 1335 and Si III λλ1206, 1892 lines which are also formed in the plateau (Lites et al. 1978; Tripp et al. 1978). These features, except Si III λ1206, can be observed even in faint stars by IUE at low dispersion, and chromospheric models of the 20,000 K region have now been constructed for several stars (e.g., Simon et al. 1979; Basri 1979; Simon and Linsky 1979).

Before discussing the general trends emerging from these models, I should bring to your attention their inherent limitations:

(1) All of the stellar models assume one-component atmospheres, whereas the Sun exhibits an embarrassingly rich variety of chromospheric structure. Inhomogeneities must also be important in many other stars as indicated by Wilson's (1978) observations of time variability in K line emission. (Such variations are likely produced in part by the appearance and temporal evolution of plage regions on the visible hemispheres of stars.) The important question is whether a one-component analysis is sufficient for an assessment of critical auxiliary quantities, for example nonradiative heating rates, or for comparison with purely theoretical chromospheric models. This question will be deferred to the next section, but it is clear that the chromosphere of at least one cool giant -- Arcturus (K2 III) -- is structured enough to make models based on diagnostics such as Ca II or Mg II, which are representative of the hotter atmospheric components, inconsistent with observations of diagnostics such as the CO fundamental vibration-rotation bands, which are sensitive to the cooler components (cf. Heasley et al. 1978).

(2) Models based on optically thick chromospheric resonance lines are uncertain to the extent that frequency redistribution in the line wings is not properly treated. The redistribution problem is most accute for estimating temperatures near the stellar temperature minimum. In these low density layers, scattering in the Ca II and Mg II resonance lines is nearly coherent at and beyond the K_1 minima, consequently the monochromatic source functions are strongly decoupled from the Planck function. For the Ca II resonance lines, radiative transitions to the $3d^2D$ metastable levels provide a lower limit of roughly 0.05 to the incoherence fraction. However, Mg II lacks analogous subordinate levels between the upper and lower states of the resonance lines, and the lower limit to the incoherence fraction Λ is only

10^{-4} (Basri 1979). Supergiants provide the most extreme examples of nearly pure coherent scattering in the Ca II and Mg II wings, owing to the low hydrogen densities and correspondingly reduced collisional redistribution. In fact, Basri (1979) finds that the largest sources of incoherence in such situations are radiative transitions to other levels. As described below, highly coherent scattering can even lead to self-reversed profiles with K_1 minimum features in isothermal models. Finally, the disagreement in solar chromospheric temperature structures based on the Ca II and the Mg II lines (Ayres and Linsky 1976) is a clear warning either that our understanding of the underlying atomic physics of the scattering process in resonance lines may be lacking in some important respect, or more likely, that single-component, homogeneous models are not a satisfactory description of the solar outer atmosphere.

(3) Baliunas et al. (1979) have shown that steep temperature gradients relatively deep in the chromospheres of active stars such as λ And (G8 III-IV) and Capella (G6 III + F9 III) can produce high pressures in the upper chromospheres and Ca II and Mg II resonance line cores with small K_2-K_3-K_2 contrasts or no central reversals at all, consistent with observations. Their approach takes advantage of properties of the line cores that were ignored in constructing earlier models (e.g. Kelch et al. 1978). While the Baliunas et al. approach may be a reasonable way to model active chromosphere stars, it is important to recognize that other processes are equally effective in filling in the line core; in particular, rotation and intermediate scale turbulence (mesoturbulence; see e.g. Shine 1975; Basri 1979). Baliunas et al. (1979) also show that the usual approach of assuming a steep temperature rise beginning at 8000 K, as appears to be valid in quiet and active regions of the Sun, is an overly restrictive assumption.

Bearing these problems in mind and the nonunique aspects of the diagnostics, it is nonetheless important to consider the basic trends that are surfacing from the modeling effort. In some cases, such trends may not be overly sensitive to the uncertainties of the modeling process:

(1) In most cases studied, temperatures in the stellar upper photosphere inferred from the K line wings are hotter than predicted by radiative equilibrium models, implying nonradiative heating in the photosphere itself. This result is sensitive to uncertainties in the PRD theory, but such uncertainties are least for the Ca II lines in dwarfs, and solar active regions show temperature enhancements of at least 400 K (Chapman 1977; Morrison and Linsky 1978) compared with quiet Sun models below the temperature minimum. If active chromosphere stars, that is stars with chromospheric line surface fluxes comparable to or brighter than solar plages, are at all analogous to solar plages, then such stars should also have photospheres with temperatures significantly hotter than radiative equilibrium models predict. In fact, enhanced photospheric temperatures are common even for nonactive chromosphere stars (Kelch 1978; Kelch et al. 1978, 1979).

(2) Kelch et al. (1979) have presented evidence that the temperature minimum in dwarfs moves outward to smaller mass column densities with decreasing nonradiative heating. (The nonradiative heating is measured by the apparent radiative cooling rate in the Ca II resonance lines.) Furthermore, there is some evidence that the T_{min}/T_{eff} ratio decreases with age among main sequence stars (Linsky et al. 1979a) as might be expected if the nonradiative heating rate decreases with age.

(3) Active chromosphere dwarf stars have larger chromospheric radiative loss rates and steeper chromospheric temperature rises than quiet chromosphere stars. The latter result is depicted in Fig. 1, where the chromospheric temperature gradients for plages (Shine and Linsky 1974; Kelch and Linsky 1978) and flares (Machado and Linsky 1975) are compared to those of 13 dwarf stars (Kelch et al. 1979). The correlation of increasing temperature gradients with increasing nonradiative heating rates is as expected. Kelch et al. (1979) and Cram and Mullan (1979) have shown that Hα emission, which distinguishes dMe stars from normal M dwarfs, can be simply explained by the steep chromospheric temperature rises implied by the bright K line emission characteristic of dMe stars.

(4) For quiet chromosphere dwarfs and giants, Kelch et al. (1979) found a correlation between the mass column density at the 8000 K level of the chromosphere (m_0) and stellar gravity (see Fig. 2). However, Baliunas et al. (1979) argue that the Kelch et al. relation is not valid for RS CVn stars. In addition, the relation does not hold for solar plages and flares, where m_0 can be several orders of magnitude larger than typical quiet Sun values. Furthermore, based on an analysis of the Mg II, Si II, Si III, and C II lines, Simon et al. (1979) find that the active

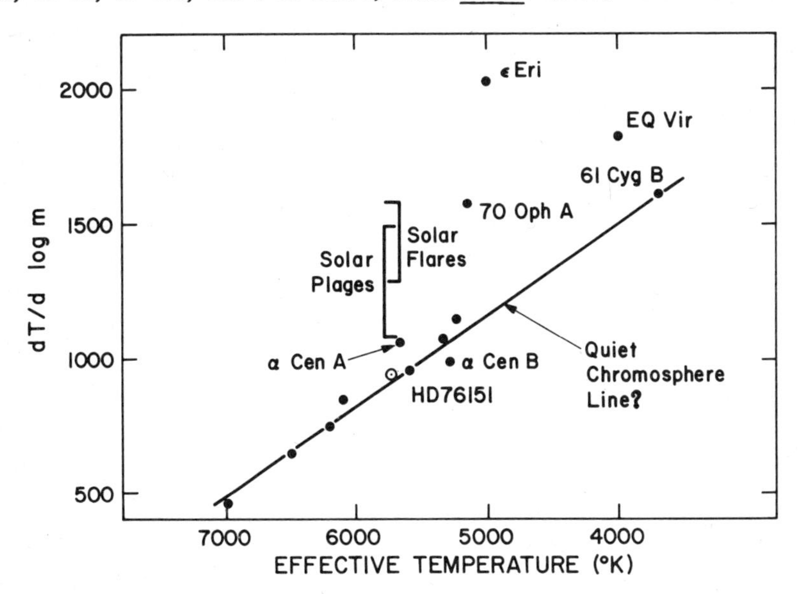

Figure 1. Temperature gradients in chromospheres of main sequence stars (Kelch et al. 1978).

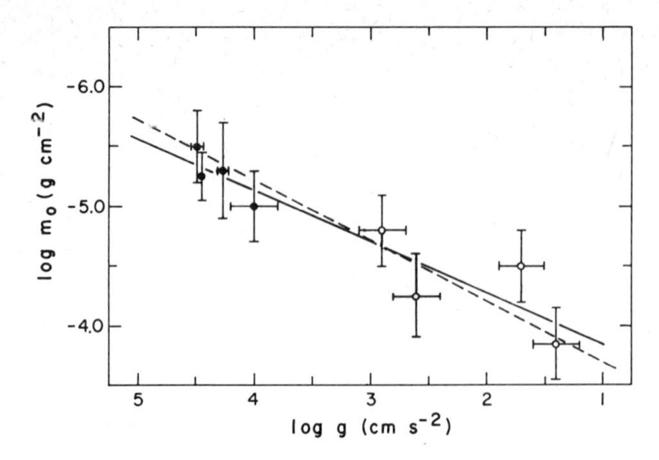

Figure 2. log m_0 versus log g (Kelch et al. 1978).

chromosphere star ε Eri (K2 V) has a value of m_0 a factor of 6 larger than predicted by the quiet chromosphere relation.

(5) A chromospheric temperature plateau near 20,000 K in the Sun (Vernazza et al. 1973), or near 16,500 K as suggested by Lites et al. (1978), may be a real feature of other stars. For example, Simon et al. (1979) propose that ε Eri (K2 V) has such a plateau. Furthermore, Basri et al. (1979) suggest that the plateau decreases in geometrical thickness with increasing brightness of Lα in the Sun.

Clearly much work remains to be done to better understand the available chromospheric diagnostics and to extend chromospheric modeling to other groups of stars. In particular, further studies of supergiants, RS CVn stars, dMe stars, F stars, and T Tauri stars are needed, as well as the extension of chromospheric models to transition regions using IUE spectra. The study of transition regions and coronae per se is beyond the scope of this review, but they are relevant to chromospheres since hydrostatic equilibrium, if valid, requires vertical pressure continuity between the top of the chromosphere and the base of the transition region. In this regard I would like to stress the following points: (1) The most reliable means of deriving pressures at the top of a chromosphere is to analyze diagnostic lines formed near 20,000 K such as the Lα core (useful primarily for the Sun), C II λλ1334, 1335 and Si III λλ1206, 1892. The Ca II and Mg II resonance lines are formed deeper in the chromosphere where the pressures are typically much larger than at the top, consequently the emission cores of these lines are not unique diagnostics. (2) Doschek et al. (1978) have proposed that the Si III λ1892/C III λ1909 line ratio, which is easily observed by IUE, is a useful diagnostic for densities at the base of the transition region. However, Simon et al. (1979) find that if a chromospheric plateau is present near 20,000 K, then Si III λ1892 is formed primarily in the plateau. Under these circumstances the Si III line is sensitive to the plateau thickness and the Si III/C III line ratio is a poor density diagnostic. Other prominent density diagnostics accessible to IUE include Si IV λ1403/C III λ1909 (Cook and Nicolas 1979) and C III λ1176/λ1909 (Raymond and Dupree 1978). However, one must keep in mind that such ratios can be sensitive to systematic flows (Raymond and Dupree 1978; Dupree et al. 1979b). Alternatively one can derive transition region pressures from emission measures, using an auxiliary relation such as constant conductive flux (e.g. Evans et al. 1975; Haisch and Linsky 1976; Brown et al. 1979). However, the validity of this approach is questionable. For example, Ayres and Linsky (1979) have argued against a conduction-heated transition region in the specific case of Capella B because that scenario requires pressures a factor of 30-50 times larger than those estimated from transition region density diagnostics, coronal emission measures, and the Mg II lines.

III. ARE THEORETICAL MODELS OF CHROMOSPHERES BECOMING REALISTIC?

Despite the rapidly growing number of spectroscopic observations of stellar chromospheres and the semiempirical models computed to match these data, we cannot claim that we understand chromospheres without first identifying the important heating mechanism(s), and second computing ab initio theoretical chromosphere models based on these heating mechanisms which accurately match the available data and semiempirical models. This particular goal of understanding stellar chromospheres has not yet been achieved, but considerable progress has been made recently.

Stein and Leibacher (1974) and Ulmschneider (1979) have described the different types of hydrodynamic and magnetohydrodynamic waves that are thought to exist in the solar atmosphere and are candidates for heating stellar chromospheres. Ulmschneider (1979) argues that magnetic modes are likely to dominate the transition region and coronae, but that short period acoustic waves are the best candidate for heating the chromospheres of the Sun and stars of spectral type A and later. His argument is based on the following elements: (1) Deubner (1976) measured the short period acoustic flux in the solar photosphere to be of order $10^8 - 10^9$ ergs cm^{-2} s^{-1}, which is ample enough to balance the total energy loss of about 6×10^6 ergs cm^{-2} s^{-1} in the outer solar atmosphere even with considerable wave damping in the upper photosphere. Other nonmagnetic modes, for example gravity waves, probably do not carry remotely comparable amounts of energy. (2) The contrast in chromospheric emission lines across the solar surface, presumably due to magnetic heating mechanisms, is not large (Ulmschneider 1974). (3) There is rather good agreement between theoretical and empirical chromospheric heating rates, based on the short period acoustic wave mechanism, for several late-type stars and the Sun.

I will consider the validity of these arguments in the context of a comparison between the predictions of the acoustic wave models and empirical data, but first it is important to state the approximations made in recent theoretical calculations.

Prior to 1977, theoretical models of chromospheres (e.g. Kuperus 1965, 1969; Ulmschneider 1967, 1971; de Loore 1970) assumed weak shock theory and time-independent solutions. In more recent calculations (e.g. Ulmschneider and Kalkofen 1977; Ulmschneider et al. 1978, 1979; Schmitz and Ulmschneider 1979a,b) shocks are treated explicitly with time-dependent hydrodynamic codes. These numerical approaches typically assume mixing-length convection, the Lighthill-Proudman theory for acoustic wave generation, a single period for the acoustic waves, and grey opacity stellar atmospheres. I now consider four comparisons between these calculations and empirical measurements of several kinds.

(1) The most basic test of the wave models is that they correctly predict the T_{eff} and gravity dependences of chromospheric radiative loss rates. This test is difficult because several important emission features (i.e. resonance lines of Ca II, Mg II and H I) must be measured to estimate the radiative loss rate in lines, and it is presently impractical to directly measure the chromospheric radiative loss rate in the H^- continuum. As a first attempt at testing the theoretical models, Linsky and Ayres (1978) estimated that Mg II h and k account for 30 percent of the chromospheric line radiative losses in the Sun and other late-type stars for which the Ca II, Mg II, and H I resonance lines are effectively thick. They then compared normalized radiative loss rates in the Mg II lines, $\mathcal{F}(\text{Mg II})/\sigma T_{eff}^4$, for 32 stars including the Sun and found a systematic trend of decreasing $\mathcal{F}(\text{Mg II})/\sigma T_{eff}^4$ with decreasing effective temperature, but essentially no dependence on stellar surface gravity. The computations by Ulmschneider et al. (1977) of the acoustic flux available to heat chromospheres exhibit the observed T_{eff} dependence, but predict an increase in chromospheric heating of 1-2 orders of magnitude between log g = 4 and log g = 2, that is certainly not seen in the data. Subsequently, Basri and Linsky (1979) determined more accurate values of $\mathcal{F}(\text{Mg II k})/\sigma T_{eff}^4$ from their IUE data and the *Copernicus* spectra of Weiler and Oegerle (1979). These data are illustrated in Fig. 3. The normalized Mg II fluxes are widely scattered, but they do show little if any dependence of $\mathcal{F}(\text{Mg II k})/\sigma T_{eff}^4$ on stellar luminosity in agreement with the previous data, and perhaps a slow decrease with decreasing T_{eff}.

An important question in this regard is the amount of chromospheric cooling in the H^- continuum and the extent to which the H^- cooling term depends on gravity. Schmitz and Ulmschneider (1979a,b) revised the previous work of Ulmschneider et al. (1977) in a way that has reduced the gravity dependence of the acoustic flux that survives radiative damping in the photosphere, and is therefore available to heat the low chromosphere. They find that the ratio of the computed H^- radiative loss rates to the empirical Mg II radiative loss rates increases with decreasing gravity in a manner consistent with their acoustic flux calculations. For example, their $\mathcal{F}(H^-)/\mathcal{F}(\text{Mg II})$ ratios range from about 2 for the Sun to roughly 20 for the giants Capella and Arcturus. Since Mg II provides roughly a third of the total line losses, $\mathcal{F}(H^-)/\mathcal{F}(\text{lines})$ ranges from order unity to about 7. However, unlike

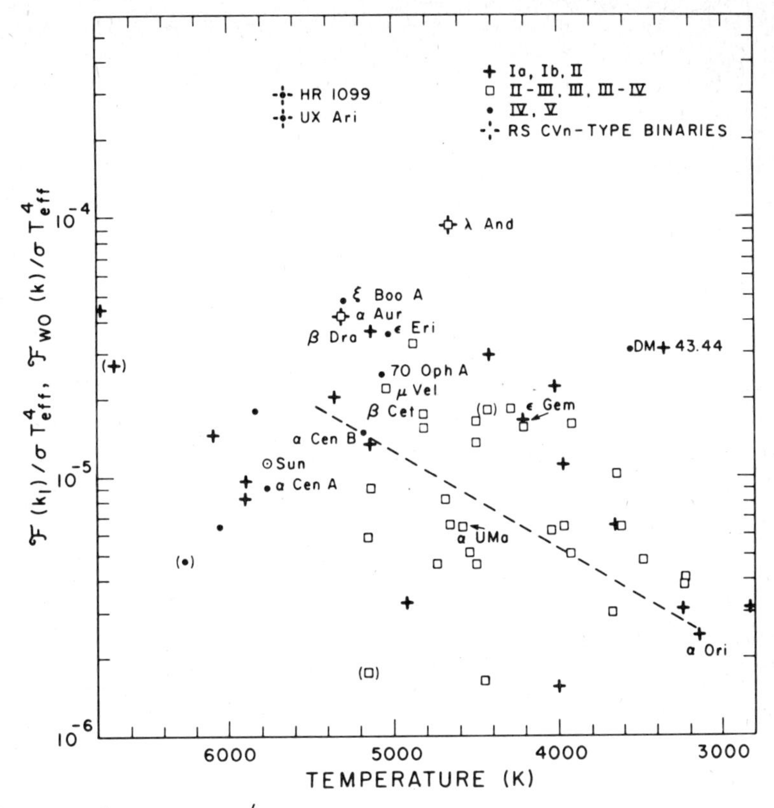

Figure 3. $\mathcal{F}$(Mg II k)/σT^4_{eff} versus T_{eff} (Basri and Linsky 1979).

radiative losses in lines, H^- cooling cannot be measured directly. Schmitz and Ulmschneider instead estimated H^- cooling indirectly using chromospheric models constructed to fit the Ca II and Mg II lines. Not only are the derived H^- radiative loss rates extremely model-dependent, but the use of models constructed to match the Ca II and Mg II lines to calculate the H^- losses is itself questionable. The reason is that the Ca II and Mg II lines average over the intrinsically inhomogeneous stellar atmosphere in a much different way than does in the H^- continuum. Since emission in the near ultraviolet resonance lines is strongly weighted toward the hotter components of an atmosphere, while the mainly visible and near infrared H^- emission is more evenly weighted over the thermal irregularities, the use of models constructed to fit the Ca II and Mg II lines will inevitably overestimate $\mathcal{F}(H^-)$. Finally there remains fundamental disagreement on the proper way to compute H^- radiative losses from a stellar chromosphere (Praderie and Thomas 1972, 1976; Kalkofen and Ulmschneider 1979; Ayres 1979b).

(2) The chromospheric k line radiative loss rates illustrated in Fig. 3 exhibit a wide range of values for stars of similar effective temperature and luminosity.

Excluding the RS CVn-type systems, the range is typically an order of magnitude. The analogous plot of normalized chromospheric radiative losses in the H and K lines (Linsky et al. 1979a) shows a comparable spread. While acoustic wave heating may explain quiet chromospheres, it cannot explain the large diversity of chromospheric radiative loss rates among single stars of similar effective temperature and gravity (same luminosity). In solar plages, chromospheric radiative losses in the Ca II and Mg II resonance lines are commonly 10 times those of the quiet Sun (Kelch and Linsky 1978; Kelch et al. 1979). Since Ca II K line intensities are well correlated with magnetic field strength in the Sun (Skumanich et al. 1975), it is generally presumed that the magnetic field plays an important role in enhancing the chromospheric nonradiative heating rates. In fact, the correlation of Ca II strength with the magnetic field must be viewed as circumstantial evidence for a hydromagnetic origin of chromospheric heating. One likely explanation, then, for the factor of 10 range in $\mathcal{F}(k)/\sigma T_{eff}^4$ at each effective temperature is that stars with low $\mathcal{F}(k)/\sigma T_{eff}^4$ ratios have few solar-type plages, while stars with large ratios are mainly covered with solar-type plages.

(3) A third test is a comparison of the theoretical acoustic wave heating in specific stars with empirically determined radiative loss rates. Ulmschneider et al. (1977) and Schmitz and Ulmschneider (1979a,b) have computed theoretical chromosphere models for the same dwarfs, subgiants, and giants that Ayres et al. (1974), Ayres and Linsky (1975), Kelch (1978), and Kelch et al. (1978, 1979) had previously computed semiempirical models based on the Ca II and Mg II lines. For many of these stars the agreement between computed acoustic fluxes and empirical radiative loss rates (including the estimated H^- contribution) is within a factor of 6, which is not a large factor when one considers the gross uncertainty in the amount of acoustic flux generated in the convective zone (see Gough 1976) and the potential importance of inhomogeneities on determining the "empirical" H^- cooling. The acoustic wave theory has the most difficulty for the cooler dwarfs such as 70 Oph A (K0 V) and EQ Vir (dK5e). For the latter star, in particular, Schmitz and Ulmschneider (1979b) estimate that a factor of 145 times more energy is needed to balance the empirical chromospheric radiative loss rate than is predicted by the acoustic wave theory. Part of the missing flux may be attributed to inadequacies in the theory, for example, the treatment of line blanketing, molecular opacity, and the effects of atmospheric stratification on wave production (Schmitz and Ulmschneider 1979a). However, I feel that the principal reason is the dominance of magnetic heating mechanisms in these active chromosphere stars. The same is almost certainly true for the RS CVn-type systems. For example, the theoretical acoustic heating rates for Capella are considerably less than the apparent chromospheric radiative losses.

(4) A final test is the location in mass column density of the temperature minimum, which is determined by the nonradiative heating rate at the base of the

stellar chromosphere. Cram and Ulmschneider (1978) called attention to inconsistencies between observed and predicted widths of the Ca II K_1 features, which should be formed in the vicinity of T_{min} (but see § VI concerning supergiants). In subsequent computations by Schmitz and Ulmschneider (1979a,b) the agreement between computed and empirical mass column densities at T_{min} has improved except for those stars (70 Oph A, EQ Vir, Capella) that show considerable deficiencies in the computed acoustic flux.

In summary, theoretical models based on the short period acoustic wave theory show promise for explaining the heating of the lower chromospheres of quiet chromosphere stars, but they are clearly inadequate to explain the heating of transition regions and coronae or active chromosphere stars and solar plages. In these "anomalous" cases, magnetic heating mechanisms are presumably dominant. Even for the quiet chromosphere stars the acoustic theory needs to be carefully examined. For example, H^- radiative losses should be calculated for the several classes of thermal inhomogeneities known in the solar case, to establish the reliability of estimating H^- cooling rates from single-component models. In addition, the acoustic wave theory should be extended to more realistic atmospheric models, including nongrey opacity sources and a more complete, nonlinear treatment of the propagation and damping of the sound waves.

IV. WHY DOES THE WILSON-BAPPU RELATION WORK?

Ever since Wilson and Bappu (1957) discovered a simple correlation between the widths of the Ca II H and K line emission cores and stellar absolute luminosity (M_v) extending over 15 magnitudes, many astronomers have expanded the data base and have attempted to explain the origin of this relation. Part of the fascination of this subject must be the inherent simplicity of the correlation and the prospect of obtaining valuable information about stellar chromospheres from simple measurements of line widths. Unfortunately, this is a topic in which one can easily be deluded into feeling that he understands something. In particular, many authors have made back-of-the-envelope calculations and arrived at relations similar in functional form to that originally proposed by Wilson and Bappu (1957), whereas the wide variety of assumptions used are often very different and even contradictory. Unfortunately, there is no substitute for careful analysis of this problem based on realistic radiative transfer calculations. Here I will briefly summarize past observational and theoretical work, and then discuss in detail two recent papers that should revolutionize our approach to the underlying physics of the Wilson-Bappu effect.

Width-luminosity relations have now been found for several lines in addition to Ca II H and K, for example the analogous Mg II h and k resonance lines, Lα, and Hα. The width that Wilson and Bappu (1957) measured, W_0 (km s^{-1}), is the separation between the outer edges of the Ca II emission features on 10Å mm^{-1} spectrograms. Lutz (1970) has shown that W_0 is very nearly the full width at half maximum (FWHM) of the emission core. Wilson and Bappu (1957) and Wilson (1959) proposed the following expression for the K line width-luminosity relation

$$M_v = 27.59 - 14.94 \log W_0 (K) \quad . \qquad (1)$$

The most extensive compilation of Mg II k line widths is that of Weiler and Oegerle (1979), based on Copernicus observations of 49 late-type stars. Their expression,

$$M_v = 34.93 - 15.15 \log W(k) \quad , \qquad (2)$$

has essentially the same slope as that for Ca II, even though the Mg II widths measured at the base of the emission feature. IUE observations will add to this data set and provide widths at different portions of the line profile. Using a small sample of Lα widths (FWHM), McClintock et al. (1975) obtained the relation

$$M_v = (40.2 \pm 4.5) - (14.7 \pm 1.6) \log W(L\alpha) \quad . \qquad (3)$$

Finally, Kraft et al. (1964), LoPresto (1971), and Fosbury (1973) have studied the dependence of the Hα line half-width, defined in several different ways, on M_v, but they have not proposed definite functional relationships.

Different characteristic features within the K line also show width-luminosity relations. For example, Ayres et al. (1975) and Engvold and Rygh (1978) find that the separation of the K_1 minimum features is correlated with M_v. Based on high resolution spectra of 26 late-type stars, Engvold and Rygh (1978) derive

$$M_v \simeq \text{const} - 15.2 \log W(K_1) \quad . \qquad (4)$$

In addition, Lutz et al. (1973) find that the entire damping wings of the H and K lines broaden with increasing stellar luminosity. Finally, Cram et al. (1979) have found in a sample of 32 stars observed at high dispersion, that both the K_2 peak separation and the K_1 minimum separation exhibit essentially the same width-luminosity slope as the FWHM.

The agreement among the slopes of the four relations above is certainly suggestive of a common, presumably simple, origin. As a first step towards understanding that origin, several authors have expressed the line widths empirically in terms of fundamental stellar parameters. For example, Lutz and Pagel (1979) have proposed the relation

$$\log W_0 = -0.22 \log g + 1.65 \log T_e + 0.10[Fe/H] - 3.69 \quad , \qquad (5)$$

based on data from 55 stars. Lutz and Pagel argue that their relation is more accurate than those derived previously without an abundance term. For the K_1 width, Cram et al. (1979) find

$$\log W(K_1) = -(0.20 \pm 0.02) \log g + (1.1 \pm 0.2) \log T_e - 3.76 \quad , \qquad (6)$$

whereas Engvold and Rygh (1978) derive a coefficient of −0.16 for the first term.

Physical interpretations of these width-luminosity relations fall into two distinct classes. The first assumes that the width, generally taken to be W_0 for Ca II K, is formed in the Doppler core of the line profile and therefore responds mainly to turbulent velocities in the chromosphere. For example, Scharmer (1976) followed Goldberg (1957) in assuming that $W_0 \simeq 6\bar{u}$, where $\bar{u}$ is mean the chromospheric turbulent velocity, and arrived at the inevitable result that such velocities are supersonic in giants and highly supersonic in supergiants. Using conservation equations and taking the mechanical energy flux proportional to $\bar{u}^{-3}$, Scharmer (1976) was able to show that $W_0 \simeq 6\bar{u} \sim g^{-1/4}$ in agreement with Eq. (5). However, the derived value of $\bar{u}$ for the Sun is 22 km s^{-1}, far larger than any estimates of turbulent velocities in the solar chromosphere. Fosbury (1973) has studied the Ca II and Hα widths together in order to better determine chromospheric turbulent velocities. He estimated chromospheric wave fluxes from the line widths and concluded that the amplitude of upward propagating acoustic waves increases with luminosity. Most recently, Lutz and Pagel (1979) have derived a relation similar to Eq. (5) assuming slab geometry with a geometrical thickness equal to the K line thermalization length, complete frequency redistribution (CRD), and Doppler control of W_0.

These three papers and previous studies of the same type share a number of difficulties: (1) They assume that W_0 is located in the Doppler core, but give no theoretical or empirical justification for this. (2) They either ignore radiative transfer altogether, despite numerous studies that show that the K and k lines are optically thick and partial redistribution effects are critically important in forming the outer portions of the emission cores, or they treat line transfer in a wholly unrealistic manner. (3) They rarely compute chromospheric densities and ionization self consistently. (4) They do not attempt to explain high resolution solar observations or turbulent velocities measured in the solar chromosphere by various techniques. The ability to arrive at expressions similar to Eq. (5) is often given as sufficient justification that the approaches are accurate enough to explain the underlying physical mechanism.

An alternative physical interpretation for the width-luminosity relations assumes that the width, generally taken to be $W(K_1)$, is formed in the damping wings of the line profile. In this scenario, the width is sensitive primarily to the mass column density above the temperature minimum and is relatively insensitive to chromospheric turbulent velocities. Engvold and Rygh (1978) have argued that $W(K_1) \simeq 7.8 \pm 1.7$ Doppler half-width units and thus the K_1 minimum features must be formed in the damping wings of the line profile. With this assumption and the approximation that the K_1 feature is formed at the temperature minimum, which models suggest occurs at roughly the same continuum optical depth in late-type stars, Ayres et al. (1975) used the pressure-squared dependence of H^- opacity to derive

$$\log W(K_1) = -0.25 \log g + 0.25 \log A_{met} + \text{const} \quad . \qquad (7)$$

The coefficient of the log g term above is close to that of Eq. (6). Thomas (1973) has proposed a somewhat different version of this explanation by showing that the location of the base of the chromosphere, which determines $W(K_1)$, may depend on the upward mass flux, which in turn is related to the chromospheric heating rate.

At this point it should be clear that a synthesis is needed to determine whether these very different explanations for W_0 and $W(K_1)$ are related, and which, if either, remains valid after a realistic study of spectral line formation in the K line. An important paper by Ayres (1979a) goes far in addressing these questions. Ayres proposed simple scaling laws for the thickness and mean electron density of stellar chromospheres as functions of surface gravity and nonradiative heating, based on hydrostatic equilibrium and the assumption that the nonradiative heating is relatively constant with height. He also took into account the influence of ionization on the general structure of a chromosphere and on the plasma cooling functions. He argued that the K_1 features are formed in the damping wings of the line profile on the grounds that PRD calculations now match the limb darkening of the solar Ca II features (Shine et al. 1975; Zirker 1968), which in itself provides strong evidence for near coherent scattering (and thus little Doppler redistribution) beyond the emission peaks. Ayres found that

$$\log W(K_1) = -1/4 \log g + (7/4 \pm 1/2) \log T_{eff} + 1/4 \log \mathcal{F}_{NR} + 1/4 \log A_{met} + \text{const} \quad , \qquad (8)$$

where $\mathcal{F}_{NR}$ is the total nonradiative heating rate and A_{met} is the metal abundance. The coefficients are close to the empirical relation [Eq. (6)] and the derived width ratio $W(k_1)/W(K_1) \cong 2.5$ is consistent with the stellar value of 2.5 ± 0.3 estimated by Ayres et al. (1975) and the solar value of $\cong 2.3$.

Ayres (1979a) further assumed that the K_2 emission peaks are formed just outside of the Doppler core and that the line source function is a maximum at one thermalization length below the top of the chromosphere. These approximations lead to

$$\log W(K_2) = -1/4 \log g - (5/4 \pm 1/2) \log T_{eff} - 1/4 \log \mathcal{F}_{NR} - 1/4 \log A_{met} + 1/2 \cdot \log \xi + \text{const} \quad , \qquad (9)$$

where ξ is the turbulent broadening velocity. The Mg II-Ca II emission peak separation ratio based on the above relation is $W(k_2)/W(K_2) \cong 0.9$.

Several important conclusions can be drawn from this work:

(1) Both $W(K_2)$ and $W(K_1)$ scale as $g^{-1/4}$ if the total nonradiative heating is independent of gravity -- the former owing to the dependence of chromospheric electron density on gravity and the latter owing to the dependence of chromospheric thickness on gravity. Because both $W(K_2)$ and $W(K_1)$ scale with gravity in the same way, it is reasonable that every width between K_1 and K_2, in particular W_0, should also scale the same way. The theoretical gravity dependence is consistent with the Lutz-Pagel scaling law $W_0 \sim g^{-0.22}$, and explains why the width-luminosity laws for W_0, $W(K_1)$ and $W(K_2)$ have essentially the same slopes (Cram et al. 1979).

(2) Ayres (1979a) found that $W(K_2)\sim\xi^{1/2}$ not $\sim\xi^1$ as generally assumed. This is a result of placing the K_2 feature just outside the Doppler core, in the Lorentzian damping wings. The assumption of Lorentzian control at K_2 is based on the argument that the solar Ca II emission peaks limb darken (Zirker 1968), which suggests significant coherent scattering, and thus little Doppler redistribution, where K_2 is formed. Engvold and Rygh (1978) estimate that on the average $W(K_2)/\Delta\lambda_D = 3.4\pm0.2$ for their sample of 26 stars, which suggests that K_2 is formed just outside the Doppler core. Consequently, the $W(K_2)$ and W_0 width-luminosity relations can be explained simply as a gravity effect without having to invoke highly supersonic turbulence in stellar chromospheres.

(3) Ayres predicts that $W(K_2)$ should decrease with $\mathcal{F}_{NR}$, while $W(K_1)$ should increase with $\mathcal{F}_{NR}$. A good test of this prediction is found by comparing solar plage profiles of Ca II (e.g., Smith 1960; Shine and Linsky 1972) with those of the mean quiet Sun. In particular, the nonradiative heating rate is large and easily estimated in plages and the stellar parameters of a plage are the same as those of the quiet Sun. The Ca II plage observations and corresponding data for the k line are consistent with the $\mathcal{F}_{NR}$ dependence of the $W(K_2)$ and $W(K_1)$ scaling laws. In addition, the K line FWHM (approximately equal to W_0) appears to be independent of $\mathcal{F}_{NR}$, which is in accord with Wilson's (1966) result that W_0 appears to be independent of the strength of the emission feature. Furthermore, Linsky et al. (1979a) find rough agreement between measured K_1 widths of a large sample of stars and the gravity and chromospheric heating rate dependences given in Eq. (8).

Finally, I comment on the calculations of Basri (1979), which cast the whole question of the interpretation of width-luminosity relations in a new light. Basri has made a number of prototype PRD calculations of line profiles for chromospheric models of late-type supergiants. His goal was to determine the factors involved in the formation of the emission widths of self-reversed chromospheric resonance lines under conditions of extreme coherency in the line wings. The PRD effects were expected to be particularly severe in supergiants owing to the very small collisional redistriubtion rates in the low density envelopes of such stars.

One set of Basri's calculations assumed a Voigt profile with parameters $a = 5 \times 10^{-4}$, $\varepsilon = 5 \times 10^{-6}$, and $r_0 = 1 \times 10^{-9}$ formed in an isothermal atmosphere. As shown in Fig. 4, he finds that the emergent line profile can have a self-reversed character for small values of the incoherence fraction Λ and r_0, the ratio of continuum to line center opacity, even though the atmosphere is isothermal and therefore has no temperature inversion. The origin of the phantom "K_1" minimum feature is as follows: Outside the Doppler core the photon scattering becomes more coherent with increasing $\Delta x = \Delta\lambda/\Delta\lambda_D$, and the monochromatic source functions rapidly decouple from the line center source function (which itself is close to the local Planck function) such that the emergent intensity decreases with increasing Δx. The intensity then rises in the far wings as the monochromatic source function begins to

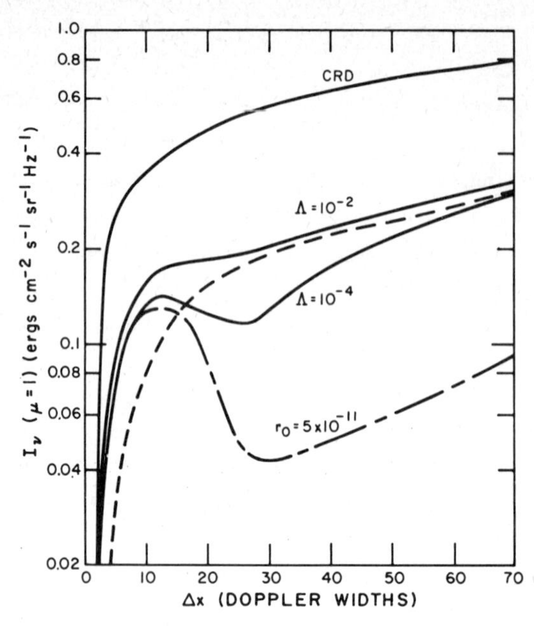

Figure 4. Line profiles for an isothermal atmosphere (Basri 1979, Fig. 4.3).

include a significant contribution from the assumed purely thermal background continuum. The extent to which the phantom K_1 feature occurs depends on the relative importance of the competing terms. Increasing coherence (decreasing Λ) tends to emphasize this Schuster-type process (Mihalas 1970), while increasing collisional redistribution (increasing Λ) or increasing continuum absorption (increasing r_0) deemphasizes it. In addition, Doppler drifting -- the frequency diffusion of photons with each "coherent" scattering owing to residual Doppler motions -- tends to increase the probability of core photons wandering into the line wings. The redistribution of core photons increases the effective noncoherent scattering term, which in turn raises the K_1 minimum over the pure coherent case. In fact, Basri found that microturbulence can influence the location of the K_1 feature even though K_1 is formed far from the Doppler core, by enhancing the Doppler drifting mechanism. Basri's calculations demonstrate the critical importance of properly treating frequency redistribution for very coherent cases and the potential dangers of naively assuming that the K_1 minimum feature is formed at the temperature minimum in extremely low gravity stars.

As a test of what mechanisms affect line profile shapes in realistic supergiant chromospheres, Basri considered the Mg II k line for the not extreme example of β Dra (G2 II), for which $W(k_1) = 2.5$ Å. Figure 5 depicts atmospheric parameters for solar-type temperature distributions (Models A and A') with T_{min} at $\log m_R = -1$ g cm^{-2} and a supergiant-type temperature distribution (Model B) with T_{min} shifted inward in mass by a factor of 100. For Model A, with a maximum turbulent velocity

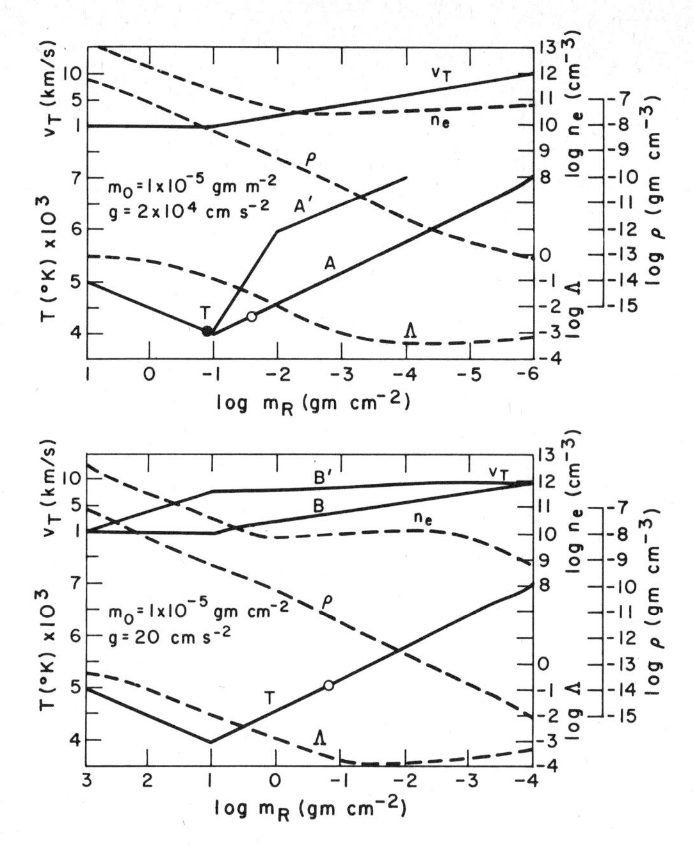

Figure 5. Atmospheric parameters for β Dra models (Basri 1979, Fig. 4.7).

in the chromosphere of 10 km s^{-1} and either the gravity of the Sun or β Dra, $W(k_1)$ occurs near 0.5 Å. By increasing the chromospheric temperature gradient (Model A') to produce a strong emission profile for β Dra, Basri finds that $W(k_1)$ is about 1.5 Å, even when the microturbulence is increased to the highly supersonic value of 30 km s^{-1}. However, the large values of microturbulence tend to wash out the k_2 emission features. Basri concludes that microturbulence by itself cannot be a viable explanation of the $W(k_1)$ - M_v or $W(K_1)$ - M_v relations in supergiants, although highly supersonic microturbulence can broaden W_0.

Alternatively, increasing the mass column density down to the temperature minimum (as is done in Model B) results in $W(k_1)$ = 1.5 Å (see Fig. 6), even when $m_{T_{min}}$ is increased without limit. The apparent lack of sensitivity of $W(k_1)$ to chromospheric column densities occurs because the k_1 minimum feature is strongly decoupled from the temperature structure (as was seen in the isothermal example). The location of k_1 is now determined by the balance of coherent and noncoherent processes. In particular, Doppler drifting is an important incoherence term which can be enhanced by increasing the microturbulence. Basri finds that $W(k_1)$ can be

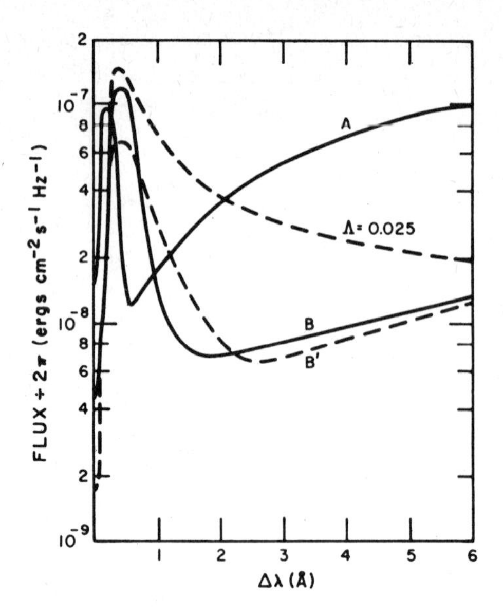

Figure 6. Mg II k lines computed for different β Dra models (Basri 1979, Fig. 4.8b).

broadened to the observed value of 2.5 Å simply by using a barely supersonic microturbulence of 8 km s^{-1} (Model B' in Figs. 5 and 6).

The prototype supergiant calculations leave us with the following unanticipated results: (1) k_1 and K_1 need not be formed at the temperature minimum. (2) Neither the turbulent velocity nor the mass column density explanation for the $W(K_1) - M_v$ relation is viable by itself. (3) Microturbulent velocities can play an important role in determining $W(k_1)$ and $W(K_1)$ via the Doppler drifting mechanism. Even though these conclusions refer to the perhaps extreme case of low density supergiant chromospheres, the most important message is that it is essential to solve the transfer equation properly before one can make meaningful statements concerning the physical basis of width-luminosity relations.

V. ARE THERE SYSTEMATIC FLOW PATTERNS IN STELLAR CHROMOSPHERES?

Until now we have been concerned with the interpretation of chromospheric line intensities and widths. These data provide valuable information on thermal structure and random nonthermal velocities in chromospheres, but they do not contain useful information on systematic flow patterns. Line profile asymmetries may provide such information with proper interpretation, and in fact they are the only means at present of studying chromospheric systematic velocity fields. Because the analysis of line asymmetries is complex, I first consider available theoretical calculations of profiles of optically thick chromospheric lines in the presence of systematic flow patterns, and then consider the extent to which solar and stellar data can be understood in terms of these models.

(1) Athay (1976) has emphasized the caution with which one should approach the question of determining velocity fields from line asymmetries. He provides examples of velocity fields for which even absorption line profiles predict the wrong sign of the flow vector. Needless to say, the analysis of optically thick self-reversed emission cores is more complex and must be undertaken with care.

(2) Using as a test case the Ca II K line formed in the solar chromosphere, Athay (1970) and Cram (1972) have shown that asymmetries provide information on velocity gradients in the line formation region but not on the magnitude or even the direction of the flow in any specific atmospheric layer. For example, Athay (1970) showed that downward motions of 10-20 km s^{-1} in the region of K_3 formation shift K_3 to the red, thereby moving absorbing material (material with a smaller source function than the underlying material producing the K_2 emission peaks) so as to partially obscure the K_{2R} emission peak and uncover the K_{2V} emission peak. The net effect is to produce brighter emission at K_{2V} and weaker emission at K_{2R}, and $\Delta\lambda_{K_3} > 0$. Unfortunately, the exact same asymmetries are produced by assuming upward motions of 3-7 km s^{-1} in the K_2-forming region and no systematic velocities where K_3 is formed. What I refer to as "blue asymmetry," $I(K_{2V}) > I(K_{2R})$ and $\Delta\lambda_{K_3} > 0$, thus only provides information on the systematic vertical velocity gradient, specifically, that $dv/dh < 0$, but no unique information on absolute vertical velocities anywhere. Similarly "red asymmetry," that is $I(K_{2V}) < I(K_{2R})$ and $\Delta\lambda_{K_3} < 0$, implies only that $dv/dh > 0$.

(3) Vertically propagating waves with scales between the microturbulent and macroturbulent limits ("mesoturbulence") and which are nonsinusoidal in character, can also produce line asymmetries. Shine (1975) has synthesized Na D profiles for a solar chromosphere model permeated by shock waves (approximated by vertically propagating sawtooth waves). He finds that the basic asymmetry of the sawtooth function produces time-averaged absorption profiles with line center shifted to the red, and the blue wing brighter than the red wing. Shine's calculations suggest that vertically propagating shock waves can produce blue asymmetries in chromospheric emission cores that might be mistaken for the symptoms of downflows where K_3 is formed.

(4) Heasley (1975) has synthesized the Ca II resonance and infrared triplet lines for models of the solar atmosphere including upward propagating acoustic pulses, perturbations in the local temperature and density induced by the pulses, and resultant changes in the line source functions, all self-consistently. He finds that the passage of a pulse through the chromosphere produces first a blue asymmetry, owing to upward motion where K_2 is formed and no motion where K_3 is formed, and then a red asymmetry, owing to upward motion where K_3 is formed and no motion where K_2 is formed. The effect of including the density and temperature perturbations associated with the pulse is to enhance hydrogen ionization. The increased electron

density in turn drives the collision-dominated line source function closer to the Planck function, thereby enhancing the K_2 emission. In fact, he notes that the dominant influence on the Ca II line source functions is not the velocity field of the pulses, but rather the atmospheric perturbations. Furthermore, the pulses do not change the wavelengths of the K_2 peaks, only their strengths.

The mean quiet Sun K line profile (e.g. White and Suemoto 1968) and the integrated solar disk K line profile (Beckers et al. 1976) both show blue asymmetry, as do the Mg II resonance lines (Lemaire and Skumanich 1973) and Lα (Basri et al. 1979). To interpret this asymmetry, we can take advantage of the high spatial resolution spectra that can be obtained from the Sun. Such spectra in the K line (e.g., Pasachoff 1970; Wilson and Evans 1972) show a variety of single and double peak profiles with red and blue asymmetry, but relatively constant K_2 peak wavelengths, consistent with Heasley's (1975) calculations. The real clue to the cause of the blue asymmetry seen in the whole Sun K line profile is provided by the high spatial resolution time-sequence spectra of Liu (1974). These data show intensity perturbations that travel from the far wings toward line center and produce a strong blue asymmetry when the disturbance reaches the line core. Liu (1974) and Liu and Skumanich (1974) have interpreted this behavior in terms of local heating of the chromosphere by upward propagating waves. The important message for us is that the solar blue asymmetry is very likely produced by something analogous to Heasley's acoustic pulses, that has the largest effect on the K line profile at those phases when the temperature and density perturbations and the upward motions are all positive in the chromospheric layers where K_2 is formed. However, the net blue asymmetry tells us nothing about the direction or magnitude of the solar wind or the nature of possible circulation patterns in the solar chromosphere.

With this background we can consider the stellar data. Profiles of the Ca II and Mg II lines in F-K main sequence stars (cf. Linsky et al. 1979a; Basri and Linsky 1979) typically have blue asymmetry, like the Sun, symptomatic of upward propagating waves in their chromospheres if our solar explanation is correct.

The G and K giants exhibit more complex behavior. Arcturus (K2 III), for example, shows Mg II line profiles with pronounced red asymmetry that did not change during 1973-1976 (McClintock et al. 1978), whereas Chiu et al. (1977) find that the K line asymmetry is variable in their data and in previous work going back to 1961, with blue asymmetry perhaps more common. Chiu et al. (1977) have modeled the red asymmetry Ca II and Mg II line profiles with a mass flux conservative stellar wind; that is, the outflow velocity is inversely proportional to the density and therefore increases rapidly with height. Such a velocity field produces red asymmetries in both the Ca II and Mg II lines, as expected from Athay's (1970) analysis for $dv/dh > 0$. The derived mass loss rate from the best fit models is $8 \times 10^{-9}\ M_{\odot}\ yr^{-1}$ and the outflow velocity is 13 km s^{-1} at $\tau(K_3) = 1$. The question remains, however, why Arcturus can have simultaneously a blue asymmetry Ca II profile and a strongly red asymmetry Mg II profile. McClintock et al. (1978) argue that the deep absorption feature at -38.3 km s^{-1} that produces or contributes to the red asymmetry of the Mg II profiles is very likely not interstellar absorption and must therefore be intrinsic to the star. Possible explanations for the opposite asymmetries include large systematic velocity gradients in the chromosphere (the Mg II lines should be formed slightly higher in the chromosphere than Ca II) or an expanding circumstellar envelope with large Mg II column density and small Ca II column density, presumably owing to ionization of Ca^{+}. The latter possibility may be correct, because the slightly cooler giant Aldebaran (K5 III) has Mg II line profiles very similar to Arcturus (van der Hucht et al. 1979), but Ca II lines which contain variable, weak circumstellar absorption features blue shifted by about 30 km s^{-1} (Reimers 1977; Kelch et al. 1978). Reimers (1977) has designated such circumstellar components "K_4."

Stencel (1978) found a statistical trend of K line blue asymmetry for giants hotter than spectral type K3 and red asymmetry for giants cooler than K4. The location of Stencel's Ca II asymmetry dividing line in the H-R diagram is depicted in Fig. 7. If blue asymmetry is symptomatic of upward propagating waves but no large wind in the chromosphere, and red asymmetry indicates the presence of a significant outward mass flux and possibly also a circumstellar shell, then the dividing line indicates the onset of massive winds in stellar chromospheres. Most recently, Stencel and Mullan (1979) (cf. Copernicus data of Weiler and Oegerle 1979) have determined a locus in the H-R diagram (see Fig. 7) where the Mg II resonance lines change their asymmetry in a manner similar to that found by Stencel (1978) for

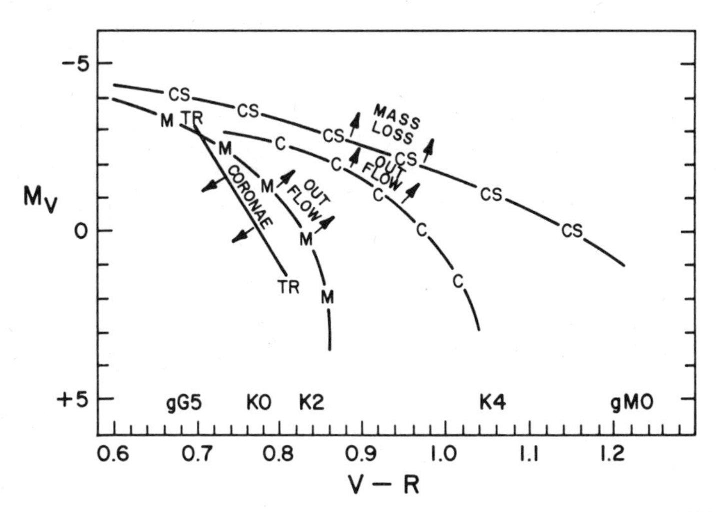

Figure 7. Transition region dividing line (Linsky and Haisch 1979), Mg II (M) asymmetry dividing line (Stencel and Mullan 1979), Ca II (C) asymmetry dividing line (Stencel 1978), and circumstellar (CS) dividing line (Reimers 1977).

H and K. The Ca II and Mg II dividing lines are located slightly to the right of that proposed by Linsky and Haisch (1979) to separate stars exhibiting high excitation emission lines characteristic of a solar-like transition region (material at 20-250 × 10^3 K) from stars showing only chromospheric emission lines (material at less than 10^4 K). Taken together these data suggest that cool stars fall into two distinct classes: those with outer atmospheres consisting of chromospheres, transition regions, and presumably also hot coronae; and those with chromospheres and massive cool winds.

Mullan (1978) has attempted to explain the apparent onset of massive winds in the early K stars. He proposes that the location of the point where the stellar wind becomes supersonic moves deeper into the stellar atmosphere as gravity and effective temperature decrease. Haisch et al. (1979) have shown that Lα radiation pressure may play an important role in initiating the cool stellar winds. The cooling effect of the wind may explain the following:

(a) For giants with color index (V-R) < 0.80 (about spectral type K0 III), the stellar wind is optically thin, but the energy associated with the mass flow is about equal to the nonradiative energy that would otherwise heat a hot corona. Such stars therefore do not have coronae and fall to the right of the Linsky-Haisch dividing line. However, the transonic point of the wind lies above the region where the Mg II lines are formed, consequently the Mg II lines are symmetric or show blue asymmetry. These stars therefore fall to the left of the Mg II asymmetry locus.

(b) For giants with (V-R) > 0.85 (about spectral type K2 III), the wind now affects the upper chromosphere where the Mg II lines are formed. The wind reverses the Mg II asymmetry and the mass loss rate rises because the transonic point has penetrated into the chromosphere where the density is high.

(c) For giants with (V-R) > 1.00 (about spectral type K4 III), the transonic point occurs well into the middle chromosphere where H and K form. The Ca II lines acquire red asymmetry and the wind now carries enough material to produce observable circumstellar features.

In cool supergiants like ε Gem (G8 Ib), ε Peg (K2 Ib), ξ Cyg (K5 Ib), and α Ori (M2 Iab), the Ca II and Mg II line asymmetries (cf. Linsky et al. 1979a; Basri and Linsky 1979) are dominated by apparent circumstellar absorption and it is difficult to determine the asymmetry of the unmutilated chromospheric line. The problem of the intrinsic asymmetries of chromospheric emission cores in supergiants is further complicated by the following points: (1) ε Gem shows apparent circumstellar absorption features at both positive and negative velocities. (2) The Mg II k profile of α Ori and presumably other stars is multilated by overlying circumstellar Fe II and Mn I absorption lines (cf. Bernat and Lambert 1976; de Jager et al. 1979). (3) Basri (1979) has constructed a chromospheric model for α Ori to match the Ca II and Mg II line profiles. He finds that broad, flat-bottomed reversals in the Ca II lines are easily predicted by the chromospheric model without including any circumstellar shell whatsoever. Consequently, our rather casual identification of circumstellar features in chromospheric lines of supergiants must be reconsidered.

Not all cool supergiants show "circumstellar" features. In particular, β Dra (G2 II) exhibits Ca II profiles with very strong blue asymmetry and Mg II profiles with less pronounced blue asymmetry (see Fig. 8). Basri (1979) has modeled these data using a comoving PRD code and a vertical velocity (see Fig. 9) that increases with height. Such a velocity field is consistent with the data, but it is perhaps unphysical and, as noted above, upward propagating waves can produce the same kind of asymmetry.

Variable asymmetry in chromospheric emission lines may turn out to be common in late-type stars as specific stars are monitored for long periods. For example, Hollars and Beebe (1976) have noted changes in the K line of α Aqr (G2 Ib), and Dravins _et al._ (1977) have found K line asymmetry changes in the δ Scuti star ρ Pup. O'Brien and Lambert (1979) have reported variable He I λ10830 emission from α^1 Her, which may result from a shock front created when high-velocity gas accretes onto the photosphere of α^1 Her. O'Brien (1979) has also reported λ10830 observations of a number of F-M stars. Many show variable absorption or emission, and α Aqr shows evidence of enormous outflow velocites of nearly 200 km s^{-1}.

The interpretation of asymmetries in chromospheric lines (Balmer series, Ca II, Na I) of T Tauri stars is a matter of dispute. Blue shifted absorption components in these lines have generally been interpreted as symptoms of a strong stellar wind (e.g. Herbig 1962). Nevertheless, Ulrich (1976) has demonstrated that nonspherically symmetric accretion can produce apparent blueshifted absorption components from an infalling postshock gas. Ulrich and Knapp (1979) find that absorption components in Hα are not reliable indicators of gas flow direction, contrary to the common presumption, but instead the Na D line absorption features are more reliable flow-vector diagnostics. They conclude that accretion occurs in the major fraction of T Tauri stars and that discrete clouds of ejected material, perhaps due to flares, continually pass through the generally infalling gas.

I wish to thank Drs. T. R. Ayres and G. S. Basri for their comments on the text, and many collegues for sending me preprints and unpublished data for inclusion in this paper. This work was supported in part by NASA through grants NAS5-23274 and NGL-06-003-057 to the University of Colorado.

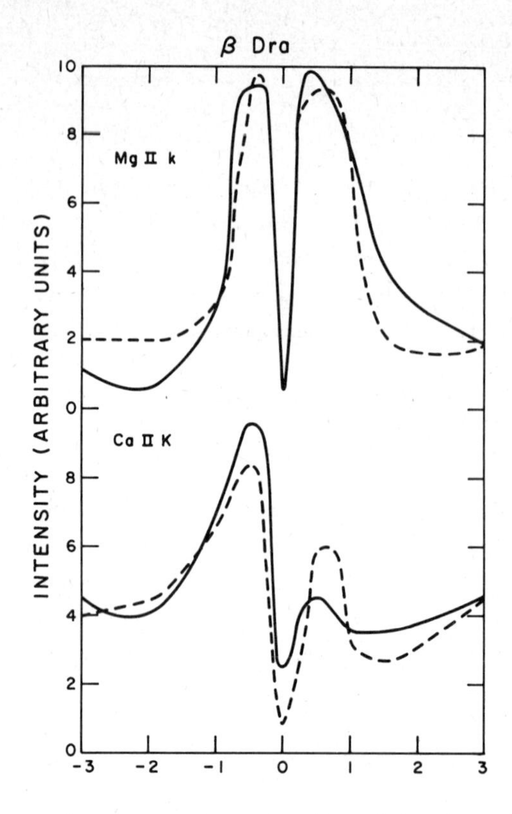

Figure 8. Computed β Dra Ca II and Mg II profiles (Basri 1979, Fig. 5.3).

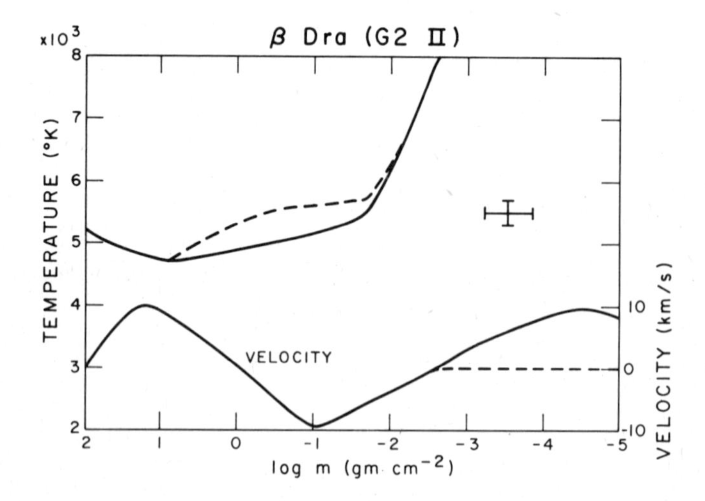

Figure 9. β Dra model and systematic velocities (Basri 1979, Fig. 5.2).

References

Anderson, C. M. 1974, Ap. J., 190, 585.
Athay, R. G. 1970, Solar Phys., 11, 347.
Athay, R. G. 1976, The Solar Chromosphere and Corona: Quiet Sun, D. Reidel, Dordrecht.
Avrett, E. H. 1979, private communication.
Ayres, T. R. 1979a, Ap. J., 228, 509.
Ayres, T. R. 1979b, private communication.
Ayres, T. R. and Linsky, J. L. 1975, Ap. J., 200, 660.
Ayres, T. R. and Linsky, J. L. 1976, Ap. J., 201, 212.
Ayres, T. R. and Linsky, J. L. 1979, in preparation.
Ayres, T. R., Linsky, J. L., Rodgers, A. W., and Kurucz, R. L., 1976, Ap. J., 210, 199.
Ayres, T. R., Linsky, J. L., and Shine, R. A. 1974, Ap. J., 192, 93 (1974).
Ayres, T. R., Linsky, J. L., and Shine, R. A. 1975, Ap. J. (Letters), 195, L121.
Baliunas, S. L., Avrett, E. H., Hartmann, L., and Dupree, A. K. 1979, Ap. J. (Letters), in press.
Baliunas, S. L. and Dupree, A. K. 1979, Ap. J., 227, 870.
Barnes, T. G. and Evans, D. S. 1976, M.N.R.A.S., 174, 489.
Basri, G. S. 1979, unpublished Ph.D. thesis, University of Colorado.
Basri, G. S. and Linsky, J. L. 1979, Ap. J., in press.
Basri, G. S., Linsky, J. L., Bartoe, J.-D. F., Brueckner, G., and Van Hoosier, M. E. 1979, Ap. J., 230, 924.
Beckers, J. M., Bridges, C. A., and Gilliam, L. B. 1976, A High Resolution Spectral Atlas of the Solar Irradiance from 380 to 700 Nanometers, Vol. 1 [AFGL-TR-76-0126(I)].
Bernat, A. P. and Lambert, D. L. 1976, Ap. J., 204, 803.
Blanco, C., Catalano, S., and Marilli, E., 1976, Astr. Ap., 48, 19.
Blanco, C., Catalano, S., and Marilli, E. 1978, Proceedings of the 4th International Colloquium on Astrophysics.
Blanco, C., Catalano, S., Marilli, E., and Rodono, M. 1974, Astr. Ap., 33, 257.
Bohm-Vitense, E. and Dettmann, T. 1979, Ap. J., in press.
Brown, A., Jordan, C., and Wilson, R. 1979, in Proceedings of the Symposium "The First Year of IUE," to appear.
Carpenter, K. G. and Wing, R. F. 1979, B.A.A.S., 11, 419.
Chapman, G. A. 1977, Ap. J. Suppl., 33, 35.
Chiu, H. Y., Adams, P. J., Linsky, J. L., Basri, G. S., Maran, S. P., and Hobbs, R. W. 1977, Ap. J., 211, 453.
Cook, J. W. and Nicolas, K. R. 1979, Ap. J., 229, 1163.
Cram, L. E. 1972, Solar Phys. 22, 375.
Cram, L. E., Krikorian, R., and Jefferies, J. T. 1979, Astr. Ap., 71, 14.
Cram, L. E. and Mullan D. J. 1979, Ap. J., to appear.
Cram, L. E. and Ulmschneider, P. 1978, Astr. Ap., 62, 239.
de Jager, C., Kondo, Y., Hockstra, R., van der Hucht, K. A., Kamperman, T., Lamers, H. J. G. L. M., Modisette, J. L., and Morgan, T. 1979, Ap. J., 230, 534.
de Loore, C. 1970, Astr. Space Sci. 6, 60.
Deubner, F.-L. 1976, Astr. Ap., 51, 189.
Doherty, L. R. 1973, in Stellar Chromospheres, ed. S. D. Jordan and E. H. Avrett, NASA SP-317, p. 99.
Doschek, G. A., Feldman, U., Mariska, J. T., and Linsky, J. L. 1978, Ap. J. (Letters), 226, L35.
Dravins, D., Lind, J., and Sarg, K. 1977, Astr. Ap., 54, 381.
Dupree, A. K., Black, J. H., Davis, R., Hartmann, L., and Raymond, J. C. 1979a, in Proceedings of the Symposium "The First Year of IUE," (in press).
Dupree, A. K., Moore, R. T., and Shapiro, P. R. 1979b, Ap. J. (Letters), 229, L101.
Evans, R. G., Jordan, C., and Wilson, R. 1975, M.N.R.A.S., 172, 585.
Engvold, O. and Rygh, B. O. 1978, Astr. Ap., 70, 399.
Fosbury, R. A. E. 1973, Astr. Ap., 27, 129.
Fosbury, R. A. E. 1974, M.N.R.A.S., 169, 147.
Giampapa, M. S. 1979, private communication.
Goldberg, L. 1957, Ap. J., 126, 318.

Gough, D. O. 1976, Proc. IAU Colloq. No. 36, ed. R. M. Bonnet and P. Delache (Clermont-Fenand: Paris), p. 3.
Haisch, B. M. and Linsky, J. L. 1976, Ap. J. (Letters), 205, L39.
Haisch, B. M., Linsky, J. L. and Basri, G. S. 1979, Ap. J., in press.
Hartmann, L., Davis, R., Dupree, A. K., Raymond, J., Schmidtke, P. C., and Wing, R. F. 1979, Ap. J. (Letters), in press.
Heasley, J. N. 1975, Solar Phys., 44, 275.
Heasley, J. N., Ridgway, S. T. Carbon, D. F., Milkey, R. W., and Hall, D. N. B. 1978, Ap. J. 219, 970.
Herbig, G. H. 1962, Adv. Astr. Ap., 1, 47.
Hollars, D. R. and Beebe, H. A. 1976, Pub. A.S.P., 88, 934.
Jordan, S. D. and Avrett, E. H. 1973, Stellar Chromospheres, NASA SP-317.
Kalkofen, W. and Ulmschneider, P. 1979, Ap. J., 227, 655.
Kelch, W. L. 1978, Ap. J., 222, 931.
Kelch, W. L. and Linsky, J. L. 1978, Solar Phys., 58, 37.
Kelch, W. L., Linsky, J. L., Basri, G. S., Chiu, H. Y., Chang, S. H., Maran, S. P. and Furenlid, I. 1978, Ap. J., 220, 962.
Kelch, W. L., Linsky, J. L., and Worden, S. P. 1979, Ap. J., 229, 700.
Kippenhahn, R. 1973, Stellar Chromospheres, ed. S. D. Jordan and E. H. Avrett, NASA SP-317, p. 265.
Kondo, Y., de Jager, C., Hoekstra, R., van der Hucht, K. A., Kamperman, T. M., Lamers, H.J.G.L.M., Modisette, J. L., and Morgan, T. H. 1979, Ap. J., 230, 533.
Kraft, R. P., Preston, G. W., and Wolf, S. C. 1964, Ap. J., 140, 237.
Kuperus, M. 1965, Recherches Astr. Obs. Utrecht, 17, 1.
Kuperus, M. 1969, Space Sci. Rev., 9, 713.
Lemaire, P. and Skumanich, A. 1973, Astr. Ap., 22, 61.
Linsky, J. L. 1977, in The Solar Output and Its Variation, ed. O. R. White (Boulder: Colorado Associated University Press), p. 477.
Linsky, J. L. 1979, in Report of Commission 36 in Transactions of the IAU, 17, in press.
Linsky, J. L. 1980, in Annual Rev. Astr. Ap., 18, in preparation.
Linsky, J. L., and Ayres, T. R. 1978, Ap. J., 220, 619.
Linsky, J. L. and Haisch, B. M. 1979, Ap. J. (Letters), 229, L27.
Linsky, J. L., Hunten, D. M., Sowell, R., Glackin, D. L., and Kelch, W. L. 1979b, Ap. J. Suppl., in press.
Linsky, J. L., Worden, S. P., McClintock, W., and Robertson, R. M. 1979a, Ap. J. Suppl., in press.
Linsky, J. L. et al. 1978, Nature, 275, 389.
Lites, B. W., Shine, R. A., and Chipman, E. G. 1978, Ap. J., 222, 333.
Liu, S. Y. 1974, Ap. J., 189, 359.
Liu, S. Y. and Skumanich, A. 1974, Solar Phys., 38, 109.
LoPresto, J. C. 1971, Pub. A.S.P., 83, 674.
Lutz, T. E. 1970, Ap. J., 75, 1007.
Lutz, T. E., Furenlid, I., and Lutz, J. H. 1973, Ap. J., 184, 787.
Lutz, T. E. and Pagel, B. E. J. 1979, M.N.R.A.S., submitted.
Machado, M. E., Emslie, A. G., and Brown, J. C. 1978, Solar Phys., 58, 363.
Machado, M. E. and Linsky, J. L., 1975, Solar Phys., 42, 395.
McClintock, W., Henry, R. C., Moos, H. W., and Linsky, J. L. 1975, Ap. J., 202, 733.
McClintock, W., Moos, H. W., Henry, R. C., Linsky, J. L., and Barker, E. S. 1978, Ap. J. Suppl., 37, 223.
Mihalas, D. 1970, Stellar Atmospheres (Freeman: San Francisco), p. 330.
Morrison, N. D. and Linsky, J. L. 1978, Ap. J., 222, 723.
Mullan, D. J. 1978, Ap. J., 226, 151.
O'Brien, G. 1979, private communication.
O'Brien, G. and Lambert, D. L. 1979, Ap. J. (Letters), 229, L33.
Pagel, B. E. J. and Wilkins, D. R. 1979, M.N.R.A.S., submitted.
Pasachoff, J. M. 1970, Solar Phys., 12, 202.
Praderie, F. 1973, Stellar Chromospheres, ed. S. D. Jordan and E. H. Avrett, NASA SP-317, p. 79.
Praderie, F. 1977, Mem. S. A. It., 553.
Praderie, F. and Thomas, R. N. 1972, Ap. J., 172, 485.

Praderie, F. and Thomas, R. N. 1976, Solar Phys., 50, 333.
Raymond, J. C. and Dupree, A. K. 1978, Ap. J., 222, 379.
Reimers, D. 1977, Astr. Ap., 57, 395.
Schmitz, F. and Ulmschneider, P. 1979a, Astr. Ap., in press.
Schmitz, F. and Ulmschneider, P. 1979b, Astr. Ap., in press.
Scharmer, G. B. 1976, Astr. Ap., 53, 341.
Shine, R. A. 1975, Ap. J., 202, 543.
Shine, R. A. and Linsky, J. L. 1972, Solar Phys., 25, 357.
Shine, R. A. and Linsky, J. L. 1974, Solar Phys., 39, 49.
Shine, R. A., Milkey, R. W., and Mihalas, D. 1975, Ap. J., 199, 724.
Skumanich, A., Smythe, C., and Frazier, E. N. 1975, Ap. J., 200, 747.
Simon, T., Kelch, W. L., and Linsky, J. L. 1979, Ap. J., submitted.
Simon, T. and Linsky, J. L. 1979, in preparation.
Smith, E., van P. 1960, Ap. J., 132, 202.
Snow, T. P. and Linsky, J. L. 1979, in Proc. Conf. on Astr. Ap. from Spacelab, ed. P. L. Bernacca and R. Ruffini (Dordrecht: Reidel), in press.
Stein, R. F. and Leibacher, J. 1974, Ann Rev. Astr. Ap., 12, 407
Stencel, R. E. 1978, Ap. J. (Letters), 223, L37.
Stencel, R. E. and Mullan, D. J. 1979, Ap. J., submitted.
Thomas, R. N. 1973, Astr. Ap., 29, 297.
Tripp, D. A., Athay, R. G., and Peterson, V. L. 1978, Ap. J., 220, 314.
Ulmschneider, P. 1967, Z. Ap., 67, 193.
Ulmschneider, P. 1971, Astr. Ap., 12, 297.
Ulmschneider, P. 1974, Solar Phys., 39, 327.
Ulmschneider, P. 1979, Space Sci. Rev., in press.
Ulmschneider, P. and Kalkofen, W. 1977, Astr. Ap., 57, 199.
Ulmschneider, P., Schmitz, F., and Hammer, R. 1979, Astr. Ap., in press.
Ulmschneider, P., Schmitz, F., Kalkofen, W., and Bohm, H. U. 1978, Astr. Ap., 70, 487.
Ulmschneider, P., Schmitz, F., Renzini, A., Cacciari, C., Kalkofen, W. and Kurucz, R. L. 1977, Astr. Ap., 61, 515.
Ulrich, R. K. 1976, Ap. J., 210, 377.
Ulrich, R. K. and Knapp, G. R. 1979, Ap. J. (Letters), 230, L99.
van der Hucht, K. A., Stencel, R. E., Haisch, B. M., and Kondo, Y. 1979, Astr. Ap., in press.
Vernazza, J. E., Avrett, E. H., and Loeser, R. 1973, Ap. J., 184, 605.
Vernazza, J. E., Avrett, E. H., and Loeser, R. 1976, Ap. J. Suppl., 30, 1.
Weiler, E. J. and Oegerle, W. R. 1979, Ap. J. Suppl., 39, 537.
White, O. R. and Suemoto, Z. 1968, Solar Phys., 3, 523.
Willstrop, R. V. 1972, private communication.
Wilson, O. C. 1959, Ap. J., 130, 499.
Wilson, O. C. 1966, Science, 151, 1487.
Wilson, O. C. 1978, Ap. J., 226, 379.
Wilson, O. C. and Bappu, M. K. V. 1957, Ap. J., 125, 661.
Wilson, P. R. and Evans, C. D. 1972, Solar Phys., 18, 29.
Withbroe, G. L. and Noyes, R. W. 1977, Ann. Rev. Astr. Ap., 15, 363.
Worden, S. P., Schneeberger, T. J., and Giampapa, M. S. 1979, Ap. J., in press.
Young, A. T. 1979, private communication.
Zirin, H. 1976, Ap. J.,
Zirker, J. B. 1968, Solar Phys., 3, 164.

OBSERVATIONS OF THE OUTER ATMOSPHERIC REGIONS OF α ORIONIS

A.P. Bernat and L. Goldberg
Kitt Peak National Observatory*
Tucson, AZ 85726, U.S.A.

Abstract

We present three separate observational studies of mass flows above the photosphere of α Ori (M2 Ia-Ib). The Ca II infrared triplet lines and Hα are asymmetric showing a systematic blue shift with decreasing residual intensity. These lines remain fixed in wavelength although the weak photospheric lines vary by ±4 km/sec. Observations of the 4.6μ vibration-rotation spectrum of CO show two sharp, cold components expanding at velocities of 10 and 17 km/sec relative to the centre of mass. Direct photographs of the shell in the light of KI λ7699 show that the cold shell is asymmetric and extends outward to at least 50".

Details of these studies are either in press or will be submitted to "The Astrophysical Journal".

* Operated by the Association of Universities for Research, Inc. under contract with the National Science Foundation.

STELLAR WINDS AND CORONAE IN COOL STARS

A.K. Dupree and L. Hartmann

Harvard-Smithsonian Center for Astrophysics
Cambridge, MA 02138/USA

ABSTRACT

Recent observational and theoretical results are reviewed that pertain to the presence and characteristics of stellar coronae and winds in late-type stars. It is found that stars - principally dwarfs - exist with "hot" coronae similar to the Sun with thermally driven winds. For stars at the lowest effective temperatures, and gravities characteristic of supergiant and giant stars, high temperature ($\sim 10^5$K) atmospheres are absent (or if present are substantially weaker than in the dwarf stars), and massive winds are present. There also exist "hybrid" examples - luminous stars possessing both a "hot" corona and a supersonic stellar wind. Constraints for theoretical models are discussed.

1. INTRODUCTION

The study of stellar chromospheres, coronae, and winds has received new impetus from recent ultraviolet and X-ray observations. A general outline of the presence and behavior of chromospheres and coronae in late-type stars is beginning to emerge from this new data. In this review, we emphasize the results of the ultraviolet observations, and make comparisons with existing theory. The X-ray data are more preliminary so these are considered briefly. Finally, we treat mass loss in the late-type stars, concentrating particularly on the relationship of gas temperature to wind structure.

Optical spectra of most cool stars ($T_{eff} \lesssim 6500$K) present an emission core in the Ca II (λ3934, λ3968) lines which has long signaled the presence of an atmospheric region hotter than the stellar photosphere, and by analogy with the Sun, called a chromosphere. The first ultraviolet measurements were principally of the strong resonance lines, Mg II, O I, and C II which gave evidence for additional chromospheric material at temperatures up to 40,000K. In two stars - Procyon (α CMi) and Capella (Alpha Aur) - species at temperatures $\sim 3 \times 10^5$K were detected (Dupree 1975; Evans, Jordan, and Wilson 1975; Vitz et al. 1976). X-ray measurements, sensitive enough to detect single "normal" stars or well-separated binaries were sparse, but a few sources-Capella (G5 III + GO III), Eta Bootis (GO IV), Vega (AO V), Alpha Cen A (G2 V)-had been detected in the soft X-ray band ∿0.2 -1 keV (Catura, Acton, and Johnson 1975; Mewe *et al.* 1975; Topka *et al.* 1979; Nugent and Garmire 1978). From this meager sampling, it was possible to say that stars had chromospheres and coronae - perhaps similar in structure to the Sun in the case of Procyon (Evans, Jordan, and Wilson 1975) but more like a solar active region in Capella (Dupree 1975).

That substantial mass loss occurs in luminous cool stars has been known for quite some time (Deutsch 1956, 1960) from optical work. More recent observations in UV lines have expanded knowledge of mass loss, including cases where optical circumstellar lines are not seen (e.g. Dupree 1976). The detection of circumstellar lines of low excitation ions clearly showed that such winds are not coronal - that is, driven by thermal expansion - but must be accelerated by some other process (Weymann 1962). Since dwarfs are presumed to have solar-type coronal mass loss, a transition between the two extreme cases - hot coronae and cool winds - must occur somewhere in the H-R diagram. Reimers (1977) suggested that the disappearance of optical circumstellar lines below an observational line in the H-R diagram was related to the increasing ionization in the circumstellar envelope.

It is clear that a knowledge of the temperature structure of late-type stellar atmospheres and envelopes is extremely important in understanding the nature of mass ejection. Such information is not readily derivable from optical data. With the advent of UV and X-ray measurements, which probe high temperature material, it is now possible to discern the general outline of the presence or absence of hot gas and its correspondence with mass loss. Such an understanding is preliminary, because of incompleteness of data and also because of observational selection effects, but our knowledge of stellar coronae and winds has advanced markedly in the last year or two.

2. ULTRAVIOLET OBSERVATIONS

a) Dwarf Stars

Ultraviolet spectra of late-type "quiet" dwarf stars appear generally similar to that of the Sun in the presence of lines typical of the solar transition region and corona (see for instance Brown, Jordan, and Wilson 1979; Dupree *et al.* 1979; Dupree 1980; Linsky and Haisch 1979; Linsky *et al.* 1978). The line ratios and surface fluxes are usually within a factor of 2 or 3 of solar values.

The active dwarfs defined as those with strong Ca II emission, and flare stars show a similar spectrum; however the surface flux of the lines is larger than solar values, by factors of 4 to 20 and is remarkably independent of photospheric temperature (Hartmann et al. 1979a). The single dwarf stars confirm the presence of chromospheres, transition regions, and coronae that are similar to the solar atmosphere although the active stars have radiative losses at a rate comparable to those found in solar active regions.

b) Giant Stars

In the more luminous stars the character of the atmosphere appears to change more dramatically from the solar prototype. Figure 1 illustrates typical spectra of single giant stars. It is important to notice that the signal-to-noise ratio is not optimum in many of these spectra and

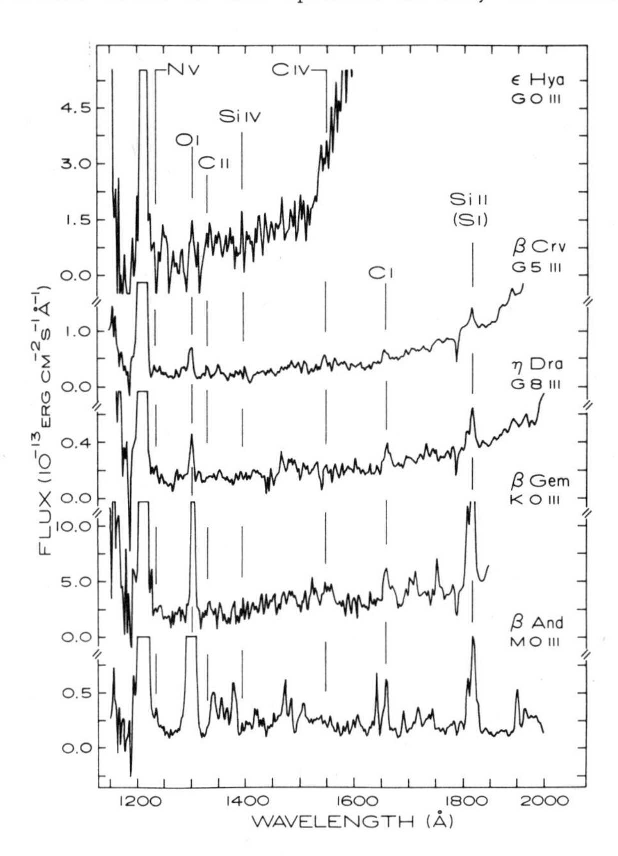

Figure 1 - IUE spectra of selected single giant stars arranged in order of decreasing effective temperature. Note the strong O I and Si II lines and the apparent presence of C IV in β Crv and β Gem. Geocoronal Ly-α has been truncated in this figure as has O I in β Gem and β And.

that an inference concerning the presence or absence of C IV as an example, is particularly difficult for many stars. The spectra are dominated by the lines of Si II and C I, all species formed at $T_c \sim 10^4$K. The strong O I multiplet at λ1302 is also indicative of these temperatures if the line formation is collisionally controlled; however, in the luminous cool stars this line may be enhanced through radiative excitation by Ly-β. High temperature species as indicated by C IV (λ1550) appear to be present in Beta Crv (G5 III) (based on 3 exposures) and in two exposures of Beta Gem (GO III) (see also Carpenter and Wing 1979) although Eta Dra (G8 III) does not show C IV - again based on multiple exposures. The character of the spectrum changes quite dramatically with β And (MO III), with only low temperature lines present. Identifications are difficult at low dispersion. However it is clear in this star that lines of Fe II and S I are present, while N V, C IV, and Si IV are not apparent.

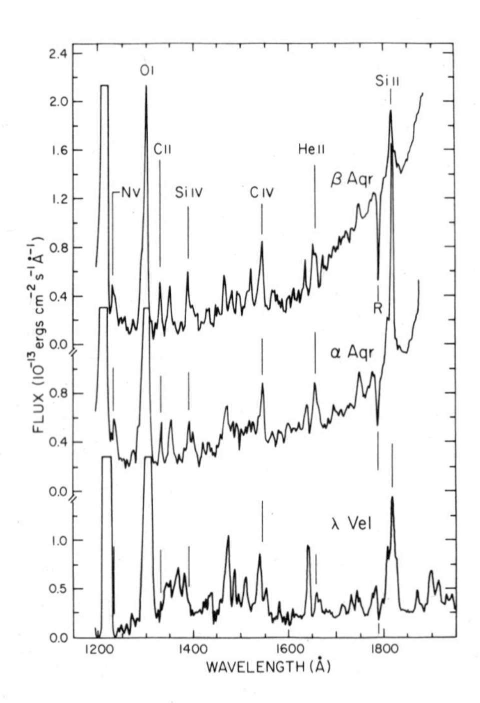

Figure 2 - IUE spectra of 3 supergiant stars: β Aqr (GO Ib); α Aqr (G2 Ib); and Vel (K5 Ib) showing low excitation lines in all and the presence of high temperature species C IV and N V in α and β Aqr only (From Hartmann, Dupree, and Raymond 1979b).

c) Supergiant Stars

IUE spectra of 3 supergiant stars (Fig. 2) show the dominance of O I and Si II as noted in the giant stars, additionally probably fluorescently excited lines of low excitation species C I, S I, Fe II near λ1450 and most importantly the clear presence of C IV and N V in the spectra of α and β Aqr (Hartmann, Dupree, and Raymond 1979b). The spectrum of λ Vel in fact is very similar to that of β And. The supergiant α Ori (M2 I) also shows no evidence of high temperature material (see for instance Linsky and Haisch 1979) but its spectrum is substantially different from λ Vel. The stellar surface fluxes of high temperature lines are approximately equal to 3 or 4 times the solar surface flux in α and β Aqr but lower by a factor of ∿100 for λ Vel (Hartmann, Dupree, and Raymond 1979b; Dupree et al. 1979).

The chromospheric lines of Mg II (see Fig. 3) for these stars show the blue-red asymmetry that has been associated with a differential chromospheric expansion. However both β and α Aqr show also the presence of narrow absorption components in the cores and wings of the lines that correspond to supersonic velocities up to ∿125 km s^{-1}. These features are found in the Ca K profiles as well (Hartmann et al. 1979b). Thus in α and β Aqr we find a corona coexisting with a massive stellar wind.

These data are remarkable for demonstrating that high-temperature gas is not inconsistent with substantial mass loss. In addition, some of the "cool" chromosphere lines in λ Vel are present in these G supergiants as well, clearly showing the hybrid nature of the spectrum. The terminal velocity of the shells at ∿-100 km s^{-1} is also significant in that they are intermediate between solar wind velocities and typical cool supergiant velocities ∿-20 km s^{-1}. Thus these objects may well represent the transition between solar-type winds and cool outflow.

3. OVERALL PRESENCE OF HOT PLASMA

The C IV doublet (λ1550) can be used to indicate hot plasma with temperatures 2 x 10^5K - and by analogy with the Sun - we infer the presence of a transition region or corona in these stars. Figure 4 contains a compilation of results from many IUE spectra (Brown, Jordan, and Wilson 1979; Carpenter and Wing 1979; Dupree et al. 1979; Hartmann et al. 1979 a, b; Linsky and Haisch 1979; Wing 1978). It is apparent that the dwarf stars (luminosity class V) consistently show evidence for C IV, but the more luminous stars (class I and II) show more varied behavior. The coolest stars (spectral type M and later) give no evidence of C IV as

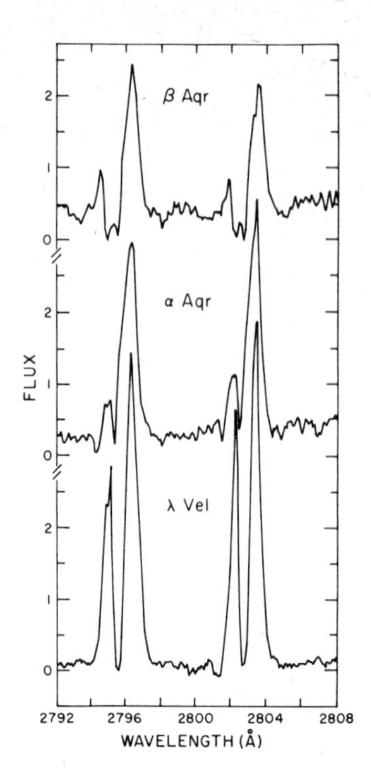

Figure 3 - Strongly asymmetric Mg II emission in 3 supergiant stars (Hartmann, Dupree, and Raymond 1979b). Note that the narrow absorption features in the core and blue wing of the lines of α and β Aqr are not as prominent in the spectrum of λ Vel.

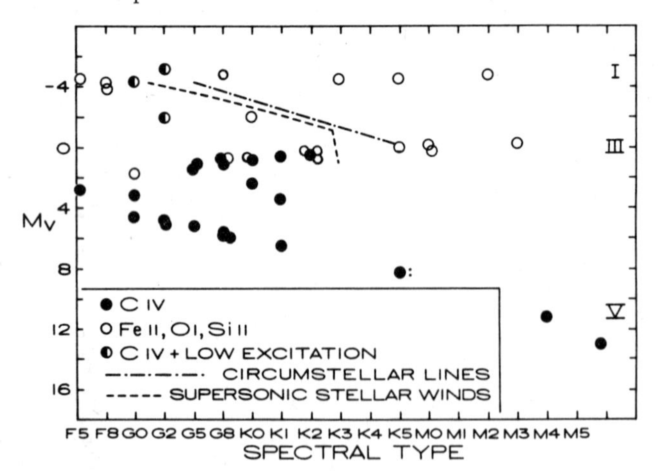

Figure 4 - The presence of emission from C IV and/or cool species in IUE spectra of late-type stars. The broken line corresponds to Mullan's (1978) position of enhanced mass loss. The dot-dash lines denote Reimers (1977) boundary for circumstellar lines. See discussion in text regarding selection effects and uncertainties in detection.

Wing (1978) first noted for γ Cru (M3.4 IIIb). At G and K spectral types there is not a clear change with effective temperature. Along the giant branch, both types occur at spectral type G8, for instance; in the supergiant stars we find the so-called hybrid atmosphere discussed earlier. The stars showing both C IV and low excitation species lie near the onset of the appearance of circumstellar lines (Reimers 1977) and near the "supersonic transition locus" proposed by Mullan (1978) that we discuss later.

Linsky and Haisch (1979) suggested that there were two types of atmospheres - solar and non-solar, that the division was sharp between these types at spectral type K0 III, and they speculated that the presence of a massive stellar wind would suppress a corona and eliminate high temperature species. The addition of more data shows that the situation is more complicated. The "division" between hot and cool atmospheres is not at all sharp. Cool atmospheres dominate at a later spectral type than shown by Linsky and Haisch, and hybrid atmospheres exist that exhibit both high temperature lines and substantial mass loss. There is other, more subtle behavior in the line ratios of C II, C IV, and O I emphasized by Brown, Jordan, and Wilson (1979) that support a smooth change in atmospheric structure with decreasing effective temperature. Generally speaking however, stars with large mass loss rates show the coolest emission features and apparently have weak or nonexistent coronae.

These differences of interpretation emphasize the problem of selection effects with small samples of stars. Observers have focussed on bright stars that in many cases possess active chromospheres; additionally, binary systems show enhanced fluxes that may be attributable to rotation. It is not easy to detect C IV in many stars. Many observations do not have sufficient signal-to-noise ratios. Spectra of stars near spectral type G0 III and earlier suffer from direct continuum emission and scattered light in the IUE instrument at short wavelengths. We cannot be sure that the open circles in Fig. 4 are stars without 10^5K gas; we can merely place upper limits on its presence.

We can infer the presence of still hotter material at $T \sim 10^6$K - from the flux in the He II, λ1640 transition. In solar active regions this transition can be enhanced by recombination following photoionization (Raymond, Noyes, and Stopa 1979). In active dwarf stars the line flux confirms the direct measurement of soft X-rays (Hartmann _et al_. 1979a). The presence of the λ1640 transition in the supergiant spectra suggests that a hot corona may well exist on these stars; detection could be difficult if soft X-ray emission is absorbed by a substantial extended atmosphere. Whereas dwarf stars show He II λ1640 (Dupree et al. 1979, Hartmann et al. 1979a), it is interesting that the spectrum of the T Tauri star RU Lupi has very weak, if any λ1640 emission, (Gahm *et al*. 1979) suggesting that relatively little coronal material is present.

X-ray measurements from the Einstein Observatory (HEAO-2) could of course give direct information on the highest temperature plasma in these stars, but the data are in a early stage of assessment. The data are consistent with the picture suggested by the ultraviolet results. It is not yet known whether cool supergiants have substantial coronae; the upper limits on coronal emission from very luminous stars can be lowered when exposure times are longer (Rosner 1979). At present, many of the objects in the stellar surveys are faint and have not been well studied optically (Vaiana *et al*. 1979). Their luminosity classes, activity, or possible binary nature are not known in most cases. The X-ray observations should shortly provide important advances in understanding the high temperature plasma in the atmospheres of cool stars.

These observations are beginning to provide some constraints for theories of coronal heating and structure. The radiative losses in a wide variety of single stars vary by a factor of $\sim$10 in Mg II (Basri and Linsky 1979) and by a factor of about 100 in higher temperature lines. The ultraviolet emission is independent of stellar effective temperature for active dwarfs, and in general appears to be more dependent on log g than on T_e. And, in several cases, stars of the same spectral type and gravity exhibit quite different ultraviolet surface fluxes. To date no single theory appears to be successful at reproducing these results.

4. MASS LOSS

Line profiles in the optical and ultraviolet have been used to determine the onset and presence of mass loss among late-type stars. Reimers (1977) investigated circumstellar lines in the optical spectra to define a boundary in the HR diagram (see Fig. 5). Additionally blue and red Mg II and Ca II asymmetries occur systematically over a large region (Stencel 1978; Mullan and Stencel 1979). The locus of the absence or weakening of C IV in the coolest stars as taken from Fig. 4 is also shown in Fig. 5.

In the coolest stars substantial mass loss occurs, and these regions correspond to a weakening of the C IV line, and by inference, a corona.

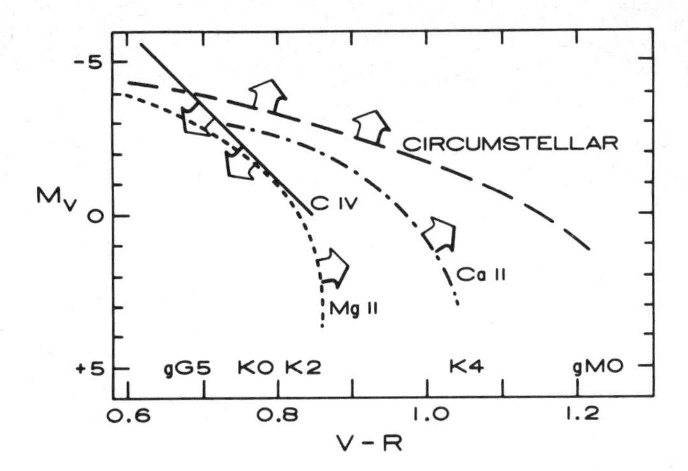

Figure 5 - The appearance of various spectral features as a function of color and luminosity. C IV emission is prominent to the left of the solid line; the ratio of red:blue emission peaks for Mg II and Ca II is greater than 1 to the right of the appropriate broken lines (Stencel 1978; Mullan and Stencel 1979). Circumstellar features are found above the long broken line (Reimers 1977).

There is not a complete anticorrelation of mass loss with high temperature gas however, as shown by the observations of the supergiants α and β Aqr. While the interpretation of Ca II and Mg II asymmetries is open to argument, this figure suggests a gradual increase in mass loss towards lower effective temperature, with the detection of mass loss in the weakest lines occurring only for the largest mass loss rates.

To date, the most popular theory of mass loss for luminous cool late-type stars is that of radiation pressure on dust (Gehrz and Woolf 1971; Kwok 1975). However, it seems very unlikely that dust forms in the vast majority of late-type stars, so that it is clearly not a universal mechanism. The original work of Durney (1973) has been extended by Mullan (1978) who recently introduced the concept of the "supersonic transition locus", where the sonic points of isothermal coronae are supposed to penetrate down to the top of the chromosphere, resulting in an abrupt enhancement of mass loss from the action of spicules ejecting matter. The semi-empirical calculations assume that Hearn's (1975) minimum-flux corona theory is applicable. The concept of a minimum flux corona is very controversial (see the review by Cassinelli 1979), and the calculation itself is based on a scaling of empirical pressures that is now known to be uncertain (Baliunas *et al*. 1979).

A suggestion due originally to Wilson (1960), that Ly-α radiation pressure can drive mass loss, was recently investigated by Haisch, Linsky, and Basri (1979) in the context of the K2 IIIp star α Boo. They calculate a Ly-α flux based on a sophisticated radiative transfer solution, but use the optically-thin approximation for the force calculation and neglect velocity gradients in the transfer. Their results show that Ly-α pressure may be larger than gravity over a small volume, but that the total momentum added is vastly insufficient to generate mass loss, so that Alfven waves are suggested as the driving mechanism. This mechanism is very sensitive to the ionization state of the gas. If the material is too cold, then Ly-α is so optically thick that the force is negligible. If the gas is too hot, insufficient neutral hydrogen is available to absorb radiation. Since the temperature distribution adopted by Haisch *et al*. is not uniquely determined by the observations, it is unclear whether these flows actually become supersonic from Ly-α radiation pressure.

One attractive possibility, suggested by observations of strong turbulence in supergiant atmospheres and chromospheres, is that the mechanical energy in various forms of the turbulence can be deposited in such a way as to drive mass loss. Hartmann and MacGregor (1979) have investigated the behavior of acoustic and magnetic wave modes as a function of stellar gravity for surface wave fluxes comparable to solar values. These waves are in principle more effective than radiation pressure in Ly-α which represents only a fraction of the mechanical energy fluxes available for driving winds. Wave modes other than light carry a factor v/c more momentum than light for a given energy flux, where v is the mode speed. This factor is typically $\sim 10^4$ for sound and magnetic waves, so that the potential for driving mass loss with such waves is considerably greater than for photons.

The calculations show that while shocks can account for chromospheric heating, their dissipation lengths are generally too short to be effective in initiating outflow. In Fig. 6 we show some static chromospheric calculations for representative wave periods. The temperature rise in units of pressure scale heights is more gradual in low gravity stars, although these waves do not explicitly account for the transition-region corona interface. A suggestive feature of these results is indicated in the temperature structures calculated for longer-period waves with larger dissipation lengths. In dwarfs, increased dissipation with height causes an increase in local gas temperature, but the opposite occurs at super-

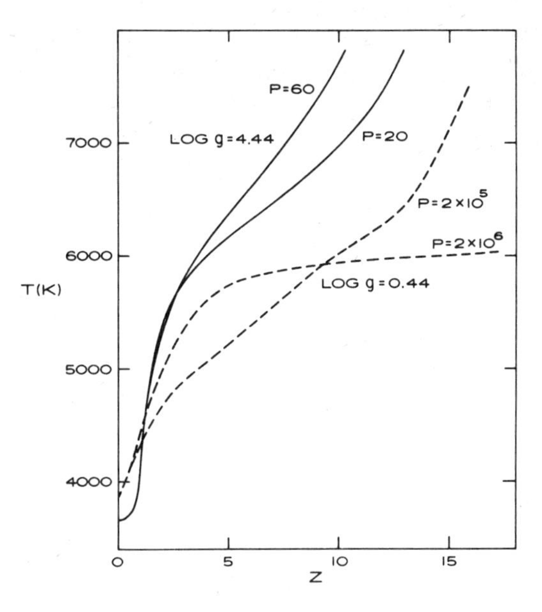

Figure 6 - Chromospheric temperature structures for dwarf (solid line) supergiant (broken line) gravities as a function of distance above the stellar surface, z, in units of base pressure scale heights. Increasing the period, $P(s^{-1})$, of the acoustic waves heating the chromosphere acts to raise the upper chromospheric temperature in dwarfs. The opposite occurs in supergiants, as the waves tend to extend the density distribution. (From Hartmann and MacGregor 1979).

giant gravities. The reason is that at low gravity the waves "push" rather than heat, and the density distribution becomes extended. Such a result is in general agreement with the suggestion of Linsky and Haisch (1979) that mass loss suppresses high-temperature gas. However, these calculations are only part of the story, for coronal formulation is not accounted for even for dwarfs. In addition, the acoustic waves do not cause mass loss, they just extend the atmosphere.

Alfven waves, on the other hand, are very efficient in driving mass loss because of their long dissipation lengths. If one assumes wave surface fluxes comparable to those adopted for solar wind models, it is possible to arrive at reasonable mass loss rates. However, such wave-driven winds generally result in very high-velocity winds, unless some damping occurs within a few stellar radii (Fig. 7). Ion-neutral damping may become very important in winds that are observationally known to be substantially neutral. Magnetic theories are unsatisfying in general because of the lack of observational quantities available for test. However, mechanical

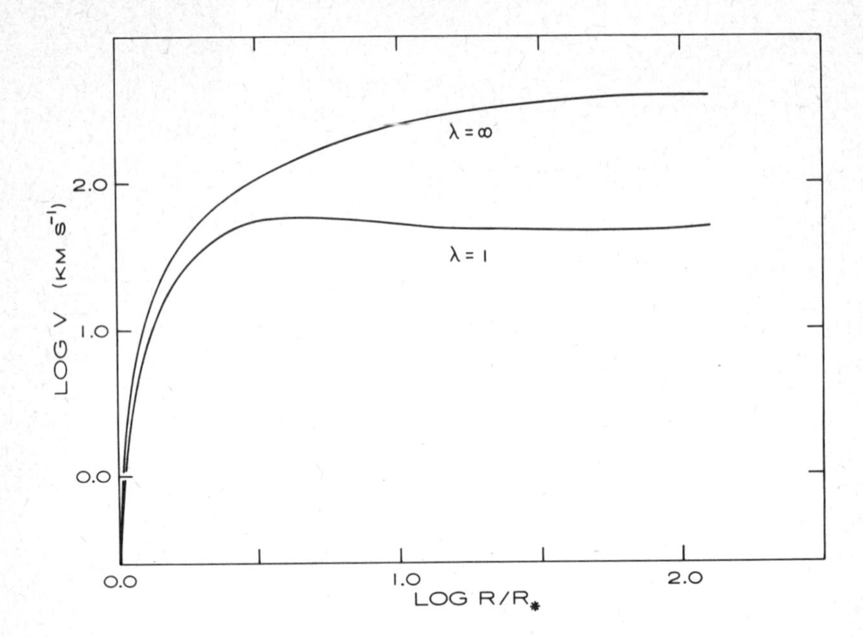

Figure 7 - Velocity structures of cool Alfven-wave driven winds for supergiants. Models in which the Alfven waves are undamped ($\lambda = \infty$) have high terminal velocities. It is necessary to damp out the wave energy within 1 R_* of the surface in order to obtain terminal velocities of tens of km s^{-1}.

energy dissipation of the magnitude required to drive observed mass loss will necessarily result in heating of the wind, probably at larger distances than acoustic waves can be effective (Hartmann and MacGregor 1979). Therefore extended warm chromospheres seem to be a necessary feature of such wind models; this should be amenable to observational test, particularly with Zeta Aur systems.

We feel that recent observational evidence, such as that of hybrid chromospheres and winds, strengthens the case for a more unified picture of solar and stellar activity than has been presented in the past. Whether mass loss as well as chromospheric and coronal heating is a result of solar-type turbulence and activity operating in lower-gravity environments is a question that the new ultraviolet and X-ray observations, probing high-temperature atmospheric regions, may be able to answer.

This work is supported in part by NASA Grant NSG 5370 to the Harvard College Observatory.

REFERENCES

Baliunas, S.L., Avrett, E.H., Hartmann, L., and Dupree, A.K. 1979, Astrophys. Journ. Letters, in press.
Basri, G.B., and Linsky, J.L. 1979, Astrophys. Journ., in press.
Brown, A., Jordan, C., and Wilson, R. 1979, in Proc. Symp. The First Year of IUE, in press.
Carpenter, K.G., and Wing, R.F. 1979, Bull. A.A.S., 11, 419.
Cassinelli, J.P. 1979, Ann. Rev. Astron. Astrophys., 17, 275.
Catura, R.C., Acton, L.W., and Johnson, H.M. 1975, Astrophys. Journ. Letters, 196, L47.
Deutsch, A.J. 1956, Astrophys. Journ., 123, 210.
Deutsch, A.J. 1960, in Stellar Atmospheres, ed. J.L. Greenstein (Chicago: Univ. of Chicago Press), p. 543.
Dupree, A.K. 1975, Astrophys. Journ. Letters, 200, L27.
Dupree, A.K. 1976, in Physique des Mouvements dans les Atmospheres Stellaires, ed. R. Cayrel and M. Steinberg, Editions du Centre National de la Recherche Scientifique, p. 439.
Dupree, A.K. 1980, in Highlights of Astronomy, Proc. of Joint Discussion of Ultraviolet Astronomy, I.A.U. Gen. Assembly, in press.
Dupree, A.K., Black, J.H., Davis, R.J., Hartmann, L., and Raymond, J.C. 1979, in Proc. Symp. The First Year of IUE, in press.
Durney, B.R. 1973, in Stellar Chromospheres, ed. S.D. Jordan and E.H. Avrett, NASA SP-317, p. 282.
Evans, R., Jordan, C., and Wilson, R. 1975, Mon. Not. Roy. Astron. Soc., 172, 585.
Gahm, G.F., Fredga, K., Liseau, R., and Dravins, D. 1979, Astron. Astrophys., 73, L4.
Gehrz, R.D., and Woolf, N.J. 1971, Astrophys, Journ., 165, 285.
Haisch, B., Linsky, J.L., and Basri, G. 1979, preprint.
Hartmann, L., Davis, R.J., Dupree, A.K., Raymond, J.C., Schmidtke, P.C., and Wing, R.F. 1979a, Astrophys. Journ. Letters, in press.
Hartmann, L., Dupree, A.K., and Raymond, J.C. 1979b, in preparation.
Hartmann, L., and MacGregor, K. 1979, in preparation.
Hearn, A.G. 1975, Astron. Astrophys., 40, 355.
Kwok, S. 1975, Astrophys. Journ., 198, 583.
Linsky, J.L. et al., 1978, Nature, 275, 389.
Linsky, J.L., and Haisch, B. 1979, Astrophys. Journ. Letters, 229, L27.
Mewe, R., Heise, J., Gronenschild, E.H.B.M., Brinkman, A.C., Schrijver, J., and den Boggende, A.J.F. 1975, Astrophys. Journ. Letters, 202, L67.
Mullan, D.J. 1978, Astrophys. Journ., 226, 151.
Mullan, D., and Stencel, R. 1979, personal communication.
Nugent, J. and Garmire, G. 1978, Astrophys. Journ. Letters, 226, L83.
Raymond, J.C., Noyes, R.W., and Stopa, M.P. 1979, Solar Phys., 61, 271.
Reimers, D. 1977, Astron. Astrophys., 57, 395.
Rosner, R. 1979, personal communication.
Stencel, R.E. 1978, Astrophys. Journ. Letters, 223, L37.
Topka, K., Fabricant, D. Harnden, F.R., Jr., Gorenstein, P., and Rosner, R. 1979, Astrophys. Journ. Letters, 229, 661.
Vaiana, G., Forman, W., Giacconi, R., Gorenstein, P., Pye, J., Rosner, R., Seward, F., and Topka, K. 1979, Bull. A.A.S., 11, 446.
Vitz, R.C., Weiser, H., Moos, H.W., Weinstein, A., and Warden, E.S. 1976, Astrophys. Journ. Letters, 205, L35.
Weymann, R. 1962, Astrophys. Journ., 136, 844.
Wilson, O.C. 1960, Astrophys. Journ., 131, 75.
Wing, R.F. 1978, in High Resolution Spectrometry (ed. M. Hack), Proceedings of the 4th International Colloquium on Astrophysics, p. 683.

RELATIONSHIP BETWEEN ENVELOPE STRUCTURE AND ENERGY SOURCE OF NON-THERMAL MOTIONS

Hiroyasu Ando
Tokyo Astronomical Observatory, University of Tokyo
Mitaka, Tokyo, Japan 181

Abstract

The ionization zone in the envelope of the late type stars is reasonably considered as a heat engine to transform some of the radiative energy into mechanical energy. This idea is suggestive for explaining Linsky and Haisch's (1979) observation, which shows the sharp division into solar-type and non-solar type stars in the outer atmosphere. Also non-thermal velocity fields in "microturbulence" and in Wilson-Bappu effect are proposed to be formed essentially from this engine. Therefore, their envelope structure dependence observationally obtained is possibly explained by the envelope parameters (g, T_e) dependence of the generated mechanical energy flux in this layer. If "microturbulence" is not contaminated by the other surface activities, it is expected to show a clear relation with envelope parameters (g, T_e) similar to Wilson-Bappu effect.

AN ANALYSIS OF MICROTURBULENCE IN THE ATMOSPHERE OF THE F-TYPE SUPERGIANT GAMMA CYGNI

A.A. Boyarchuk and L.S. Lyubimkov
The Crimean Astrophysical Observatory
P/O Nauchny, Crimea, 334413, U.S.S.R.

Abstract

We have analysed high dispersion spectra of the supergiant γ Cyg (F8 Ib). On the basis of the curve of growth method it has been shown that there is no dependence of microturbulent velocity ξ_t on excitation potentials of spectral lines. Using model atmospheres we considered about 100 Fe I lines and found that no constant value ξ_t makes possible to remove the systematic discrepancy in iron abundance between groups of lines with different equivalent widths. The depth dependence of microturbulence in the atmosphere is investigated. It is shown that parameter ξ_t increases outwards from about 0-1 km/s at the optical depth τ_{5000} = 0.2 to 10 km/s at the depth $\tau_{5000} = 10^{-3}$. Deduced function $\xi_t(\tau_{5000})$ gives the same iron abundance log ε(Fe) = 7.45 ± 0.05 for all groups of Fe I lines. The detailed analysis will be published in Izv. Crimean Astrophys. Obs.

THE SOLAR CHROMOSPHERIC MICROTURBULENCE AND THE EMISSION OBSERVED AT ECLIPSE

Y. Cuny
Observatoire de Paris
92190 Meudon, France

Abstract

For a long time there was not a clear interpretation of the fact that the emission observed at eclipse is systematically too high compared with predictions based on disk observations. This discrepancy has been attributed to the hydrogen density, the atomic abundances, or the multiplet excitation temperatures, the physical cause being assumed to be some inhomogeneities. We have shown (Liege 1978, under press) that it can be attributed to the microturbulence. The radiative energy absorbed in the doppler width of the absorption profiles of lines, produces high excitation temperatures. The relative abundances of Ba II, Sr II, Ti II with these excitation temperatures are in agreement with the photospheric abundances.

Assuming photospheric abundances, we get the hydrogen density versus the altitude while the observations give $n_e n_p/T_e^{3/2}$. The solution of these two equations, taking account of non-LTE for hydrogen but without the assumption of hydrostatic equilibrium, gives a chromospheric model. The model obtained is very near the usual quiet chromospheric models based on disk data.

This work confirms the validity of our assumption of the influence of microturbulence on the chromospheric emission observed at eclipse.

EXCITATION DEPENDENT GF-VALUES AND DEPTH DEPENDENT MICROTURBULENCES

Toshio Hasegawa
Hokkaido University of Education
Asahikawa, 070, Japan

Abstract

The velocity of microturbulence is frequently determined from Fe I lines. Unfortunately classical gf-values of this element have excitation dependent errors. In the absolute curve of growth analysis of some F-type stars with new gf-values, the author found that a great part of the depth dependence (i.e. excitation dependence) of microturbulent velocity, which had been derived by many authors with old gf-values, is the consequence of these errors.

The errors of old gf-values increased with excitation potential and the errors were compensated by adopting velocities of microturbulence decreasing with excitation potential. On the other hand, gf-values of ionized elements (e.g. Ti II and Fe II) are not changed as much as those of Fe I, accordingly the ionization dependence can also be eliminated.

ON THE STRUCTURAL AND STOCHASTIC MOTIONS IN THE SOLAR AND STELLAR ATMOSPHERES

E.I. Mogilevsky
Izmiran, USSR, Academy of Sciences
Moscow, U.S.S.R.

Abstract

Existence of discrete structures of velocity $\overline{V}$ and magnetic field $\overline{B}$ in stellar atmospheres follows from the Vlasov's integral equation

$$\int \mathfrak{F} d\tau = F(v\ \overline{B},\ T...) \qquad (1)$$

where $\mathfrak{F}(u,\dot{v},r)$ is a statistical function of the distribution of the elements of matter having masses $m = \frac{\rho}{\dot{V}}$ and velocities, v, $d\tau$ is elementary phase volume, F is functional which arrange the ties between macrovelocity, magnetic field, temperature, etc., providing self containment of cosmic plasma. The self containment of cosmic plasma leads to some specific fundamental properties of solar and stellar atmospheres.

Equation (1) has discrete solutions, implying the existence of the structural hierarchy, with definite characteristics for the density, velocity, magnetic field, etc. As a result, "strange" phenomena take place in the solar and stellar atmospheres, for example: Magnetization, non-balancing of the magnetic fluxes in N and S polarities, surprising fast dynamics of the active phenomena, the non-classical values of electro and conductivity, etc.

The predominance of filamentary structures, which are well observed in the solar atmosphere, leads to the stochasticity and quasiperiodicity of motions.

I U E OBSERVATIONS OF CIRCUMSTELLAR LINES AND MASS LOSS FROM B-STAR

S.P. Tarafdar, K.S. Krishna Swamy and M.S. Vardya
Tata Institute of Fundamental Research
Homi Bhabha Road, Bombay 400 005, India

Abstract

The circumstellar lines due to A $^1\Pi$ - $X^1\Sigma$ transition of CO has been identified for the first time in the spectra of an early type star. The spectrum, taken from the International Ultraviolet Explorer between 1150 - 1900 Å, of 9 Cep, a B2Ib star with E(B-V) = 0.47, shows not only the interstellar absorption lines of CO but also absorption features shifted towards short wavelength relative to the interstellar lines. The amount of shifts towards short wavelength from the 1 - 0, 2 - 0, 3 - 0, 4 - 0, & 6 - 0 interstellar absorption bands of CO has been found to be fairly close and corresponds to a velocity of 450 km sec^{-1}. All attempts to identify these set of lines in a consistent way with atomic or ionic lines have yielded essentially negative results. This has led us to tentatively identify these lines due to circumstellar CO around 9 Cep. An estimate of rate of mass loss from the circumstellar CO-column density leads to a value of $\dot{m} \gtrsim 10^{-11}$ $M_\odot$ yr^{-1}.

ON THE ESTABLISHMENT OF INTERNALLY CONSISTENT ABUNDANCE-OSCILLATOR STRENGTH SCALES

E.A. Gurtovenko and R.I. Kostik
Main Astronomical Observatory of the Ukrainian Academy of Sciences
Kiev 127, U.S.S.R.

Abstract

The method of establishing of internally consistent abundance-oscillator strength scales by using solar fraunhofer lines is elaborated and investigated.

The error of internal accuracy should not exceed 0.05 - 0.06 dex. The absolute accuracy depends on the accuracy of "reference" gf-values.

The oscillator strengths for about 800 Fe I lines are obtained. The comparison of the results for 19 lines common in our and Blackwell et al. (1976) investigations gives the difference $\log gf_{Black} - \log gf_{auth} = \Delta = -0.044 \pm 0.010$. The accidental part of the difference actually determines the internal accuracy of the obtained oscillator strengths.

The Kurucz and Peytremann (1975) oscillator strengths for Fe I lines are analysed. Large systematic errors depending on gf and excitation potential are revealed and investigated. For some lines those errors may change the true values of gf by two orders of magnitude.

DIFFERENTIAL ROTATION AND MAGNETIC ACTIVITY OF THE LOWER MAIN SEQUENCE STARS

G. Belvedere
Istituto d'Astronomica dell'Universita di Catania
1-95125, Catania, Italy
L. Paterno
Osservatorio Astrofisico di Catania
1-95125, Catania, Italy
M. Stix
Kiepenheuer Institut für Sonnenphysik
D-7800 Freiburg, Germany

Abstract

We extend to the lower main sequence stars the analysis of convection interacting with rotation in a compressible spherical shell, already applied to the solar case (Belvedere and Paterno, 1977; Belvedere et al. 1979a). We assume that the coupling constant ε between convection and rotation, does not depend on the spectral type. Therefore we take ε determined from the observed differential rotation of the Sun, and compute differential rotation and magnetic cycles for stars ranging from F5 to M0, namely for those stars which are supposed to possess surface convection zones (Belvedere et al. 1979b, c, d). The results show that the strength of differential rotation decreases from a maximum at F5 down to a minimum at G5 and then increases towards later spectral types. The computations of the magnetic cycles based on the $\alpha\omega$-dynamo theory show that dynamo instability decreases from F5 to G5, and then increases towards the later spectral types reaching a maximum at M0. The period of the magnetic cycles increases from a few years at F5 to about 100 years at M0. Also the extension of the surface magnetic activity increases substantially towards the later spectral types. The results are discussed in the framework of Wilson's (1978) observations.

References

Belvedere, G. and Paterno, L. 1977 Solar Phys. 54, 289.
Belvedere, G., Paterno, L. and Stix, M. 1979a Geophys. Astrophys. Fluid Dyn. (in press).
Belvedere, G., Paterno, L. and Stix, M. 1979b Astron. Astrophys. submitted.
Belvedere, G., Paterno, L. and Stix, M. 1979c Astron. Astrophys. submitted.
Belvedere, G., Paterno, L. and Stix, M. 1979d Astron. Astrophys. to be submitted.
Wilson, O.C. 1978, Astrophys. J. 226, 379.

CHANGES OF PHOTOSPHERIC LINE ASYMMETRIES WITH EFFECTIVE TEMPERATURE

David F. Gray
Department of Astronomy, University of Western Ontario
London, Canada

Abstract

Asymmetries in photospheric lines were described briefly in my earlier contribution to these proceedings and in more detail in the January 15, 1980 issue of the Ap. J. The cores of the stronger lines in α Boo (K2III) show a blue shift, quite opposite to the effect seen in the solar lines. Suspecting this difference to be due to the shift in ionization with effective temperature, two other stars were measured, one slightly hotter than α Boo (β Gem K0III) and one slightly cooler (β UMi K4IV). The preliminary results seem to confirm the effective temperature dependence since no asymmetries were found for β Gem and stronger blue-shifted core asymmetries were found for β UMi.

SMALL-SCALE VERSUS LARGE SCALE MOTIONS IN THE SOLAR ATMOSPHERE DERIVED FROM A NON-LTE CALCULATION OF MULTIPLET 38 OF Ti I

R. Cayrel
CFHT Corporation, Waimea Office
Kamuela HA 96743, U.S.A.
S. Dumont and P. Martin
Observatoire de Meudon
F-92190 Meudon, France

Abstract

A non-LTE computation of multiplet 38 of Ti I (11 lines) has been undertaken in order to determine small and large scale unresolved motions contributing to the Doppler broadening of solar lines at the centre of the disk (vertical motions) and at the edge of the disk (horizontal motions).

The abundance of Titanium and the total Doppler velocity (all scales) are determined by fitting observed and computed profiles of weak unsaturated lines ($W < 12$ mÅ) of the multiplet. Then saturated lines having the same lower level as the weak lines are computed for a variety of partitions of the total kinetic energy between the small scale and the large scale modes going from 0% small scale to 100% small scale. Oscillator strengths with internal accuracies of about 2% from Wahling (1977) have been used. The location of the observed profile (taken from Delbouille et al. for the centre of the disk and from Brault and Testerman, KPNO for $\mu = 0.2$) among the computed profiles yields the partition of the energy between the two modes.

The computations done so far with 4 levels and continuum give less than 20% of the energy in the small scale mode. Further computations with more levels are needed to establish this ratio with better accuracy. The total energy for vertical motions has a root square velocity of about 1.4 km/sec whereas the same quantity is 2.2 km/sec for horizontal motions.

EFFECTS OF FLUX TUBES ON CONVENTIONAL CHROMOSPHERIC DIAGNOSTICS

Thomas R. Ayres
JILA and University of Colorado
Boulder CO 80309

Abstract

Magnetic flux tubes are usually envisioned as small discrete structures sparsely distributed throughout an otherwise uniform "intertube" medium. An important distinction between the flux tubes and the surrounding atmosphere is the presence of a strong chromospheric temperature inversion at high pressures in the flux tubes, while the intertube component has only a mild, low pressure chromosphere, if any at all. This implies that the flux tubes will be enormously brighter in conventional chromospheric diagnostics than the intertube component. However, the unresolved magnetic elements cover perhaps only 10% of the "quiet" Sun at chromospheric heights. Consequently the intense K and k emission cores are severely diluted. The net result is a weak emission reversal that is not characteristic of either the flux tube or intertube chromosphere. Even in the thermal microwave continuum longward of 100 μm, the flux tubes can contribute significantly to low spatial resolution spectra. Consequently, the spatially averaged microwave emission is also not characteristic of either of the distinct components.

If magnetic flux tubes are indeed the dominant class of atmospheric inhomogeneity in the Sun and other cool stars, then single-component interpretations of spatially unresolved data can be completely misleading, especially for inferring important auxiliary quantities, for example chemical abundances, line broadening parameters and chromospheric energy budgets. In the latter case, chromospheric radiative cooling rates derived from empirical mean models could be overestimated by up to an order of magnitude.

SUMMARY

Erika Böhm-Vitense
University of Washington
Seattle, Washington 98195 USA

Let me start by saying that we have heard some excellent review papers and I doubt very much that I can contribute much of importance in addition to what we have heard already.

I also wish to apologize to all those who have made important contributions during this meeting and who will not be mentioned in this 30-minute summary. If I wanted to mention everybody I would just have enough time to read the program.

I. Origin of Non Thermal Motions

The possible origins which we discussed the first day are summarized briefly in Table I. We did not come up with an explanation for the observed turbulence in B and O stars. G. Nelson confirmed that convection cannot be it, even though radiation pressure has some enhancing effects. K. Kodaira pointed out that circulation induced by rotation is unlikely to be the origin. However, S.R. Sreenevasan informed us that shear turbulence originating from inward increasing rotation might be a likely explanation.

For late type stars the general consensus seems to be that convection or at least the convection zone is the generator for all observed motions--except, of course, rotation.- J.P. Zahn pointed out the difficulties which the theoreticians encounter when trying to solve the highly nonlinear hydrodynamic equations in extended convection zones. So far it has not been possible to derive theoretically the expected velocity field. Therefore, I believe, the observers have to go to work and help the theoreticians. G. Nelson (1978) has pointed out ways how to do it: The maximum velocity in the convection zone is mainly determined by the turbulent exchange or the drag length while the overshoot or penetration into the stable zone is mainly determined by the horizontal scale. Therefore, I think such measurements as reported here by A. Nesis are very important. Progress in convection theory has been made by the inclusion of the pressure fluctuations which can be fairly large near the boundaries and lead to such interesting effects as antibuoyancy. The inclusion of the pressure fluctuations enables us to understand the observed scales of the granules (Nelson 1978) and the exploding granules as Nordlund has shown in his movie which was one of the highlights of this meeting, and we certainly would like to know more about the physics, numerical methods and boundary conditions that went into this numerical simulation of solar granulation.

There was some suggestion that in the mixing length theory of convection the mixing length ℓ to be used should be two scale heights H instead of one in order to increase the thickness of the convection zone as seems to be required by P. Gilman's theory of differential rotation and apparently also by the observed oscillation modes of the sun. I would like to point out, however, that the possible increase in the extent of the convection zone is limited by the observed solar lithium abundance which I think does not permit the depth obtained for $\ell = 2H$.

I was quite surprised to hear a hydrodynamicist suggest to calibrate the elaborate hydrodynamic theory with the mixing length theory. I always thought it should go the other way.

Y. Osaki clearly pointed out the instabilities and the possible waves and oscillations in the upper layers of the sun and late type stars, most of which are actually observed in the sun as J. Becker showed. If observable in other stars they could be used to probe the deeper invisible atmospheric layers as is done for the sun. To the observers of special interest is the fact that the expected motions are anisotropic, mainly vertical, while the convective motions are mainly horizontal in the stable surface layers, as we saw from A. Nordlund's velocity fields. Y. Osaki also pointed out that G and K stars are expected to be overstable to many nonradial p modes. If modern studies of turbulence confirm earlier observations indicating an increase in microturbulence from G to K main sequence stars (Chaffee et al. 1971), these instabilities could be the explanation. Convective motions are not expected to increase towards cooler stars.

Since for the sun, for which we have high spatial and time resolution, we can disentangle the observed velocity fields in the k-ω plane as J. Beckers demonstrated, we might hope that solar observations can in the future give us the velocity distributions for the different velocity fields. J. Beckers pointed out to me, however, that the observational integration over height would wipe out the information about the velocity amplitudes. Fortunately L.E. Cram told us later that there may still be some hope, though the relaxation times assumed to be negligible in his computations, will have to be checked.

II. Observations of Velocity Fields in the Sun and Stars

After having reviewed the complicated fields of motion in the sun we turned to the observed velocity fields in stars and the situation looks much simpler,at least from the observer's point of view, clearly only an effect of aspect as R. Glebocki pointed out nicely.

There was a divergence of opinions between theoreticians, who want to know the origin of the observed velocity fields,while the observers can reasonably only investigate what can be measured.

At present there appear to be three ways to measure non-thermal motion fields. They are briefly summarized in Table 2.

(1) We can measure line shifts or asymmetries which occur if velocity fluctuations are correlated with intensity fluctuations or if asymmetries in the velocity field are present. We have seen theoretical line profiles for acoustic waves and for oscillations which beautifully demonstrated this. These asymmetries may help to separate such velocity fields from others where the velocity fluctuations are not coupled with intensity fluctuations. D. Dravins' studies may prove to be very important in this respect. The dependence on the excitation potential which gives information about the depth dependence of the generating velocity field may be another tool to separate different origins. Clearly theory and observation have to work in close collaboration to extract this information.

(2) We can measure the increase in the equivalent widths which will occur if velocity gradients over scales smaller than $\Delta\tau_\nu = 1$ are present. This leads to the concept of microturbulence. For different frequencies within the line the condition $\Delta\tau_\nu < 1$ refers to different distances leading to some conceptual difficulties. A Gaussian velocity distribution is assumed for the microturbulence field which, if wrong, may lead to errors that have not yet been studied. The depth dependence of the microturbulence can be studied by the investigation of lines with different excitation potentials or of lines in spectral regions with large differences in the continuous κ. The Goldberg-Unno method (1958, 1959) using different depth points in line profiles of one multiplet suffers from the fact that the observations always integrate over the whole line forming region and one cannot decide whether the larger velocities are at the top or at the bottom of this layer.

(3) Line profiles can be measured. They reflect all velocity fields including rotation and therefore contain all the information but provide the largest difficulties in separating the different fields. After correcting for instrumental broadening, for thermal broadening and microturbulence, for rotation and for observational noise, the broadening that is left over is called macroturbulence, again assumed to have a Gaussian velocity distribution which, if wrong, can cause large errors in the separation of rotation and macroturbulence. Obtainable only after so many deconvolutions it will generally not be determined accurately. Since this turbulence does not increase the line strength it must refer to scales $\Delta\tau_\nu > 1$. Large scale velocity changes occurring horizontally will be observed as macroturbulence.

The concepts of micro and macroturbulence have one property in common with the mixing length theory: for many years they have been criticized strongly but are still widely used for lack of knowing anything better. In the turbulence case the new concept of mesoturbulence is clearly a step forward since it does not rely on the assumption of either large or small scale velocity variations. It can be used for any scale of velocity variations. In the approach actually used in the numerical work shown to us by E. Sedlmayr it still uses one correlation length only which can have any size. However, the formalism presented to us by H.-R. Gail can well be used for fields with several correlation lengths. A correlation length of the order of $\Delta\tau_\nu=1$ will influence the equivalent width of the line, i.e. will manifest itself as microturbulence, but will also show up in additional line broadening, interpreted as macroturbulence. We have heard that the carefully determined values for microturbulence,--i.e. using the correct model, the correct oscillator strengths f and abundances Z as well as the correct damping constant γ--and macroturbulence can be used directly to determine the correlation length and the amplitude of the velocity field.

The concept of mesoturbulence clearly is not the ultimate solution, but, as G. Traving pointed out privately, is still a method suggested due to our ignorance. Ultimately we will have to determine the velocity distributions for the different velocity fields and see if and how the line profiles differ for different fields. The solar observations may provide some guidance. Theory will have to help. Hopefully different velocity fields will lead to measurably different line profiles or show different dependences on excitation energies and wavelengths, which may be used to distinguish different origins.

For the deconvolutions of the different contributions to the line profile the Fourier transformation, explored in this context especially by D. Gray, promises to be very helpful provided the profiles from the different velocity fields are measurably different in the frequency domaine. As D. Gray pointed out we cannot expect miracles from the Fourier transformation. Uncertain differences of line profiles in the frequency domaine will remain uncertain in the Fourier domaine even though the differences may appear amplified.

III. Measured Values of Micro-and Macroturbulence and their Origin

The modern carefully-determined micro-and macroturbulence velocities in F and G stars increase with increasing luminosity and with increasing effective temperature in accordance with the variations expected for the maximum convective velocities as T. Gehren pointed out. This confirms the suspicion that for the F and G stars micro-and macroturbulence have their origin in the convection zone. The measured macroturbulence values are generally larger than the microturbulence velocities but vary proportionately. In the concept of mesoturbulence as outlined by E. Sedlmayr this indicates a correlation length somewhat larger than $\Delta\tau_\nu=1$,again in qualitative agreement with mixing length theory expectations.

Earlier measurements of microturbulence with wrong f values showed high microturbulence values also for late A stars. There are, however, several indications--see below-- that convection stops at F0. If microturbulence persists to higher temperatures we probably have to come back to the suggestion of Baschek and Reimers 1969 that for these stars microturbulence may be due to a superposition of many pulsation modes similar to the ones observed by L.B. Lucy for α Cyg.

IV. Effects of Velocity Fields on the Outer Layers of the Stars

R. Stein in his very clear review convinced us that none of the observed velocity fields and also magnetohydrodynamic waves could be responsible for the heating of the upper chromosphere and corona. Acoustic waves are dissipated in the lower chromosphere and can therefore well heat the lower chromosphere but they cannot penetrate to higher layers. Gravity waves travel mainly horizontally and for Alfvèn waves the energy distribution over large volumes in the corona seems to be a problem. Again the observers have to assist the theoreticians.

If acoustic waves are indeed responsible for the heating of the lower chromosphere then the energy input into these layers should be correlated with the acoustic energy generation in the convection zones. R. Stein pointed out that along the main sequence the acoustic energy generation in the convection zone increases roughly proportional to T_{eff}^{16}. Most of this energy is absorbed already in the photosphere but roughly 10% may heat the lower chromosphere. We should then expect chromospheres for all stars with convection zones. Indeed modern observations especially by the International Ultraviolet Explorer, IUE, reveals chromospheric emission line spectra for most late type stars as J. Linsky has reviewed here. Our own observations (Böhm-Vitense and Dettmann 1979) show that for luminous stars chromospheric emission stops at the Cepheid instability strip and on the main sequence for B-V = 0.3. This is also the color for which the average rotation for stars begins to decrease and is also the red edge of the gap in the two color diagram (Böhm-Vitense and Canterna 1974) which can also be attributed to the abrupt onset of convection. As G. Nelson (1978) has pointed out inhomogeneous photospheres of convective stars look more red than homogeneous ones.

Convective stars generally appear to have chromospheres--except perhaps old stars--. If acoustic heating is responsible for the energy input into the lower chromosphere then the energy loss of these layers should increase with increasing convective velocities, i.e. with increasing T_{eff} and luminosity. As J. Linsky pointed out, the energy loss in the lower chromosphere can be measured by the MgII h and k line emission Ulmschneider claimed that a steep increase with T_{eff} is indeed observed while R. Stein expressed some doubt. Whether the MgII emission increases with increasing luminosity is still debated. In J. Linsky's graph most of the G and K supergiants appear to have a larger emission than most of the G and K giants. There are, however, a few supergiants which show very small MgII emission (for instance ɩ Aur and ξ Cyg). It will have to be checked whether interstellar absorption might have reduced their MgII intensity. Measuring uncertainties also appear to be very large for these stars, (see Weiler and Oegerle 1979). In the discussion of the luminosity dependence of the energy input varying amounts of absorption in the photosphere may also turn out to be important.

Additional information about the heating of the lower layers may be obtained from the Wilson Bappu effect. There are mainly two suggestions to explain the emission line width luminosity correlation. One group wants to relate the increasing width to the increasing "turbulent" velocities, the other group wants to explain it by an increasing optical depth effect. I always found it very important that the width luminosity effect holds independently of the emission line strength which obviously can be different for stars of the same luminosity. I wonder whether this can be understood if the width is determined by the optical thickness in the emission lines. This has not been discussed here.

Information about the transition layers can be obtained from the CIV emission lines. The situation is, however, somewhat unclear since, as J. Linsky pointed out, based on Mullan's study, the possibility exists that stellar winds may be an important energy sink for the transition region and may even eliminate it. In fact the possibility has been discussed here that for cool and luminous stars observable winds may reach down to the MgII and CaII emitting regions. For these stars the CIV lines become invisible. It is still debated whether they disappear abruptly or continuously. The early G supergiants observed by us (Böhm-Vitense and Dettmann 1979) show CIV emission.

Additional information may be obtained from the observations of old and metal poor stars. T. Gehren pointed out that, as expected from convection theory, they have the same observed microturbulence as young stars. The acoustic heating should therefore be the same. Wilson (1966) and Kraft (1967) found, however, that the CaII K_2 emission decreases with increasing age. From a few IUE observations obtained so far I also find a decrease in MgII emission. This line, however, is still observable in a few cases, but the far UV emission lines are completely invisible. Clearly more observations are needed before a final conclusion can be drawn. But based on those few observations we may perhaps speculate that some acoustic heating does occur in the layers which emit the CaII and MgII lines but that in young star the layers which emit the higher excitation lines are heated by another mechanism which decays with increasing age and is probably connected with rotation and magnetic fields.

It was very interesting to me that Y. Cuny pointed out that the absorption of photospheric light, which may increase considerably with increasing turbulence, should not be neglected in the energy balance of chromospheric lines.

V. Major Open Questions

Let me conclude in summary by listing the major open questions discussed at this meeting:

1) Contributions of different velocity fields to the line broadening;
2) Origin of line broadening in A, B and O stars;
3) Origin of "chromospheres" in O and B stars;
4) The heating mechanism for the upper chromospheres, transition regions and coronae in convective stars;
5) The explanation of the Wilson Bappu Effect,
   a) is it due to an optical depth effect, or
   b) is it due to the velocity field?
6) Can coronae, transition layers and upper chromospheres in cool luminous stars be extinguished by stellar winds?

References *

Baschek, B. and Reimer, D. 1969, Astron. & Astrophys. 2, 240
Böhm-Vitense, E. and Cantera, R. 1974, Astrophys. J. 194, 629
Böhm-Vitense, E. and Dettmann, T. 1979, Astrophys. J. in press
Chaffee, F.H., Carbon, D.F. and Strom, S.E. 1971, Astrophys. J., 166, 593
Goldberg, L. 1958, Astrophys. J., 127, 308.
Kraft, R.P. 1967, Astrophys. J., 150, 551.
Nelson, G. 1978: Thesis, University of Washington, Seattle
Unno, W. 1959, Astrophys. J., 129, 375
Weiler, E.J. and Oegerle, W.R. 1979, Astrophys. J. Suppl., 39, 537
Wilson, O.C. 1966, Astrophys. J., 144, 695

* Only references not contained in this volume are given

TABLE 1

Origin of Non-Thermal Motions

	F Stars and Later			A Stars and Earlier	
SPEAKER:	Zahn	Osaki	Gilman, Durney	Nelson	Kodaira
	Convective Velocities due to buoyancy	*Waves*	*Differential Rotation*		
	Theory does not yet give v(x,y,z,t) because of highly non linear equations.	acoustic, gravity	for deep convection zones	not due to convection	probably not due to circulation
	(My question: Can observations provide v(x,y,z,t) to be put into theory?	*Pulsations* radial, non-radial oscillations	No influence of rotation on convection in early F stars	A Stars: pulsations? (Baschek& Reimers, Lucy)	
		excitation mainly by κ mechanism seated in convection zone		O, B stars: shear turbulence from rotation? (Sreenivasan)	
	Δp leads to anti= buoyancy. Numerical work explains observed solar Δt, ξt, exploding granules size of granules (Nordlund, Nelson 1978)	motions are anisotropic mainly vertical		Radiation driven instabilities? (Hearn& Nelson)	
		G K stars are overstable for many non-radial p modes			

TABLE II

Observations of Velocity Fields in Stars

Asymmetric Line Shifts	*Microturbulence*	*Macroturbulence*	*Rotation*
if ΔV_r is coupled with ΔI	gives increase in A_λ	additional line broadening	additional line broadening
or if	implies $\frac{dV_r}{d\tau_\nu} \neq 0$	$\frac{dv_r}{d\theta}$, $\frac{dv_r}{d\phi} \neq 0$	$\frac{dv_r}{d\theta}$, $\frac{dv_r}{d\phi} \neq 0$
asymmetric velocity distribution	for $\Delta\tau_\nu < 1$	for $\Delta\tau_\nu > 1$	characteristic velocity distribution
or if	may occur for expansion: global or local contraction: global or local pulsations waves small scale turbulence	possibilities: pulsations waves expansion contraction differential rotation large scale turbulence	
asymmetric areas with opposite V_r			

Mesoturbulence (Traving) (between Microturbulence and Macroturbulence)

$$\frac{dv_r}{dt} \neq 0$$

for any scale s
statistical description with
correlation lengths (Gail)
s can be determined from ratio
of apparent microturbulence
to apparent macroturbulence
(Sedlmayr)

Selected Issues from

Lecture Notes in

Vol. 594: Singular Perturbations and Boundary Layer Theory, Lyon 1976. Edited by C. M. Brauner, B. Gay, and J. Mathieu. VIII, 539 pages. 1977.

Vol. 596: K. Deimling, Ordinary Differential Equations in Banach Spaces. VI, 137 pages. 1977.

Vol. 605: Sario et al., Classification Theory of Riemannian Manifolds. XX, 498 pages. 1977.

Vol. 606: Mathematical Aspects of Finite Element Methods. Proceedings 1975. Edited by I. Galligani and E. Magenes. VI, 362 pages. 1977.

Vol. 607: M. Métivier, Reelle und Vektorwertige Quasimartingale und die Theorie der Stochastischen Integration. X, 310 Seiten. 1977.

Vol. 615: Turbulence Seminar, Proceedings 1976/77. Edited by P. Bernard and T. Ratiu. VI, 155 pages. 1977.

Vol. 618: I. I. Hirschman, Jr. and D. E. Hughes, Extreme Eigen Values of Toeplitz Operators. VI, 145 pages. 1977.

Vol. 623: I. Erdelyi and R. Lange, Spectral Decompositions on Banach Spaces. VIII, 122 pages. 1977.

Vol. 628: H. J. Baues, Obstruction Theory on the Homotopy Classification of Maps. XII, 387 pages. 1977.

Vol. 629: W.A. Coppel, Dichotomies in Stability Theory. VI, 98 pages. 1978.

Vol. 630: Numerical Analysis, Proceedings, Biennial Conference, Dundee 1977. Edited by G. A. Watson. XII, 199 pages. 1978.

Vol. 636: Journées de Statistique des Processus Stochastiques, Grenoble 1977, Proceedings. Edité par Didier Dacunha-Castelle et Bernard Van Cutsem. VII, 202 pages. 1978.

Vol. 638: P. Shanahan, The Atiyah-Singer Index Theorem, An Introduction. V, 224 pages. 1978.

Vol. 648: Nonlinear Partial Differential Equations and Applications, Proceedings, Indiana 1976–1977. Edited by J. M. Chadam. VI, 206 pages. 1978.

Vol. 650: C*-Algebras and Applications to Physics. Proceedings 1977. Edited by R. V. Kadison. V, 192 pages. 1978.

Vol. 656: Probability Theory on Vector Spaces. Proceedings, 1977. Edited by A. Weron. VIII, 274 pages. 1978.

Vol. 662: Akin, The Metric Theory of Banach Manifolds. XIX, 306 pages. 1978.

Vol. 665: Journées d'Analyse Non Linéaire. Proceedings, 1977. Edité par P. Bénilan et J. Robert. VIII, 256 pages. 1978.

Vol. 667: J. Gilewicz, Approximants de Padé. XIV, 511 pages. 1978.

Vol. 668: The Structure of Attractors in Dynamical Systems. Proceedings, 1977. Edited by J. C. Martin, N. G. Markley and W. Perrizo. VI, 264 pages. 1978.

Vol. 675: J. Galambos and S. Kotz, Characterizations of Probability Distributions. VIII, 169 pages. 1978.

Vol. 676: Differential Geometrical Methods in Mathematical Physics II, Proceedings, 1977. Edited by K. Bleuler, H. R. Petry and A. Reetz. VI, 626 pages. 1978.

Vol. 678: D. Dacunha-Castelle, H. Heyer et B. Roynette. Ecole d'Eté de Probabilités de Saint-Flour. VII-1977. Edité par P. L. Hennequin. IX, 379 pages. 1978.

Vol. 679: Numerical Treatment of Differential Equations in Applications, Proceedings, 1977. Edited by R. Ansorge and W. Törnig. IX, 163 pages. 1978.

Vol. 681: Séminaire de Théorie du Potentiel Paris, No. 3, Directeurs: M. Brelot, G. Choquet et J. Deny. Rédacteurs: F. Hirsch et G. Mokobodzki. VII, 294 pages. 1978.

Vol. 682: G. D. James, The Representation Theory of the Symmetric Groups. V, 156 pages. 1978.

Vol. 684: E. E. Rosinger, Distributions and Nonlinear Partial Differential Equations. XI, 146 pages. 1978.

Vol. 690: W. J. J. Rey, Robust Statistical Methods. VI. 128 pages. 1978.

Vol. 691: G. Viennot, Algèbres de Lie Libres et Monoïdes Libres. III, 124 pages. 1978.

Vol. 693: Hilbert Space Operators, Proceedings, 1977. Edited by J. M. Bachar Jr. and D. W. Hadwin. VIII, 184 pages. 1978.

Vol. 696: P. J. Feinsilver, Special Functions, Probability Semigroups, and Hamiltonian Flows. VI, 112 pages. 1978.

Vol. 702: Yuri N. Bibikov, Local Theory of Nonlinear Analytic Ordinary Differential Equations. IX, 147 pages. 1979.

Vol. 704: Computing Methods in Applied Sciences and Engineering, 1977, I. Proceedings, 1977. Edited by R. Glowinski and J. L. Lions. VI, 391 pages. 1979.

Vol. 710: Séminaire Bourbaki vol. 1977/78, Exposés 507–524. IV, 328 pages. 1979.

Vol. 711: Asymptotic Analysis. Edited by F. Verhulst. V, 240 pages. 1979.

Vol. 712: Equations Différentielles et Systèmes de Pfaff dans le Champ Complexe. Edité par R. Gérard et J.-P. Ramis. V, 364 pages. 1979.

Vol. 716: M. A. Scheunert, The Theory of Lie Superalgebras. X, 271 pages. 1979.

Vol. 720: E. Dubinsky, The Structure of Nuclear Fréchet Spaces. V, 187 pages. 1979.

Vol. 724: D. Griffeath, Additive and Cancellative Interacting Particle Systems. V, 108 pages. 1979.

Vol. 725: Algèbres d'Opérateurs. Proceedings, 1978. Edité par P. de la Harpe. VII, 309 pages. 1979.

Vol. 726: Y.-C. Wong, Schwartz Spaces, Nuclear Spaces and Tensor Products. VI, 418 pages. 1979.

Vol. 727: Y. Saito, Spectral Representations for Schrödinger Operators With Long-Range Potentials. V, 149 pages. 1979.

Vol. 728: Non-Commutative Harmonic Analysis. Proceedings, 1978. Edited by J. Carmona and M. Vergne. V, 244 pages. 1979.

Vol. 729: Ergodic Theory. Proceedings 1978. Edited by M. Denker and K. Jacobs. XII, 209 pages. 1979.

Vol. 730: Functional Differential Equations and Approximation of Fixed Points. Proceedings, 1978. Edited by H.-O. Peitgen and H.-O. Walther. XV, 503 pages. 1979.